Energy, Environment, and Sustainability

Series Editor

Avinash Kumar Agarwal, Department of Mechanical Engineering, Indian Institute of Technology Kanpur, Kanpur, Uttar Pradesh, India

AIMS AND SCOPE

This books series publishes cutting edge monographs and professional books focused on all aspects of energy and environmental sustainability, especially as it relates to energy concerns. The Series is published in partnership with the International Society for Energy, Environment, and Sustainability. The books in these series are edited or authored by top researchers and professional across the globe. The series aims at publishing state-of-the-art research and development in areas including, but not limited to:

- Renewable Energy
- Alternative Fuels
- Engines and Locomotives
- Combustion and Propulsion
- Fossil Fuels
- Carbon Capture
- Control and Automation for Energy
- Environmental Pollution
- Waste Management
- Transportation Sustainability

Review Process

The proposal for each volume is reviewed by the main editor and/or the advisory board. The chapters in each volume are individually reviewed single blind by expert reviewers (at least four reviews per chapter) and the main editor.

Ethics Statement for this series can be found in the Springer standard guidelines here https://www.springer.com/us/authors-editors/journal-author/journal-author-helpdesk/before-you-start/before-you-start/1330#c14214

More information about this series at https://link.springer.com/bookseries/15901

Avinash Kumar Agarwal ·
Antonio García Martínez · Ankur Kalwar ·
Hardikk Valera
Editors

Advanced Combustion for Sustainable Transport

Springer

Editors
Avinash Kumar Agarwal
Department of Mechanical Engineering
Indian Institute of Technology Kanpur
Kanpur, Uttar Pradesh, India

Ankur Kalwar
Department of Mechanical Engineering
Indian Institute of Technology Kanpur
Kanpur, Uttar Pradesh, India

Antonio García Martínez
Department of Thermal and Reciprocating Engines
Universitat Politècnica de València
Valencia, Spain

Hardikk Valera
Department of Mechanical Engineering
Indian Institute of Technology Kanpur
Kanpur, Uttar Pradesh, India

ISSN 2522-8366 ISSN 2522-8374 (electronic)
Energy, Environment, and Sustainability
ISBN 978-981-16-8417-3 ISBN 978-981-16-8418-0 (eBook)
https://doi.org/10.1007/978-981-16-8418-0

This Springer imprint is published by the registered company Springer Nature Singapore Pte Ltd.
The registered company address is: 152 Beach Road, #21-01/04 Gateway East, Singapore 189721, Singapore

Preface

Energy demand has been rising remarkably due to the increasing population and urbanization. The global economy and society significantly depend on energy availability because it touches every facet of human life and activities. Transportation and power generation are two major sectors which are extensively dependent on energy. Today's requirement is to manage efficient utilization of energy in a sustainable manner which can satisfy the global increasing demands. Further, exploration of new energy resources is equally important.

The International Society for Energy, Environment and Sustainability (ISEES) was founded at the Indian Institute of Technology Kanpur (IIT Kanpur), India, in January 2014 to spread knowledge/ awareness and catalyze research activities in the fields of Energy, Environment, Sustainability, and Combustion. Society's goal is to contribute to the development of clean, affordable, and secure energy resources and a sustainable environment for society and spread knowledge in the areas mentioned above, and create awareness about the environmental challenges the world is facing today. The unique way adopted by ISEES was to break the conventional silos of specializations (engineering, science, environment, agriculture, biotechnology, materials, fuels, etc.) to tackle the problems related to energy, environment, and sustainability in a holistic manner. This is quite evident by the participation of experts from all fields to resolve these issues. The ISEES is involved in various activities such as conducting workshops, seminars, conferences, etc., in the domains of its interests. Society also recognizes the outstanding works of young scientists, professionals, and engineers for their contributions in these fields by conferring them awards under various categories.

The Fifth International Conference on 'Sustainable Energy and Environmental Challenges' (V-SEEC) was organized under the auspices of ISEES from December 19–21, 2020, in virtual mode due to restrictions on travel because of the ongoing COVID-19 pandemic situation. This conference provided a platform for discussions between eminent scientists and engineers from various countries, including India, Spain, Austria, Bangladesh, Mexico, USA, Malaysia, China, UK, the Netherlands, Germany, Israel, and Saudi Arabia. At this conference, eminent international speakers presented their views on energy, combustion, emissions, and alternative energy resources for sustainable development and a cleaner environment. The conference presented two high-voltage plenary talks by Dr. V. K. Saraswat, Honorable Member, NITI Ayog, on 'Technologies for Energy Security and Sustainability' and Prof. Sandeep Verma, Secretary, SERB, on 'New and Equitable R&D Funding Opportunities at SERB.'

The conference included nine technical sessions on topics related to energy and environmental sustainability. Each session had 6–7 eminent scientists from all over the world, who shared their opinion and discussed the trends for the future. The technical sessions in the conference included Emerging Contaminants: Monitoring and Degradation Challenges; Advanced Engine Technologies and Alternative Transportation Fuels; Future Fuels for Sustainable Transport; Sustainable Bioprocessing for Biofuel/ Non-biofuel Production by Carbon Emission Reduction; Future of Solar Energy; Desalination and Wastewater Treatment by Membrane Technology; Biotechnology in Sustainable Development; Emerging Solutions for Environmental Applications' and Challenges and Opportunities for Electric Vehicle Adoption. Five hundred plus participants and speakers from all over the world attended this three days conference.

The conference concluded with a high-voltage panel discussion on 'Challenges and Opportunities for Electric Vehicle Adoption,' where the panelists were Prof. Gautam Kalghatgi (University of Oxford), Prof. Ashok Jhunjhunwala (IIT Madras), Dr. Kelly Senecal (Convergent Science), Dr. Amir Abdul Manan (Saudi Aramco) and Dr. Sayan Biswas (University of Minnesota, USA), Prof. Avinash K. Agarwal, ISEES, moderated the panel discussion. This conference laid out the roadmap for technology development, opportunities, and challenges in energy, environment, and sustainability domain. All these topics are very relevant for the country and the world in the present context. We acknowledge the support received from various agencies and organizations for the successful conduct of the Fifth ISEES conference V-SEEC, where these books germinated. We want to acknowledge SERB (special thanks to Dr. Sandeep Verma, Secretary) and our publishing partner Springer (special thanks to Ms. Swati Meherishi).

The editors would like to express their sincere gratitude to a large number of authors from all over the world for submitting their high-quality work on time and revising it appropriately at short notice. We would like to express our special gratitude to our prolific set of reviewers, Dr. Gan Xiao, Dr. Krishnamoorthi, Dr. Dimitris Assanis, Ashutosh, Dr. Yao Mingfa, Dr. D. Deshmukh, Dr. Nikhil Sharma, Sam Joe, Dr. Kohei Kuzuoka, Dr. Chetan Patel, Dr. Chao Han, Utkarsha, Dr. Shubhankar Dev, Dr. Peter Kucher, Dr. Burak Zincir, Dr. Weiqiang Han, Dr. S. Thirumalini, Harsimran, Dr. B. Ashok, Dr. Maciej Mikulski, Dhananjay, who reviewed various chapters of this monograph and provided their valuable suggestions to improve the manuscripts.

This book is based on advanced combustion technologies, which are currently used in internal combustion engines. This book describes about different strategies for improving conventional diesel combustion. This book includes few chapters on low-temperature combustion techniques of compression–ignition engines which result in significant reduction of NOx and soot emissions. Few chapters based on newly evolved gasoline compression technology and optical techniques in advanced gasoline direct injection engines are another important aspect of this book. The chapters include recent results and are focused on current trends of automotive sector. In this book, readers will get the idea about the advancements in combustion technologies, which will help them in analyzing various issues related to in-cylinder combustion, pollutant formation, and alternative fuels. Few chapters of this book are based on dual-fuel combustion technology for the adoption of biofuels or cleaner alternative fuels along with conventional fuels. We hope that the book would greatly interest the professionals, post-graduate students involved in fuels, IC engines, engine combustion research.

Kanpur, India — Avinash Kumar Agarwal
Valencia, Spain — Antonio García Martínez
Kanpur, India — Ankur Kalwar
Kanpur, India — Hardikk Valera

Contents

Editors and Contributors

About the Editors

Avinash Kumar Agarwal joined IIT Kanpur in 2001. He worked at the Engine Research Center, UW@Madison, the USA as a Post-Doctoral Fellow (1999–2001). His interests are IC engines, combustion, alternate and conventional fuels, lubricating oil tribology, optical diagnostics, laser ignition, HCCI, emissions, and particulate control, 1D and 3D Simulations of engine processes, and large-bore engines. Prof. Agarwal has published 435+ peer-reviewed international journal and conference papers, 70 edited books, 92 books chapters, and 12200+ Scopus and 19000+ Google Scholar citations. He is the associate principal editor of FUEL. He has edited "Handbook of Combustion" (5 Volumes; 3168 pages), published by Wiley VCH, Germany. Prof. Agarwal is a Fellow of SAE (2012), Fellow of ASME (2013), Fellow of ISEES (2015), Fellow of INAE (2015), Fellow of NASI (2018), Fellow of Royal Society of Chemistry (2018), and a Fellow of American Association of Advancement in Science (2020). He is the recipient of several prestigious awards such as Clarivate Analytics India Citation Award-2017 in Engineering and Technology, NASI-Reliance Industries Platinum Jubilee Award-2012; INAE Silver Jubilee Young Engineer Award-2012; Dr. C. V. Raman Young Teachers Award: 2011; SAE Ralph R. Teetor Educational Award -2008; INSA Young Scientist Award-2007; UICT Young Scientist Award-2007; INAE Young Engineer Award-2005. Prof.

Agarwal received Prestigious CSIR Shanti Swarup Bhatnagar Award-2016 in Engineering Sciences. Prof. Agarwal is conferred upon Sir J C Bose National Fellowship (2019) by SERB for his outstanding contributions. Prof. Agarwal was a highly cited researcher (2018) and was in the top ten HCR from India among 4000 HCR researchers globally in 22 fields of inquiry.

Prof. Antonio García Martínez is an Associate Professor in the Department of Thermal and Reciprocating Engines at the Universitat Politècnica de Valencia, Spain, where he develops his teaching responsibilities in the framework of combustion fundamentals. His research area focuses on low-temperature combustion. He has published more than 140 peer-reviewed articles, and is an active member in different Engineering International Organizations such as SAE or IEES, acting as session organizer, reviewer and author at different events. He received his M.S. and Ph.D. in Mechanical Engineering from the Universitat Politècnica de València, Spain. Professor Garcia has been a visiting professor at the Combustion Engines division at Lund University, as well as a visiting researcher at RWTH Aachen University, where he developed relevant works on the implementation of advanced combustion systems on CI engines.

Mr. Ankur Kalwar is currently a Ph.D. research scholar at Indian Institute of Technology (IIT) Kanpur. He received his Bachelor's and Master's degree in Mechanical Engineering from Lovely Professional University, India in 2016. His Master's work focused on developing the liquid desiccant—vapour compression hybrid refrigeration system. Currently, his areas of research include gasoline direct injection technology, fuel spray analysis, optical diagnostics, combustion diagnostics, engine emissions measurement, particulate characterization and their control, and alternative fuels. He has authored several book chapters and research articles in international journals and conferences.

Mr. Hardikk Valera is pursuing his Ph.D. from Engine Research Laboratory (ERL), Department of Mechanical Engineering, Indian Institute of Technology (IIT) Kanpur. He has completed his M.Tech. and B.Tech. from National Institute of Technology (NIT) Jalandhar, India, and Ganpat University, respectively. His research interests include methanol fuelled SI engines, methanol fuelled CI engines, optical diagnostics, fuel spray characterization and emission control from engines.

Contributors

Avinash Kumar Agarwal Engine Research Laboratory, Department of Mechanical Engineering, Indian Institute of Technology Kanpur, Kanpur, India

Sadaf Batool Michigan Technological University, Houghton, MI, USA

Sawan Bharti Invertis University, Bareilly, India

Krishn Chandra Department of Engineering, University of Campania Luigi, Vanvitelli, Italy;
Engine Research Laboratory, Department of Mechanical Engineering, Indian Institute of Technology Kanpur, Kanpur, India

Balendra V. S. Chauhan Institute of Fundamental Technological Research, Polish Academy of Sciences, Warsaw, Poland;
Invertis University, Bareilly, India

Sam Joe Chintagunti Engine Research Laboratory, Department of Mechanical Engineering, Indian Institute of Technology Kanpur, Kanpur, India

Antonio García Martínez Department of Thermal and Reciprocating Engines, Universitat Politècnica de Valencia, Valencia, Spain

Akshay Garg Department of Mechanical Engineering, College of Engineering Roorkee, Roorkee, India

Ashutosh Jena Engine Research Laboratory, Department of Mechanical Engineering, Indian Institute of Technology Kanpur, Kanpur, India

Sulav Kafle Engine Research Laboratory, Department of Mechanical Engineering, Indian Institute of Technology Kanpur, Kanpur, India

Ankur Kalwar Engine Research Laboratory, Department of Mechanical Engineering, Indian Institute of Technology Kanpur, Kanpur, India

Jinsu Kim West Virginia University, Morgantown, WV, USA

Niraj Kumar Department of Mechanical Engineering, ASET, Amity University, Noida, India

Oronzio Manca Department of Engineering, University of Campania Luigi, Vanvitelli, Italy

Jeffrey D. Naber Michigan Technological University, Houghton, MI, USA

Babalola Aisosa Oni Department of Chemical Engineering, Covenant University, Ota, Ogun State, Nigeria;
Department of Chemical Engineering, China University of Petroleum, Changping, Beijing, People's Republic of China

Dosawada Pavan Sai Department of Mechanical Engineering, ASET, Amity University, Noida, India

Samuel Eshorame Sanni Department of Chemical Engineering, Covenant University, Ota, Ogun State, Nigeria

Vishal Saxena Department of Fuel (FMME) Engineering, I.I.T (ISM), Dhanbad, Jharkhand, India

Imran Sayyed Panjab University, Chandigarh, India

Mahdi Shahbakhti University of Alberta, Edmonton, AB, Canada

Mritunjay Shukla Automotive Fuels and Lubricants Application Division, CSIR-Indian Institute of Petroleum, Dehradun, Uttarakhand, India

Harold Sun West Virginia University, Morgantown, WV, USA

Andrea Unich Department of Engineering, University of Campania Luigi, Vanvitelli, Italy

Hardikk Valera Engine Research Laboratory, Department of Mechanical Engineering, Indian Institute of Technology Kanpur, Kanpur, India

Ajitanshu Vedrantam Universidad de Alcalá, Alcala de Henares, Spain

Yu Zhang Aramco Americas, Aramco Research Center- Detroit, Novi, MI, USA

Part I
General

Chapter 1
Introduction to Advanced Combustion for Sustainable Transport

Avinash Kumar Agarwal, Antonio García Martínez, Ankur Kalwar, and Hardikk Valera

Abstract Research in the IC Engines domain is directed towards novel combustion technologies to improve engine efficiency and reduce harmful pollutants. Technologies are also being developed to adopt alternative, cleaner fuels that meet stricter emission norms globally and reduce dependence on conventional fuels. This book covers various engine combustion technologies under development globally. The first section of the book introduces the combustion technologies being developed for internal combustion engines and has one chapter. In second section, strategies for clean diesel combustion are covered, which would help automotive OEMs continuously improve existing engine technologies. Recent advances in low-temperature combustion (LTC) techniques in CI engines such as Reactivity controlled compression ignition (RCCI), Premixed charge compression ignition (PCCI), Homogeneous Charge Compression Ignition (HCCI), Gasoline compression ignition (GCI) are comprehensively covered. For improving the understanding and control of these LTC techniques, different stages of combustion in LTC regimes are also discussed threadbare. The third section comprehensively discusses optimisation for various control parameters in gasoline direct injection (GDI) engine development. Experimental optical investigations of in-cylinder processes are covered for a superior understanding of fundamental processes. In the fourth section, dual-fuel combustion technology in IC engines is covered since the research interest of the global research community is converging towards fuel diversification. The fifth section includes miscellaneous areas. It discusses the different aspects of metal hydride technology for hydrogen storage, which will help realise hydrogen-fueled vehicles in future. Also, a description of waste heat recovery potential from IC engines using the organic Rankine cycles (ORC) is covered. The information covered in this book would be useful for students, researchers, academicians and the automotive researchers in industry alike.

A. K. Agarwal (✉) · A. Kalwar · H. Valera
Department of Mechanical Engineering, Indian Institute of Technology Kanpur, Kanpur, India
e-mail: akag@iitk.ac.in

A. G. Martínez
Department of Thermal and Reciprocating Engines, Universitat Politècnica de Valencia, Valencia, Spain

A. K. Agarwal et al. (eds.), *Advanced Combustion for Sustainable Transport*, Energy, Environment, and Sustainability, https://doi.org/10.1007/978-981-16-8418-0_1

Keywords Low-temperature combustion (LTC) · Mode-switching · Gasoline compression ignition (GCI) · Gasoline direct injection (GDI) · Optical diagnostics · Dual-fuel combustion · Compressed natural gas (CNG) · Metal hydrides · Waste heat recovery

The chapters in this book are organised in five different sections: (i) General, (ii) Advanced Engine Technologies for CI Engines, (iii) Advanced Engine Technologies for SI Engines, (iv) Dual-fuel Combustion Technology, and (v) Miscellaneous.

IC Engines are the prime movers of the global automotive industry. This technology is continuously evolving in the last 140 years for achieving sustainable transport. As the emission norms are getting stricter, it is becoming more and more difficult for the automotive industry to comply with the legislative emission norms using existing engine technologies (Payri et al. 2015; Reitz et al. 2020). Hence, researchers focus on developing advanced combustion technologies having lower emissions and improved fuel economy. Extensive research is underway to develop technologies to use cleaner alternative fuels more efficiently. This book includes these aspects of advanced combustion technologies for achieving sustainable transport.

The first section of the book has one chapter, introducing different aspects of the book's contents.

The second section of the book has four chapters that cover strategies for developing cleaner diesel engines. Detailed discussions of various LTC techniques such as RCCI, PCCI, HCCI and GCI in CI engines is covered in this section. The first chapter explains the fundamental causes of in-efficiency, pollutant formation, and possible approaches to tackle them in conventional diesel combustion. The focus on the evolution of diesel engines, modern designs, and strategic optimisation trends is also comprehensively presented. The next chapter discusses the state-of-the-art LTC engine control, including dynamic modelling, model predictive combustion control, experimentation, and implementation of real-time closed-loop combustion controllers (Agarwal et al. 2017). It discusses the limitation of the constrained optimal operating range of LTC engines and explains controller development for these engines. Another chapter thoroughly reviews the emission performance, combustion characteristics, and thermal efficiency of HCCI, RCCI, and PCCI technologies and compares their applications for dual-fuel and single fueled engines. The understandings obtained by researchers about the two-stage combustion occurring in PPCI conditions using optical diagnostics are also incorporated in yet another chapter. In the last chapter of this section, gasoline compression ignition (GCI) has been shown as a potential cost-effective engine technology to achieve higher fuel efficiency with lower engine-out NO_x and soot emissions. This chapter presents an overview of gasoline's autoignition behaviour, fundamental gasoline combustion and basic principles, benefits, and challenges of varying GCI combustion strategies.

The third section of the book discusses the optical diagnostics investigations for GDI engine development in one chapter. Current generation GDI engines offer higher power output and superior fuel economy (Kalwar and Agarwal 2020). The chapter broadly covers studies related to in-cylinder flow characteristics, sprays characterisation, fuel spray-air interactions, fuel evaporation, mixture formation, flame propagation characteristics, and pollutant formation, corresponding to different GDI engine operating parameters. It includes a detailed perspective that helps optimise engine design and control parameters adapted to conventional and alternative fuels. The rich information derived from the optical diagnostics helps develop new-generation GDI engines capable of meeting market requirements and current and future emission compliances.

The fourth section of the book has two chapters, which deals with dual-fuel combustion technology to utilise different alternative fuels and conventional fuels. In the first chapter, the properties of different dual-fuel systems are discussed, along with the degree of homogenisation that can improve fuels' combustion/atomisation potential towards attaining higher BTE, higher engine power output, and moderate heat release rate and good air–fuel ratios. The second chapter of this section discusses CNG utilisation in a dual-fuel mode in conventional CI and SI engines. Dual-fuel operation using CNG is a good option for emission compliance while improving the overall fuel economy (Wei and Geng 2016). Different strategies such as advanced injection timings, oxygenated pilot fuels, EGR, piston bowl geometry optimisation, and injector nozzles optimisation are discussed to improve engine performance and reduce emissions in dual-fuel mode engines.

The fifth section of the book discusses miscellaneous topics. The potential of metal hydrides for hydrogen storage in a hydrogen-fueled vehicle is discussed. Metal hydrides can overcome the problems associated with hydrogen storage and offer advantages such as thermal stability, greater safety, no fuel loss, long-term storage, etc. Thermodynamic and kinetics of different metal hydrides are mathematically analysed based on pore size and surface area. By knowing the thermodynamic and kinetic requirements of metal hydride reactions, suitable materials for automotive applications can be predicted. The technologies required for the appropriate release of hydrogen can be suggested. The second chapter deals with waste heat recovery potential from IC engines using an organic Rankine cycle (ORC). The ORC is a promising technology for recovering waste heat from various low-temperature sources and potentially enhancing IC engines' overall performance. Several optimisation methods that could potentially improve engine performance and decide optimum operating conditions for the ORC systems are discussed in this chapter.

This monograph presents different technologies to increase IC engine efficiency and lower exhaust emissions. Specific topics covered in the monograph include:

- Introduction to Advanced Combustion for Sustainable Transport
- Strategical Evolution of Clean Diesel Combustion
- Multi-mode low-temperature combustion (LTC) and mode switching control
- State of the Art in Low-Temperature Combustion Technologies: HCCI, PCCI, and RCCI
- Combustion in Diesel Fuelled Partially Premixed Compression Ignition Engines
- Gasoline Compression Ignition Combustion Strategies and Recent Engine System Developments for Commercial and Passenger Transport Applications

- Optical Diagnostics for Gasoline Direct Injection Engines
- Dual fuel internal combustion engines for sustainable transport fuels
- Compressed Natural Gas Utilization in Dual-Fuel Internal Combustion Engines
- Analysis of the potential metal hydrides for hydrogen storage in automobile applications
- Waste Heat Recovery Potential from Internal Combustion Engines Using Organic Rankine Cycle.

References

Agarwal AK, Singh AP, Maurya RK (2017) Evolution, challenges and path forward for low-temperature combustion engines. Prog Energy Combust Sci 61:1–56

Kalwar A, Agarwal AK (2020) Overview, advancements and challenges in gasoline direct injection engine technology. In: Advanced combustion techniques and engine technologies for the automotive sector. Springer, Singapore, pp 111–147

Payri F, Luján JM, Guardiola C, Pla B (2015) A challenging future for the IC engine: new technologies and the control role. Oil Gas Sci Technol Rev IFP Energ Nouv 70(1):15–30

Reitz RD, Ogawa H, Payri R, Fansler T, Kokjohn S, Moriyoshi Y, Agarwal AK, Arcoumanis D, Assanis D, Bae C, Boulouchos K, Canakci M, Curran S, Denbratt I, Gavaises M, Guenthner M, Hasse C, Huang Z, Ishiyama T, Johansson B, Johnson TV, Kalghatgi G, Koike M (2020) IJER editorial: the future of the internal combustion engine

Wei L, Geng P (2016) A review on natural gas/diesel dual-fuel combustion, emissions and performance. Fuel Process Technol 142:264–278

Part II
Advanced Combustion Technologies for CI Engines

Chapter 2
Strategical Evolution of Clean Diesel Combustion

Ashutosh Jena and Avinash Kumar Agarwal

Abstract Diesel engines have been dominant powerplants for transporting goods and public transport, powering agricultural machines as well as industrial machines. After hundred years of evolution, these machines continue to be the most efficient among their counterparts. A practical vision and unbiased analysis would conclusively prove that diesel engines are here to stay for long. Over the years, technology has evolved to meet the demands of customers and policymakers. Fuel efficiency has improved significantly along with the reduction in noise to near-zero levels. An order of magnitude reduction in NO_x and PM has been achieved in the last couple of decades. Future fuels and advanced combustion technology are pioneering the clean diesel revolution. However, the efficiency of conventional diesel combustion (CDC) engines should not be compromised. Therefore, modern-day engine developers need to appreciate the challenges faced in the past and how they have been resolved. This chapter summarises the fundamental causes of in-efficiency, pollutant formation, and possible approaches to tackle them. The chapter also focuses on the technological evolution of diesel engines, modern designs, and strategic optimization trends.

Keywords CDC · Bowl-geometry · Pollutants · NO_x · PM

Abbreviations

aTDC	After Top Dead Centre
bTDC	Before Top Dead Centre
BTE	Brake Thermal Efficiency
CARB	California Air Resource Board
CDC	Conventional Diesel Combustion
CH_2O	Formaldehyde
CO	Carbon Monoxide

A. Jena · A. K. Agarwal (✉)
Engine Research Laboratory, Department of Mechanical Engineering, Indian Institute of Technology Kanpur, Kanpur 208016, India
e-mail: akag@iitk.ac.in

A. K. Agarwal et al. (eds.), *Advanced Combustion for Sustainable Transport*, Energy, Environment, and Sustainability, https://doi.org/10.1007/978-981-16-8418-0_2

CO_2	Carbon Dioxide
CVCC	Constant Volume Combustion Chamber
DoE	Department of Energy
EGR	Exhaust Gas Recirculation
EV	Electric Vehicle
FIP	Fuel Injection Pressure
H_2O_2	Hydrogen Peroxide
HACA	Hydrogen Abstraction Acetylene Addition
HEV	Hybrid Electric Vehicle
HRR	Heat Release Rate
HSDI	High-Speed Direct Injection
ICE	Internal Combustion Engine
KHP	Ketohydroperoxide
LTC	Low-Temperature Combustion
NO_x	Oxides of Nitrogen
PAH	Polycyclic Aromatic Hydrocarbons
PM	Particulate Matter
RO_2	Peroxy Radicals
RoPR	Rate of Pressure Rise
SiRPA	Silica Reinforced Porous Anodized Aluminium
TDC	Top Dead Centre
UHC	Unburnt Hydrocarbon
US	United State

2.1 Introduction

2.1.1 Future of Diesel Engine

Ever since the start of civilization, the development of society has been governed by power and propulsion science. Even today, when power production is discussed, combustion is the first thought that comes to mind. Combustion has immense contribution in the success story of humans on earth. It would not be justified to give a logic that we are going to get rid of combustion anytime soon. However, the time has come to move towards cleaner combustion for the sake of environmental preservation. It is not that we realized this very late. The efforts towards cleaner combustion started way back in the late twentieth century and have made significant advances ever since. Combustion is a complicated thermochemistry process, and several power plants have been devised to control and utilize it for power generation. The internal combustion (IC) engine is one such power plant. This propulsion technology, which is more than 125 years old, powers the most efficient and reliable prime movers for the automotive, marine, and power generation sectors. Several alternative fuels and technologies have emerged as options for powering society, along with IC engines. Therefore, it is essential to determine the energy mix by considering all possible sources of power in an unbiased manner. While the personal transport sector has relied upon both SI and CI (diesel) engines, the heavy-duty transport in marine, construction, agriculture, and decentralized power sectors have betted on diesel engines for decades.

Higher fuel economy is the main reason behind a significantly higher share of diesel-powered vehicles in light-duty transports in Europe, India, and South Korea, which stands at 53%, 51%, and 30%, respectively (Bosch 2014). Electric vehicles (EVs) and hybrid-electrical vehicles (HEVs) have emerged as cleaner alternatives to IC engines in the passenger car segment. However, for worldwide market acceptance, EV and HEV technologies need to be economically comparable as well. The economy becomes more important when considering the public transport, agriculture, marine, and industrial sectors, specifically in developing countries. Despite the ongoing efforts for the globalization of e-mobility, several developed countries continue to rely heavily on diesel engines. As an example, 76% of commercial vehicles in the US are powered by diesel engines, which registered a 6% surge since last year with more than 11 million vehicles on the road (Fig. 2.1). Moreover, 84% of transit buses and 95% of school buses are not yet electrified (Markit 2020).

California Air Resource Board (CARB) stated that the introduction of biodiesel and renewable diesel fuels under the state's 'low carbon fuel standard program' has eliminated 25 M tons of CO_2. Since 2019, BEVs could achieve only 1/3rd reduction in CO_2 emission compared to diesel counterparts running on renewable fuels (Diesel-Powertrains 2021). The US DOE super-truck program promises to better the standard set by policymakers by as early as 2025. The team has already demonstrated >50% BTE, which is projected to be implemented largely by the 2030 time frame (Lutsey 2018).

These figures, predictions, and analyses are in no way advocating in favor of any particular technology. However, it is obvious that a significant amount of research effort for more than a century now benefits the IC engines in improving faster and meeting the expectations. Therefore, it is essential to focus on steady development and implementation of proven technologies before realizing the 'promised' zero-emission future. The journey and perspective should be very clear among engineers and researchers in the field for continuous development. This chapter makes an effort to summarize the technological evolution that has helped diesel engines adapt to the changes. In subsequent sections, conventional and advanced combustion technologies are discussed. The understanding behind the formation of major pollutant species is described in detail. Further, the strategical development based on the understanding of pollutant formation and losses is presented.

2.1.2 CDC and LTC

The fuel (typically higher cetane) is sprayed into the hot ambient air in the CDC, closer to the compression stroke TDC. The high velocity of the fuel spray and high

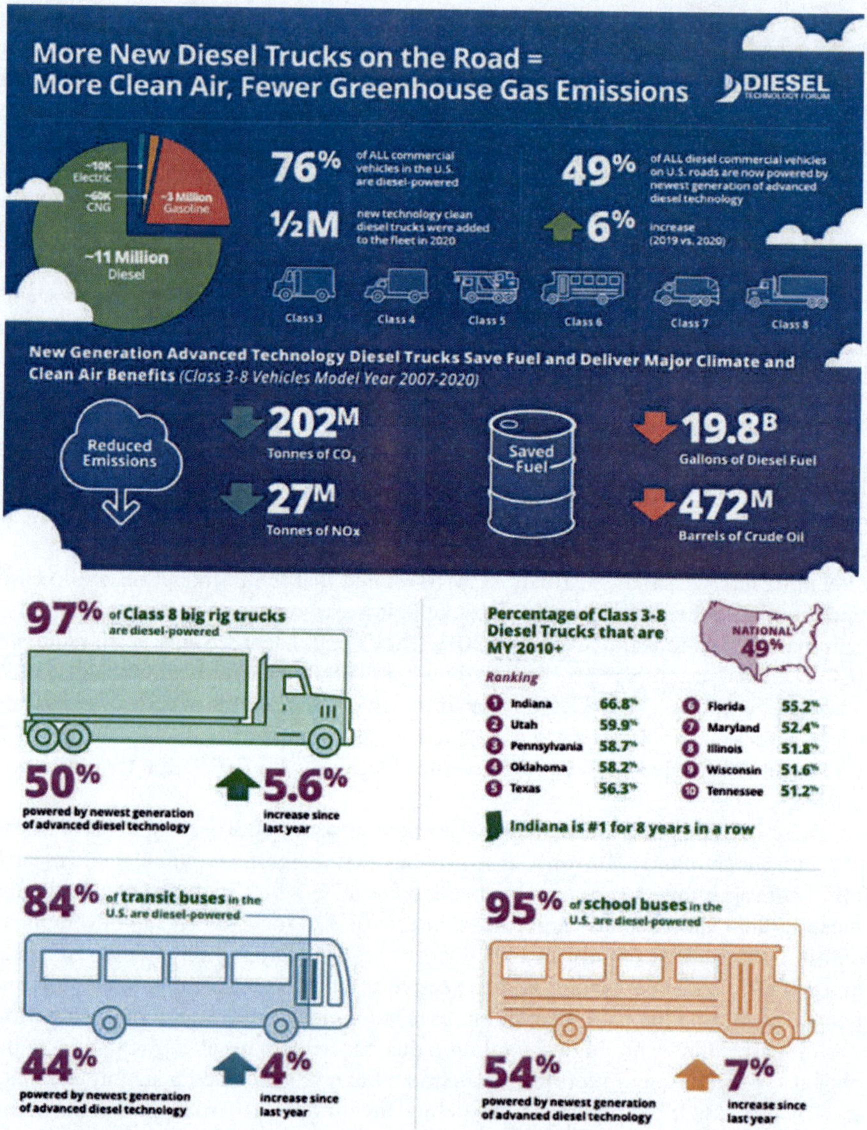

Fig. 2.1 Share of diesel engines in medium-duty and heavy-duty transport sector in the US (Diesel Technology Forum 2021)

ambient air density promotes the breakup of spray jets into fine droplets. The droplets then evaporate and mix with the ambient air to form a combustible mixture. The time elapsed between the fuel injection to ignition is termed 'ignition delay.' Therefore, it is evident that the spray jet momentum and its interaction with in-cylinder flow govern the ignition delay.

Furthermore, the charge distribution, injection timing, and ambient conditions are vital parameters controlling combustion. During the initial days of development, diesel engines had relatively lower fuel injection pressure (FIP). Hence, by the onset of ignition, only about 20% of the injected fuel mass formed a localized premixed charge. Rapid combustion of this premixed charge resulted in a higher rate of pressure rise (RoPR), giving rise to characteristic diesel noise common to conventional engines. This noise is still common to some old tractors. Over time, higher FIP accompanied by increased flow turbulence has provided better control on the combustion. The recent evolution of laser diagnostics for IC engines has provided significant insights into the complicated processes during the combustion in diesel engines. Several models have been proposed for the depiction of the CDC. The most common model developed by Dec (1997) for combusting diesel spray is applicable to free spray combustion, which is likely to happen in large-bore diesel engines. However, for modern-day high-speed direct injection (HSDI) engines, spray impingement on the piston/cylinder walls is inevitable. Figure 2.2 depicts the combustion of impinging spray inside the piston-bowl based on imaging studies conducted in IC engines and in the constant volume combustion chamber (CVCC).

The minimum flame luminosity near the wall resulted from incomplete combustion and higher heat losses in the vicinity of the cylinder liner. Spray then re-enters into the bowl, where it finds sufficient oxygen for burning. Soot oxidation results in high flame luminosity. The high soot emissions from the CDC mode are the result of significantly rich mixture pockets. Also, the higher adiabatic flame temperature results in a significant increase in NO_x emissions. This forms the basis for the low-temperature combustion (LTC) concept. In LTC, fuel is premixed to avoid rich pockets, and the ambient is highly diluted to restrict higher temperatures during

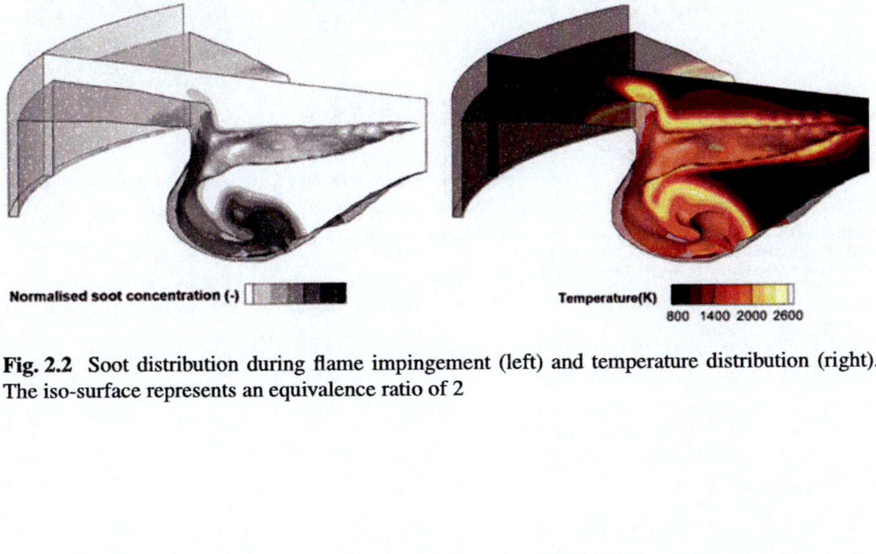

Fig. 2.2 Soot distribution during flame impingement (left) and temperature distribution (right). The iso-surface represents an equivalence ratio of 2

combustion. Premixing of the charge can be achieved by advancing the injection timing. For advanced injection timing, the in-cylinder temperature is lower which prolongs the ignition delay allowing the injected mass to form lean mixture before onset of ignition. The ignition delay can also be increased by decreasing the oxygen concentration (charge dilution by EGR) or decreasing the fuel reactivity. The conceptual model for LTC in large-bore engines has been shown in Fig. 2.3, which compares the CDC mode spray flames with the LTC mode spray flames in large-bore engines (Musculus et al. 2013).

For initial crank angles, the spray evolves identically for both modes, and the high-temperature ambient results in earlier ignition in the CDC mode. The fuel mass is continuously injected into the hot cloud of burning fuel, resulting in rich mixture

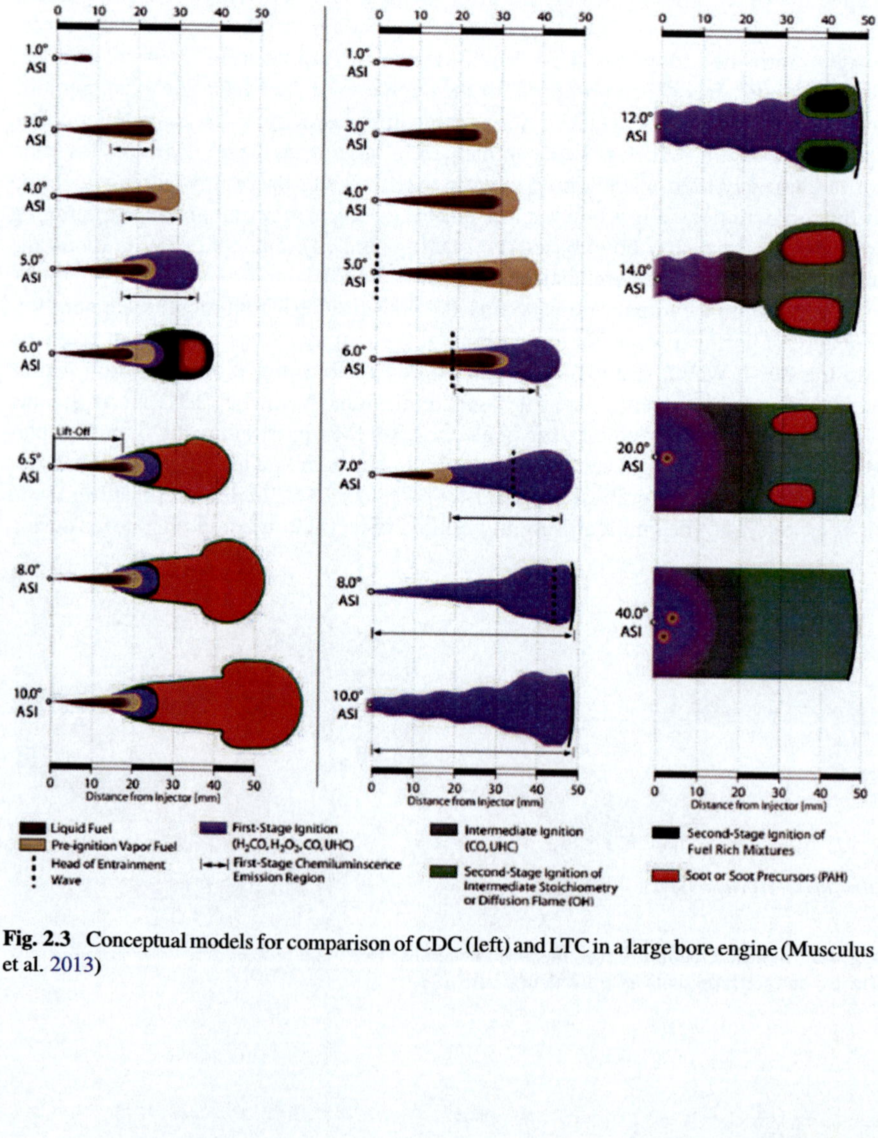

Fig. 2.3 Conceptual models for comparison of CDC (left) and LTC in a large bore engine (Musculus et al. 2013)

combustion and a significantly higher soot level. Unlike CDC, in the LTC mode, the first stage ignition and the second stage ignition are decoupled. The first stage ignition results in formaldehyde formation (blue region), which is consumed in the second stage (green region). The premixed charge reaches the bowl periphery, and small soot pockets are formed after the spray impingement, which disappears in the later crank angles, as seen in Fig. 2.3.

2.2 Practical Limit of the Efficiency

Diesel engines are regarded as heat engines. Hence it is quite obvious to think that the efficiency limits must be governed by Carnot efficiency. However, there are certain differences between the two. Carnot engine is a conceptual engine that is far from being a practical device. In the Carnot cycle, energy is added to the system as heat through temperature differences in an isothermal process. The process is infinitely slow, which is impractical in real engines. In a diesel engine, energy is transferred into the combustion chamber by mass transfer (fuel injection), incurring certain irreversibility. However, unlike the Carnot engine, the diesel engine works on an open cycle with no constraint of heat rejection to the sink. Theoretically, the efficiency of the diesel engine can reach close to 100% (Edwards et al. 2011). Despite the higher combustion efficiency of diesel (~98%), the maximum efficiency of the engine is limited to 60% (reported for heavy-duty marine engines). The immediate cause of this reduction can be attributed to the exergy efficiency of the system. The combustion irreversibility is the prime cause of loss of efficiency by availability analysis (Rakopoulos and Giakoumis 2006; Edwards et al. 2010). Here, one must appreciate the difference between combustion efficiency and exergy efficiency. Combustion efficiency is defined as the ratio of energy released at the end of combustion to that of the chemical energy content of the fuel mass under consideration.

On the other hand, exergy efficiency talks about the work potential of the energy released. For example, 10 mg fuel undergoing complete combustion under free atmosphere and inside the combustion chamber at high pressure–temperature ambient has different work potentials. The combustion irreversibilities account for nearly 25% of exergy destruction. Heat transfer to the surfaces evolved as the second major factor contributing 17% to the net exergy destruction. Hence, it is important to understand the constraints that limit the efficiency of a practical engine.

2.2.1 *Constraints for Optimisation*

Theoretically, efficiency can be increased by increasing the compression ratio hence the peak cylinder pressure and temperature. However, certain process constraints

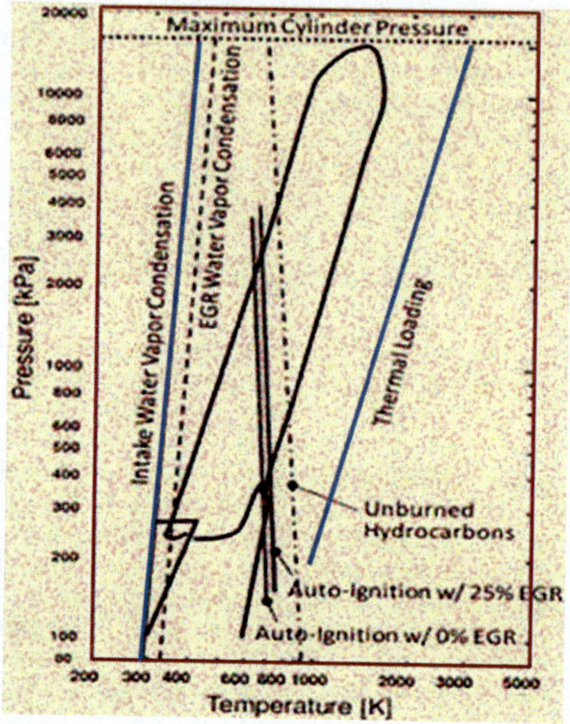

Fig. 2.4 Factors limiting the practical efficiency of the diesel engine

should be taken care of for the engine's unhindered operation in real-world applications. Figure 2.4 depicts the typical P versus T curve for a diesel engine with various constraints limiting the operational regime.

Peak Cylinder Pressure and Pressure Rise Rate: The maximum allowable pressure in the cylinder is limited by the material limits (dashed horizontal line in Fig. 2.4). The position of the heat release governs the peak cylinder pressure during the combustion event. The centroid of the heat release rate closer to the TDC with minimum area results in a higher degree of constant volume combustion. The degree of constant volume combustion (dCVC) is given by Bergstrand and Denbratt (2001)

$$\mathrm{dCVC} = \frac{1}{\eta_{otto} Q_{hr}} \int \left(1 - \left(\frac{V_h + V_c}{V(\theta)} \right)^{1-\gamma} \frac{dQ_{hr}}{d\theta} \right) d\theta$$

where V_h is the swept volume, Vc is the clearance volume, V (θ) is the instantaneous volume, η_{otto} is the otto cycle efficiency, and Q_{hr} is the total heat release.

A sharp heat release rate translates to a higher RoPR. RoPR is a determinate factor for the dynamic load on the engine connecting rod and crankshaft. It should be limited to 10 bar/deg for acceptable cylinder noise and ensuring component durability (Curran et al. 2013).

Thermal Loading: Thermal loading resulting from service temperatures leads to undue thermal stresses in the engine components. Piston crown, valve seats, and port junctions are most susceptible to thermal failure. The peak cylinder temperature should be limited to the left of the 'thermal loading line' for all operating conditions.

Vapor Condensation: Cooled EGR has been standard for most modern-day diesel engines. This brings an additional issue of vapor condensation into the picture. Lower intake temperature is desired for higher air density and hence higher specific power. However, EGR puts further constraints on the minimum charge temperature, as seen in Fig. 2.4. The introduction of the EGR circuit also puts an extra burden on the turbocharger. Before introducing EGR (before 2002), the engine intake was maintained at a steady boost of 0.5 to 1 bar. The unconstrained turbocharger operation helps in reducing back pressure on the engine. The introduction of EGR in 2002 resulted in a sudden drop in engine efficiency (Stanton 2013).

2.2.2 *Heat Loss*

The optimal HRR peak is instrumental in determining engine efficiency. Ideally, the timing hovers around 7°–9° aTDC for all diesel engines, irrespective of their design (Stanton 2009; Charlton 2005). Therefore, advancing the centroid of HRR would theoretically increase engine efficiency. However, in-cylinder heat losses increase as the centroid is advanced towards the TDC. Therefore, the heat release timing is a trade-off between the effective expansion ratio and the heat loss.

Several factors govern the wall heat transfer in diesel engines. In light and medium-duty engines, wall impingement of flames contributes to a significant heat loss (Jena 2020). Therefore, providing thermal insulation on the critical component (e.g., piston cavity for diesel engines) has emerged as an important area of research. This led to the concept of 'Low Heat Rejection engines' in the 1980s. TACOM and Cummins Inc. conceptualized the novel engine concept, known as an adiabatic engine, in 1975 (Bryzik and Kamo 1983). This engine design used ceramic coating to reduce the heat transfer through the in-cylinder surfaces. However, the outcomes of these studies were somewhat mixed. Several issues related to lubrication and side-wall thrust discouraged further developing and commercializing these engines (Jaichandar and Tamilporai 2003). These drawbacks encouraged thinner coatings called 'Thermal Barrier coating.' Recently, Toyota came up with a solution to match the surface temperature with local gas temperature to reduce thermal gradient and heat losses. The Japanese auto giant developed 'Temperature swing insulation coating (TSWIN)' to achieve higher engine efficiency through reduced heat losses (Kosaka et al. 2013). In 2016, Toyota introduced a new material called 'Silica Reinforced Porous Anodized Aluminium' (SiRPA) for improved heat loss characteristics (Kogo et al. 2016). In the current state of the art, low thermal conductivity and low heat capacity materials have been explored to instantly attain and follow the gas temperature to minimize thermal gradients. In addition to the wall surface temperature, the wall surface area exposed to high-temperature gases is also critical from a heat loss perspective. Indeed, the low surface-to-volume ratio of large-bore engines gives lower heat loss and higher efficiency (Edwards et al. 2011; Reitz and Duraisamy 2015). Several other factors, such as FIP, fuel injection strategy, and in-cylinder flow, affect the engine's heat transfer characteristics.

2.3 Mechanisms of Pollutant Formation

Stringent emission norms have inspired several after-treatment solutions for cleaning the tailpipe emissions of diesel engines. However, the efficiency of these systems relies upon the mode of engine operation and exhaust gas condition (Sonawane and Agarwal 2021). In-situ pollutant control is an attractive way to reduce the burden on these devices. Therefore, the formation of these species needs to be understood. Soot and NO_x are the primary pollutant species from diesel engines. However, advanced combustion technologies have increased hydrocarbon emission concerns. NO_x formation is not directly related to the fuel conversion efficiency, but it depends on the peak in-cylinder temperature. The optical diagnostics of the pollutant species is difficult. The formation of NO_x could be explained and predicted by the Zeldovich mechanism (Ya 1946). This section explores soot and unburned hydrocarbons (UHC) formation using optical diagnostics and numerical investigations.

2.3.1 Soot Formation

Conventional diesel combustion is characterized by two phases of heat release after the main injection. The first phase of heat release consumes the premixed charge prepared during the delay period. The second phase is mixing controlled diffusion combustion, and the fuel injection rate governs that. The fuel is diffused from the high concentration region to the outer spray region, where it gets oxidized. Due to the limited time, soot formation takes place in the fuel-rich envelope. Soot formation is higher during the diffusion combustion phase (Pickett and Siebers 2004a). The combustion becomes predominantly mixing controlled (diffusion) at higher loads (because of higher injected fuel mass), leading to a significant increase in soot mass in the exhaust. One immediate thought is to shift the heat release to the premixing side; however, premixed combustion is uncontrolled and depends on several operating parameters, thus restricting the operational range (Dec 2003). Therefore, diffusion combustion is inevitable at higher loads. The final soot mass of the cycle is believed to be governed by soot inception and soot oxidation. Soot inception takes place downstream of the nozzle in the fuel-rich vapor core. A higher concentration of aliphatic and aromatics is the main reason for soot inception (Velásquez et al. 2013).

Naphthalene (A2) has been identified as a dominating contributor for soot inception and growth through the condensation of hydrocarbons, as shown in Fig. 2.5 (Attili et al. 2014; Zhou et al. 2014; Mao et al. 2017).

Acetylene and H contribute to the growth by the HACA mechanism (Frenklach and Wang 1991). OH radicals help in the oxidation of the soot. Figure 2.6 shows the concentration distribution across a turbulent non-premixed flame (Attili et al. 2014). The x-axis represents the mixture fraction (Z), while the normalised concentration of the species are shown along the y-axis. The peak of O_2 and OH radicals are present in the range of $Z < 0.25$, where the soot oxidation probability is higher (Attili et al. 2014).

Acetylene and A2 are distributed towards a higher mixture fraction, suggesting soot inception in those regions. Typically, the size of incepted soot particles lies in the range of 1–6 nm, while the soot mass density of ~1 g/cm^3 can be assumed for

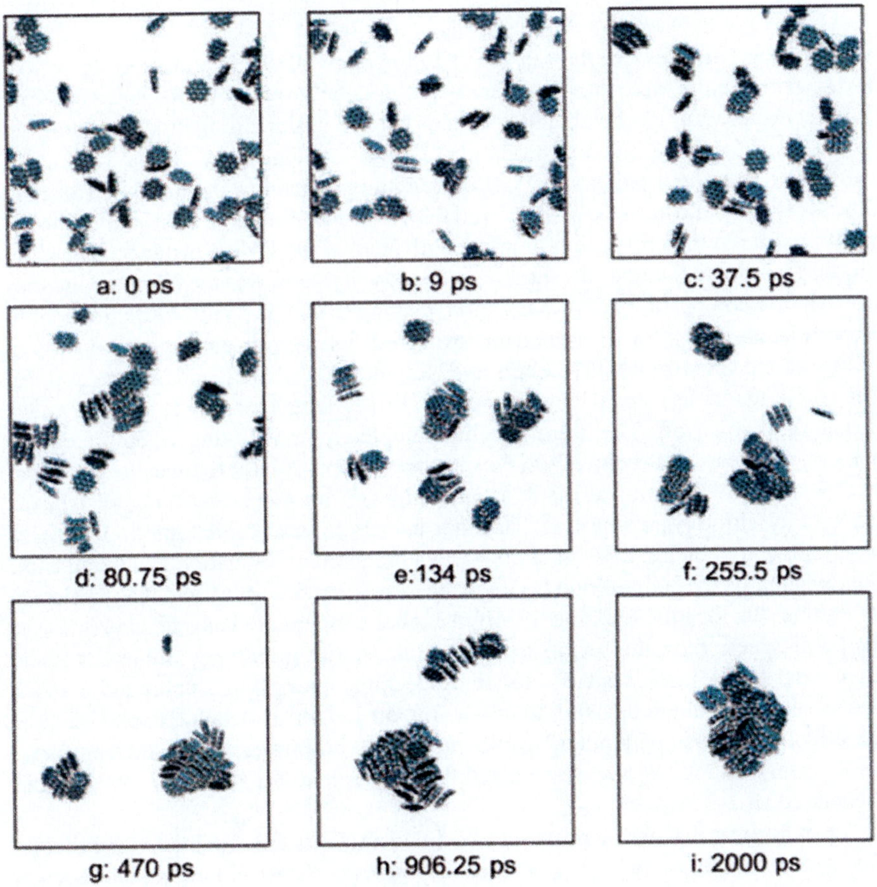

Fig. 2.5 Evolution of soot particle (Mao et al. 2017)

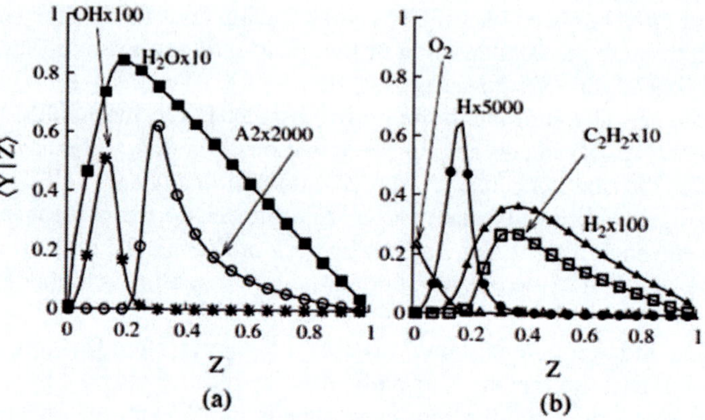

Fig. 2.6 Distribution of species concentration across turbulent premixed flame (Attili et al. 2014)

all modeling purposes (Michelsen 2017; Totton et al. 2010). The number density of PAHs, temperature, size are vital parameters for the growth of soot particles. The local concentration of the species depends on the scalar dissipation rate and the turbulence. Therefore, the air-entrainment and lift-off length are crucial factors for the formation of soot precursors. Lift-off length is defined as the distance from the nozzle orifice to the location of stabilized flame during combustion. A smaller lift-off length results in the presence of liquid length beyond the lift-off distance. Thus, the mixture fraction (Z) across the jet is significantly higher due to the complex interactions during combustion and vaporization, leading to a higher rate of soot inception. Experimental studies have revealed the strong link between air-entrainment upstream of lift-off and soot formation (Siebers and Higgins 2001; Pickett and Siebers 2004b). Air-entrainment increases with an increase in lift-off length leading to a lower equivalence ratio at lift-off. The flame became completely non-sooting when the equivalence ratio decreased below 2.0 (Pickett and Siebers 2004a). However, the flame was observed to be non-sooting at a much higher 'cross-section averaged' equivalence ratio with oxygenated fuel. The soot inception was avoided due to the lower jet mixing temperature. Several factors affect the flame stabilization in a combustor. Fundamentally, the competition between the advancing fuel spray and flame velocity determines the location of flame lift-off. A higher temperature leads to an increase in the flame speed; thus, the lift-off length decreases with increasing ambient temperature. Higher FIP increases the velocity of the spray, pushing the stabilization point downstream, thereby increasing the flame lift-off length. A detailed explanation of the effect of various parameters on lift-off length and limits of non-sooting flame can be found in Pickett and Siebers 2004a; Siebers and Higgins 2001; Pickett and Siebers 2004b).

The preceding discussion provides a useful insight into the soot inception process. However, in modern-day high-speed diesel engines, jet-wall and jet–jet interactions further complicate the soot formation process. Under these conditions, late-cycle soot oxidation governs the final engine-out soot emission. In-cylinder flows play a defining role in turbulence generation and species transport in the engine. The flow fluctuations may lead to cyclic variation in the lift-off length location, leading to varying soot mass for different engine cycles (Xuan et al. 2019). An experimental study conducted in the heavy-duty diesel engine revealed that increased inter plume spacing reduced the cyclic fluctuations of lift-off length (Lequien et al. 2015). The lift off-length was increased with spacing between the plumes. However, the in-cylinder swirl showed a minimal effect. Plume-to-plume interactions and swirl-plume interactions are more complex in light-duty small-bore engines due to smaller space and higher swirl operations. The spray plumes spread on the piston-bowl walls after impingement and form fuel-rich regions. Soot inception in this region results in forming immature soot particles that agglomerate and grow as the spray travels along with the bowl. Further downstream, continual surface growth and increased oxidation restrict the overall size of the soot particles (Zhang et al. 2017). For increased injected mass at higher loads, the spreading plumes interact and form overlapped regions of a higher equivalence ratio. The soot formed from the overlap region shows a higher growth rate and persists for a longer duration due to limited interaction with OH than a single plume/no-overlap case (Le et al. 2017). The spreading of the impinged spray plumes is affected by the direction of the swirl. In the up-swirl direction, the mixing rate increases due to interactions between the spreading lamina of spray and the wall-flow of the swirl. The increased turbulence enhances the oxidation of soot in this location than the down-swirl direction resulting in compact soot particles (Zhang et al. 2019) as shown in Fig. 2.7.

2.3.2 CO and UHC Formation

The HC and CO emissions typically form due to combustion inefficiency. These emissions are not too much problematic for CDC mode, though. UHC emissions are anyway insignificant compared to other LTC modes. Therefore, most research studies related to HC and CO emissions have been carried out for the LTC mode of operation. Diesel is a complex mixture of several hydrocarbon species, including saturated hydrocarbons and aromatics. A large fraction of this mixture consists of straight-chain alkanes. n-Heptane (also a straight-chain alkane with seven carbon atoms) shows ignition properties similar to diesel. Therefore, understanding the chemical kinetic of n-heptane combustion helps understand diesel combustion. Figure 2.8 shows the numerical simulation results of n-heptane combustion in a well-stirred reactor (Curran et al. 1998).

The initial condition of simulations was similar to LTC under constant pressure conditions. These results form a basis for interpreting engine-out emissions and in-cylinder imaging results despite significantly different conditions and in-cylinder dynamics. The simulation was initialized at 787 K. It was observed from Fig. 2.8 that before 1st stage ignition, ketohydroperoxide (KHP) concentration increased rapidly to attain a peak value. KHP is formed from the isomerized peroxy radicals (RO_2). Further decomposition of RO_2 and KHP yields OH, and this reaction accelerates

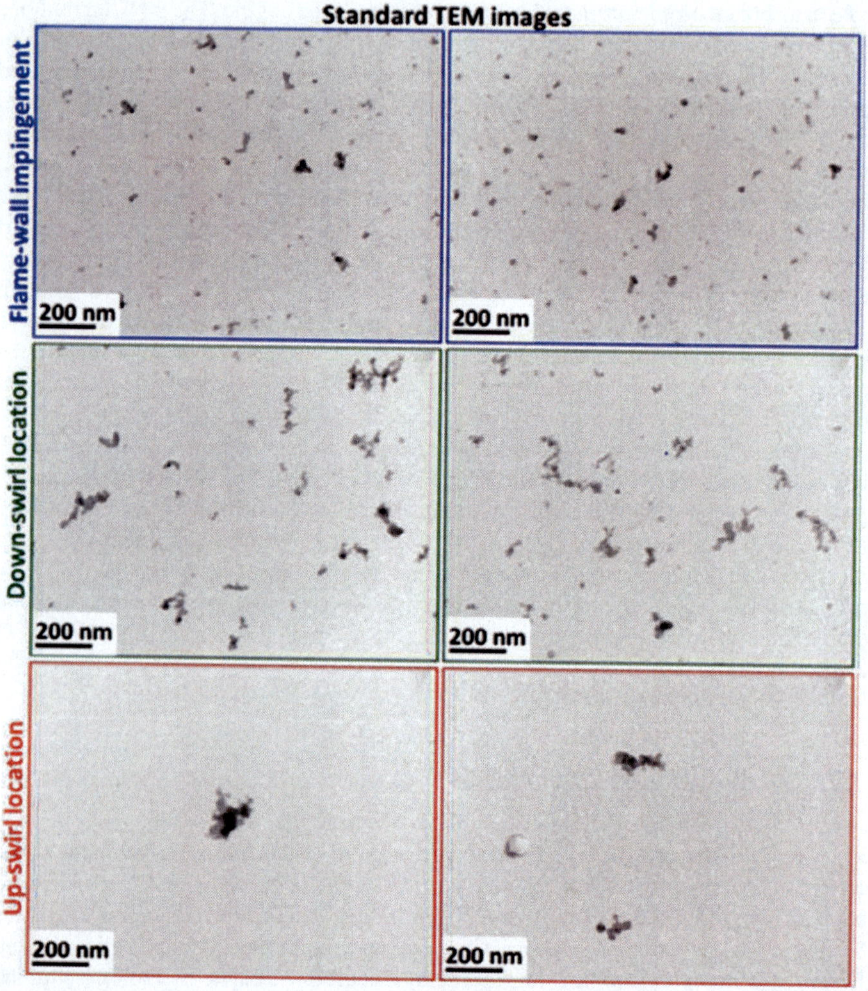

Fig. 2.7 Morphology of soot particles probed at different locations in swirl supported light-duty engine (Zhang et al. 2019)

the onset of ignition. The rise in temperature results in a sharp reduction in KHP. The reactions slow down, and the 1st stage ignition concludes. KHP decomposes to formaldehyde (CH_2O). As can be observed from the figure, the concentration variation of CH_2O is similar to UHC; therefore, it is used as a UHC marker in engines and fundamental studies. Formaldehyde then further oxidizes to hydrogen peroxide (H_2O_2). Above 1000 K, H_2O_2 decomposes to OH radicals, which again accelerates the reaction rate for the onset of second stage ignition (high-temperature heat release). It is evident from these results that as OH radicals peak, UHC and CO are completely oxidized.

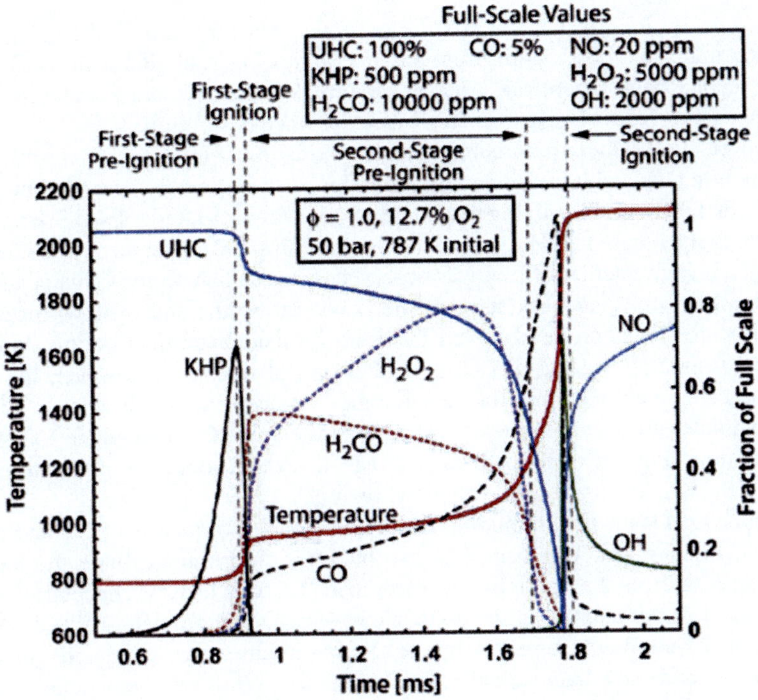

Fig. 2.8 Chemical kinetics of two-stage ignition of n-heptane in diesel-like combustion conditions (Curran et al. 1998)

Further simulations were carried out at lean and rich mixture conditions. The higher fuel mass in a richer mixture result in a lower adiabatic temperature due to increased latent heat absorption resulting from evaporation. The low temperature significantly affects the ignition events. Thus, the first stage ignition occurs earlier in the leaner mixture relative to the richer mixture, which is counter-intuitive. However, the richer mixture attains higher temperature after the first-stage ignition, resulting in a faster depletion of H_2O_2 and a shorter dwell before the second stage.

The above discussion highlights the importance of temperature, equivalence ratio, and interdependency in the combustion chamber. During real engine operation, temperature distribution depends upon several factors. Temperature rise due to adiabatic compression is one key factor. On the other hand, charge stratification (equivalence ratio distribution) depends on the injection and flow parameters. For early injection conditions, the mixture gets sufficient time under high-temperature ambient due to compression heating. Therefore, a slightly leaner mixture can also enter the second stage of ignition.

On the other hand, for late injection, the ignition shifts to the expansion stroke. The expansion cooling restricts the second stage of ignition, leading to UHC emissions. Deteriorated ambient conditions, further increase the mixing time due to a longer ignition delay. As a result, the mixture becomes over lean and remains unburned. Another important observation from the kinetic model simulation is that CO forms in the second stage. Therefore, a leaner mixture contributes more towards the UHC emissions as it is less likely to enter the second stage of ignition.

Conversely, a richer mixture may promote oxidation of UHC to CO, resulting in a reduction in UHC with increasing CO emission. A significantly advanced injection may further increase the mixing time. The mixture strength attains $\phi < 0.3$, leading to a substantial increase in UHC (Lachaux and Musculus 2007). For advanced injection timing, a higher fraction of fuel was observed in the squish region. This penetrated fuel fraction is pushed back towards the bowl rim by the squish flow, forming a lean mixture. This can be observed from the formaldehyde distribution shown in Fig. 2.9 (top, SoI: −23.1° aTDC, 5.9° after end of injection).However, the mass fraction penetrated deep into the squish region attains near stoichiometric mixture and completes the second stage of combustion as evident from the plane 1 shown in Fig. 2.9 (top). For late injection (SoI: 1.2° aTDC, 11.25° after start of injection), the formaldehyde is confined inside the bowl region.

Engine load is the next important factor affecting the formation of CO and UHC. Figure 2.10 shows the UHC and CO distribution in the engine cylinder for various loads. As seen from the figure for the lower load (1.5 bar), UHC is concentrated near the cylinder center and the squish zone. However, CO predominantly appeared near the cylinder wall. As discussed earlier, the leaner mixture near the injector could not enter the second stage and resulted in UHC formation. While relatively richer mixture in squish undergoes second stage combustion, forming CO. With an increase in the engine load, both temperature and equivalence ratio in the squish zone promote UHC oxidation. As a result, UHC started decreasing in the squish region. The increased temperature also helped in the oxidation of CO. As the load increased, CO started to build up towards the center of the cylinder. The near injector UHC gets partially oxidized at an intermediate load, adding to CO. At a higher load, the CO also gets

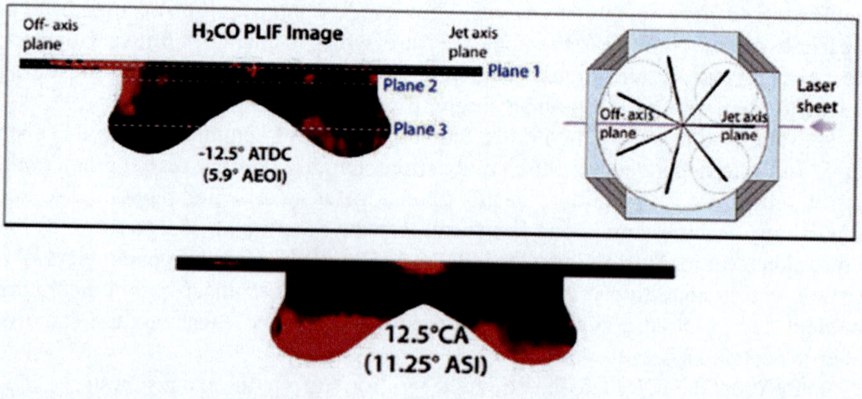

Fig. 2.9 Formaldehyde distribution in the diesel engine for early (top) and late injections under LTC condition (Musculus et al. 2013)

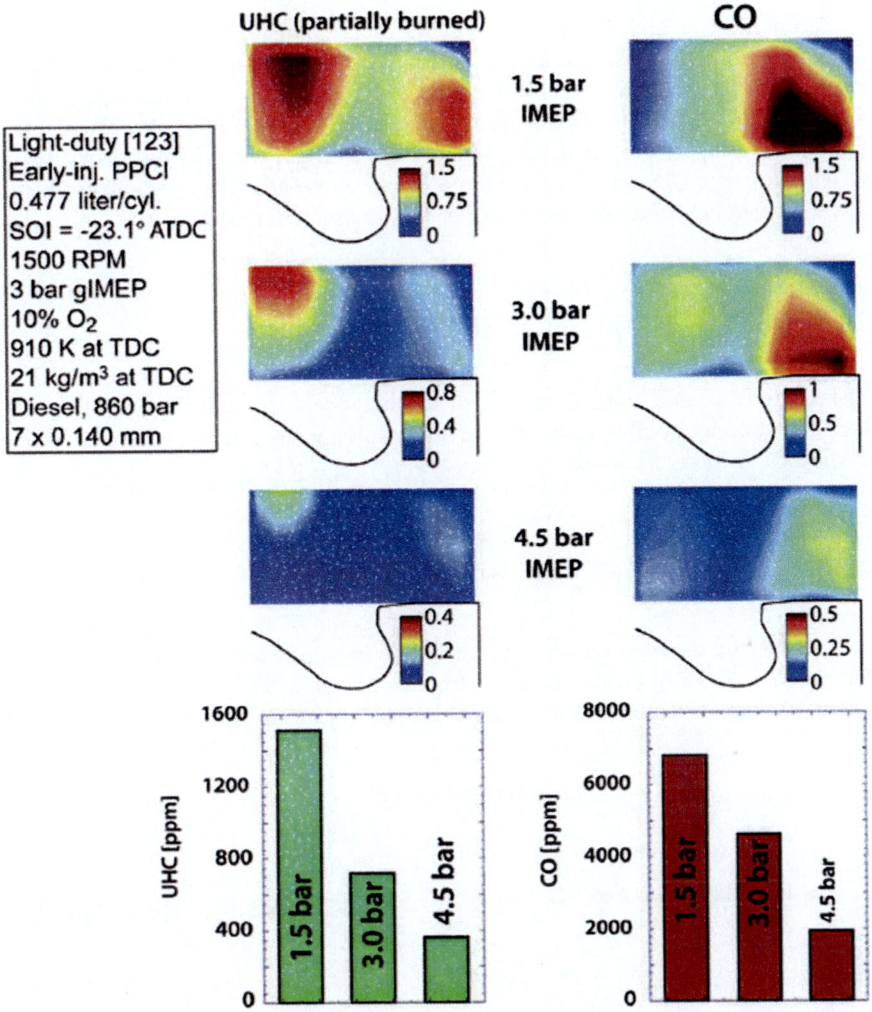

Fig. 2.10 In-cylinder distribution of HC and CO for different engine loads (Musculus et al. 2013)

oxidized due to increased temperature. Therefore, both UHC and CO diminish at a higher load.

Near injector fuel introduced due to the injector dribbling is a major contributor to the UHC formation near the cylinder center. This trend has been observed in CDC as well (Greeves et al. 1977). In CDC, the combustion is predominantly diffusion combustion and fuel-rich. The CO resulted from the rich combustion gets oxidized in the post-combustion phase, leading to lower CO emission. However, in LTC, the premixed charge further lean-out during the post-combustion phase in low load conditions, leading to higher CO emission.

Furthermore, at higher loads, the cylinder wall impingement and slower evaporation of the liquid fuel film are the major sources of UHC and CO emissions (Kanda et al. 2005; Kashdan et al. 2007; Hotta et al. 2005). Charge dilution also plays a vital role in the emission of both species. Increased charge dilution lowers the peak in-cylinder temperature, increasing the mixture strength required for the oxidation UHC. Similarly, CO oxidation would be suppressed due to the combined effect of lower oxygen concentration and lower peak in-cylinder temperatures.

2.4 Strategic Evolution of CDC

The previous section highlighted the background chemistry and physics behind the formation of major pollutant species. In this section, various aspects of the evolution of the CDC will be discussed.

2.4.1 Injection Strategies

In-cylinder fuel distribution plays a central role in engine combustion and pollutant formation. FIP, SoI, injection dwell control the fuel–air mixing in a DI diesel engine. These parameters taken together determine the injection rate shape. Figure 2.11

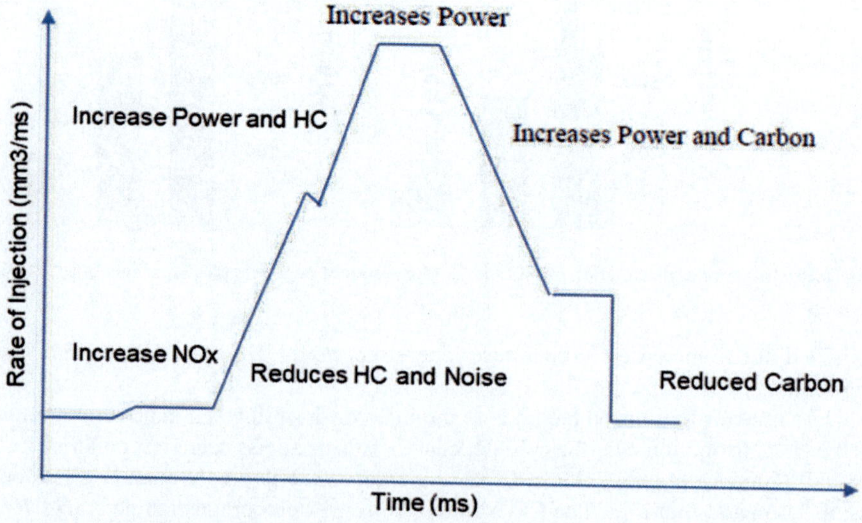

Fig. 2.11 Stages of injection rate shape in diesel engine

depicts typical injection rate shape features in a DI diesel engine with special relevance to pollutant formation, noise, and performance. Optimal values of the injection parameters, rate shape, and placement are known as 'injection strategy.'

Since their invention, mechanical fuel injection systems were an integral part of diesel engine architecture. Late in the twentieth century, environmental whistleblowers forced the development and introduction of electronic fuel injection (EFI) systems in the ICEs. In the late 1980s, the first-generation electronic control system was introduced. The first generation EFI system had a limited number of sensors and actuators. The system worked on an open-loop control architecture. With the growing demand for improved performance, fuel economy, and emission reduction, second-generation systems evolved with limited closed-loop features. The modern-day third-generation control systems use an increased number of sensors and actuators to precisely implement optimized injection strategies. These systems incorporate several physical models (called virtual sensors) for efficient performance and faster response under different conditions. However, several light-duty agricultural equipment to heavy-duty earthmoving equipment still rely on mechanical fuel injection systems due to their simplicity and low cost. In the following sub-section, the impacts of fuel injection strategy on noise, performance, and emissions are discussed.

2.4.1.1 In-Cylinder Noise Reduction

Fuel injection systems have evolved with diesel engines since their invention. The first significant development was to enable multiple injections for noise reduction. Injection rate shaping was instrumental in achieving smooth engine operation. Pilot injection provided control over the shape of the heat release curve, leading to a smooth pressure rise rate and a significant reduction in noise at part load conditions (Rinolfi 1995; Imarisio and Rossi Sebastiano 2001). The pilot-main injection strategy has been used for noise reduction since the early 90s. In-cylinder noise control continues to be a hot area of research even today (Busch et al. 2014; Denny et al. 2020). However, post the 90s, the injection strategy has been optimized primarily to reduce engine-out emissions. Optimization of injection strategy can successfully reduce the engine noise without any penalty in emissions. Destructive interference between the pressure signatures of the pilot and the main injections is the main reason for combustion noise reduction. RoPR, combustion phasing, etc., are not always correct indicators for the combustion noise (Busch et al. 2016). Advanced acoustic analysis is required for sustainable improvement in noise. Noise, vibration, and harshness (NVH) management are the keys to commercialize advanced combustion technologies. Linearity of engine noise with engine speed is of utmost importance. Ford's 6.7 l turbocharged engine claims to successfully reduce NVH by twin pilot injection strategy, optimal fuel pump placement, and improved manifold design (Deraad et al. 2010). Indeed, there are several variables other than combustion parameters that affect the NVH.

2.4.1.2 Performance and Emissions

Injection rate shaping is crucial for engine performance and emissions. A slower rate shape expands the combustion duration and allows more time for soot oxidation; however, the reduction in dCVC results in reduced engine efficiency (Nehmer and Reitz 1994). Injection rate shaping improves post-development of the common rail direct injection system significantly. Centro Ricerche Fiat and Fiat Auto led the technological development in Turin in the early 90s. Alfa Romeo 156 JTD and the Mercedes-Benz 220 CDI were the first vehicles to feature advanced injection systems in 1997, designed by Bosch in association with Fiat. In those days, the injection pressure of the CR system was limited to 1350 bar only.

Only two injections per engine cycle were possible with the existing system due to the constraints of minimum dwell (1.8 ms) between successive injections (Ganser 2000; Flaig et al. 1999). The modern piezo-injectors can achieve a faster rate shape (square rate shape) than its solenoid counterpart and can deliver up to eight injections in a single-engine cycle. Split armature modules have achieved near-zero injection dwell, contributing significantly to the injection rate shaping. The piezo-injectors have a minimum dwell of 0.1 ms compared to 0.4 ms in solenoid injectors. The dwell between successive injections is crucial in the optimization of noise and pollution. Higher injection pressure increases the chance of injector dribbling, leading to deteriorated engine performance and emissions.

Cummins XPI fuel system operates at over 2400 bar FIP with 5 injection events in an engine cycle. The system claims to have eliminated injector leakage by developing a centralized pressurization cavity in the XPI injector (Stanton 2013). The development of advanced fuel injection systems has given flexibility for innovative injection strategies. The pilot-main injection strategy came into existence for mitigating the characteristic diesel noise. However, the strategy showed great potential in managing well-known NO_x-soot trade-off of diesel engines. Faster rate shapes have resulted in increased efficiency and power with a penalty in NO_x emissions and noise. The higher pressure levels boosted the torque level by 20–30%, improving the drivability of engines, which is comparable to gasoline engines (Stanton 2013). The flexibility of multiple injections gave freehand for the optimization. This resulted in significant attention in split injection strategies for emission reduction. Numerous researchers have explored the split injection strategy in the last two decades. Initial studies focused on equally splitting (50:50) the fuel mass/stroke over the pilot and main injections (Tow et al. 1994; Minami et al. 1995; Pierpont et al. 1995). A higher amount of fuel injected in the first injection improves the engine-out soot emission (Zhang et al. 2001; Zhang and Nishida 2003; Shayler and Ng 2004). Spray-induced turbulence of the first injection affects the spray penetration and mixing characteristics of the fuel injected in the second injection. The spray momentum and hence the injected mass determines the intensity of turbulence and vorticity. The residual flow field of the first injection dissipates quickly. Therefore, dwell time between the closely coupled injections is key for the outcomes along with split ratio (Parrish et al. 2012). Optical investigations have shown that the interaction of the flames of the first

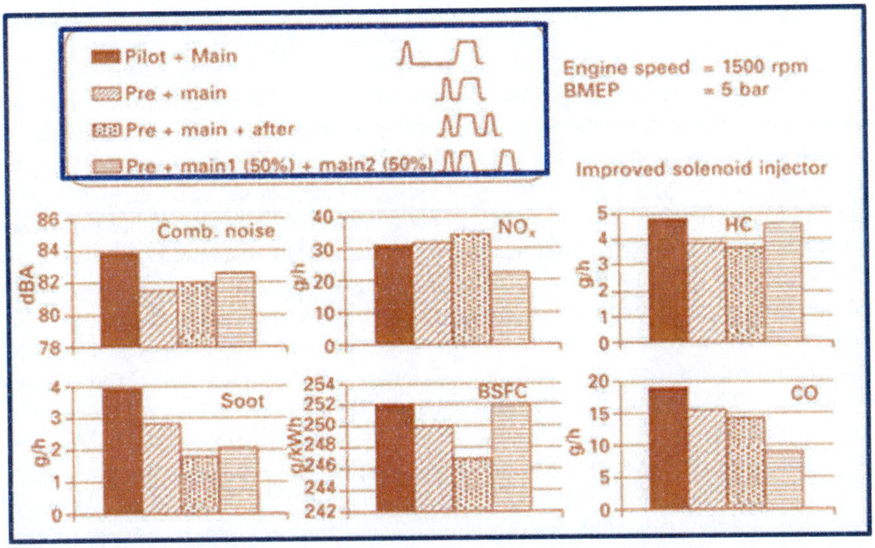

Fig. 2.12 Typical results show the impact of various injection strategies on the engine performance, emissions, and noise (Imarisio and Rossi Sebastiano 2001)

injection with the fuel injected in the second injection may lead to higher soot mass (Herfatmanesh and Zhao 2013; Yang et al. 2018; Denny et al. 2019) (Fig. 2.12).

Optimization of injection dwell, injection timing with ultra-high injection pressure are the current trends in the design of advanced diesel engines. Multiple injection strategies and injection rate shaping are instrumental in the transition of CDC mode to LTC mode. While the timing and pressure optimizations do not add to the vehicle cost, high-pressure fuel injection equipment constitutes a big share of the total vehicle cost. A detailed analysis of the evolution of multiple injection strategies, fuel injection equipment, and emission trends has been provided by Vaglieco (2010).

2.4.2 *Swirl and Intake Geometry*

In-cylinder flow dynamics is an important factor for charge motion and heat transfer characteristics of the engine. The intake geometry design determines the evolution of flow structures in diesel engines.

2.4.2.1 Variable Swirl Control

Most modern-day light-duty diesel engines are swirl supported. Swirl significantly reduces the spray-wall impingement in small-bore engines, which is otherwise inevitable (Jena, 2020; Sahoo et al. 2012; Hotta et al. 2002). However, most engine designs are characterized by a fixed swirl due to scarce understanding of the in-cylinder flow and its impact on combustion. The rate of mixing and heat transfer are largely dependent on the swirl ratio of the engine. Therefore, it is essential to tailor the swirl according to the engine's combustion mode and operating regime (load-speed). The typical swirl ratio of a diesel engine lies in the range of 1–5. The in-cylinder swirl scales with the engine speed under the assumption that the intake runner and manifold wave dynamics are insignificant. The reader may refer to Miles (2009) for deep learning on flow scaling of IC engines. The two intake valve configuration is a common feature these days. Out of the two, one port is designed so that the flow enters tangentially into the cylinder (called a tangential port). The tangential jet strikes the cylinder wall and generates a large-scale flow structure of the order of engine bore. The other port is shaped like a helix (helical port or fill port) to generate local vortical structures. The helical port typically has a higher flow coefficient. One of the ports is throttled to achieve higher momentum through the other to achieve variable swirl operation. Restricting the flow through the helical port leads to a stronger momentum of the tangential jet. Therefore, the angular momentum associated with the large-scale structure increases, contributing to a higher global swirl. The upper range of swirl falls in the range of 3–5 with the helical port throttle condition. Similarly, the swirl can be decreased by throttling the tangential port. Figure 2.13 depicts the swirl control valves (often called the swirl flap) and the flow velocity distribution in the ports.

The flow restrictions do not have a significant effect on the volumetric efficiency at a lower speed. Higher swirls are preferred during low-speed operation to achieve improved soot oxidation (Cursente et al. 2009; Chi et al. 2008). However, the filling efficiency starts to fall beyond a certain level of throttling. Recent head designs have used chamfer to achieve a higher swirl at lower valve lift (Eidenböck et al. 2012; Bauder et al. 2010; Lee et al. 2009). The combined effect of pumping losses, heat transfer losses, and improved combustion determines the engine's thermal efficiency (Olmeda et al. 2017; Li et al. 2016). Therefore, the swirl can be optimized to achieve a simultaneous reduction of emissions with improved engine performance across the entire operating range (Fig. 2.14).

Engines with injectors having an increased number of holes can be optimized at significantly lower swirl levels (Fasolo et al. 2005; Heuvel et al. 2006; Kurtz and Styron 2012; Akihama et al. 2009). However, jet–jet interactions become crucial as the hole number increases.

2.4.2.2 Variable Valve Actuation Mechanism

Variable Valve Actuation (VVA) helps optimize the valve opening and closing time and the valve lift, similar to the variable swirl system. However, the concept is still new for the automotive industry and has garnered a lot of research interest (Srivastava et al. 2018). Most on-road vehicles use fixed valve-lift profiles for intake and exhaust valve operations. The cam-driven valve mechanism includes several

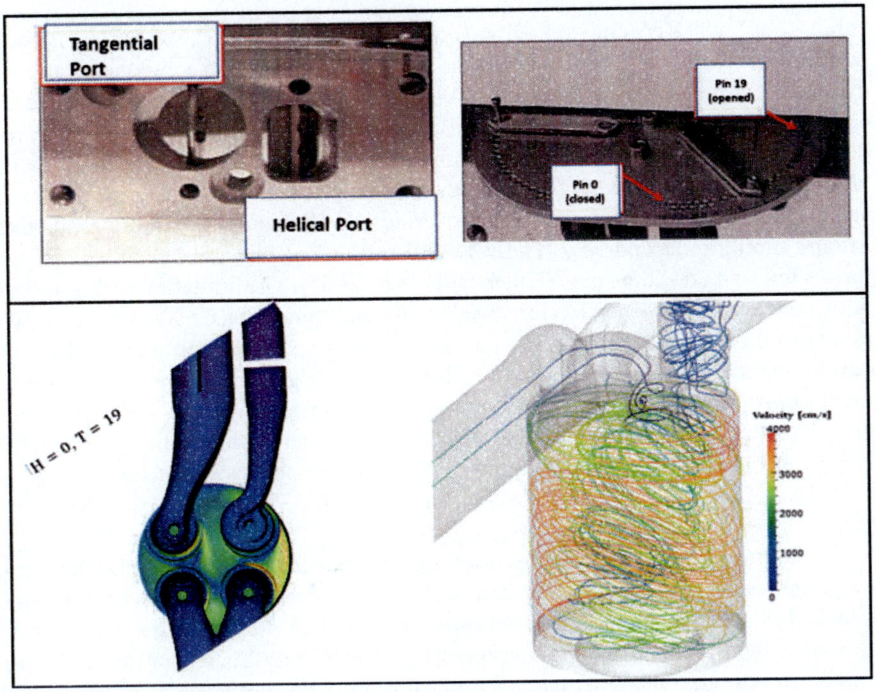

Fig. 2.13 Swirl control system in a light-duty engine. Swirl flaps (top-left), control mechanism (top-right), CFD model for maximum swirl case (bottom-left), velocity magnitude for maximum swirl case (bottom-right) (Perini et al. 2014)

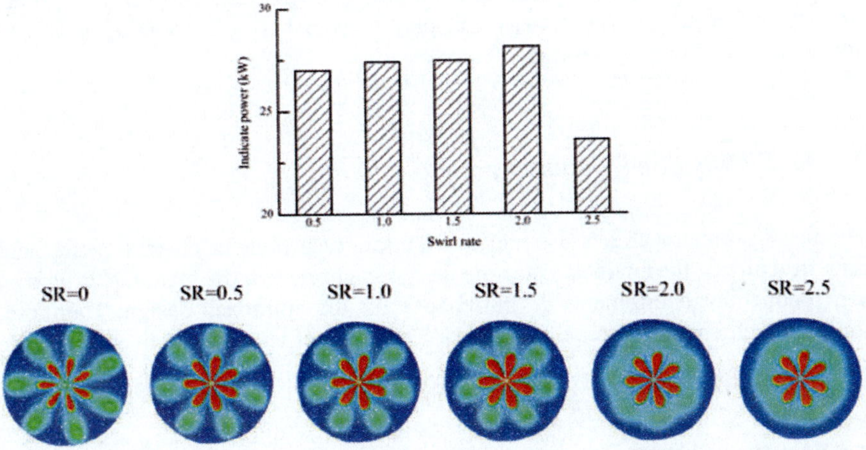

Fig. 2.14 Effect of swirl on performance (top) and equivalence ratio distribution (bottom) (Li et al. 2016)

components, which are constrained and are properly lubricated. However, for engines requiring very high-speed operations, the inertial effect of these components affects the overall system performance and sometimes leads to catastrophic failure. Diesel engines are inherently low-speed engines, compared to gasoline engines. Therefore, the high-speed operation-related concerns may not be as severe as in the case of high-performance gas engines. However, optimization of turbocharger operation, advanced concepts like negative valve overlap (NVO) require flexibility to control lift and timing independently. The valve lift profile can be varied seamlessly by a three-dimensional sliding cam (Fallahzadeh et al. 2005). The limitation of this design lies in the interdependency of cam profile and dwell time. Thus, valve lift and valve timing cannot be independent. The necessity to decouple both events has led to the development of cam-less engines. Cam-less valve designs are independent of kinematic constraints. Therefore much aggressive lift profiles can be adapted to meet the demand of high filling efficiency and variable lift. The valve lift can be controlled by electro-mechanical, electro-hydraulic, or electro-pneumatic mechanism (Sugimoto et al. 2004; Cope and Wright 2006; Nam and Choi 2012; Zhang et al. 2016; Ma et al. 2006).

The cam-less engine concept provides several other advantages other than its main purpose. The flexibility of timing can be used for internal-EGR by synchronizing inlet and exhaust valve timings. The trapped mass in the cylinder can be controlled to alter the engine's effective compression ratio as per the requirement of the particular combustion mode. Valve lift has a significant effect on the flow characteristics in diesel engines; therefore, variable flow control can also be realized. As discussed in the previous section, a particular valve can be de-activated completely to achieve higher or lower swirl levels. The VVA concept was originally developed to reduce throttle losses and increase high-speed operation in SI engines. However, with proper strategy and research, the strategy can play a crucial role in adopting advanced combustion technology in diesel engines.

2.4.3 *Piston Bowl Geometry*

The bowl geometry affects the fuel distribution, heat transfer characteristics, and flow structure of the engines. Once the design is optimized, the manufacturing cost and production requirements do not change for the optimized design. Therefore, this approach continues to be the first choice for the efficiency improvement of the engines.

2.4.3.1 Geometric Parameters of Piston-Bowl

Piston geometry plays a key role in amplifying the swirl and turbulence during compression. The small-scale structures formed during induction dissipate by the time piston reaches the BDC. During compression, the upward motion of the motion pushes the fluid mass from the peripheral regions of the cylinder towards the center into the piston bowl. Due to the conservation of angular momentum, the fluid forced into the bowl moves with higher angular velocity. The squish flow increases significantly as the piston nears the TDC. Thus, the instantaneous swirl ratio gets amplified (Jena et al. 2021). The amplification is proportional to the volume dilation hence the compression ratio (r_c). The r_c can be altered by changing the k_{factor} keeping the bore diameter constant

$$k_{factor} = V_{bowl}/V_{total}$$

where V_{bowl} is the bowl volume, and V_{total} is the total volume at TDC.

Reducing k_{factor} increases r_c, thereby lowering the air utilization. However, the increased squish area helps in higher amplification of the in-cylinder swirl and turbulence (Jena et al. 2021). An increased air utilization results in higher specific power output. Hence modern diesel engines are considering decreasing r_c. It has several disadvantages, such as higher soot emissions at mid-load, increased noise level, etc. (Cursente et al. 2009; Terazawa et al. 2011; Catania et al. 2009). Therefore, the design must be optimized carefully, along with injection strategy and geometry. The other important geometric feature of the piston is the 'central projection inside the bowl,' often called the 'piston pip.' The central region of the piston bowl remains mostly unutilized throughout the cycle. The idea behind the origin of piston pip is to fill this area with metal to increase air utilization. It also helps confine the squish flow to the piston bowl's peripheral region, leading to higher angular momentum.

Bore to stroke (B/S) ratio is yet another factor affecting the piston design. The piston assembly friction is inversely proportional to B/S. Therefore, oversquare (B/S > 1) engines offer lower frictional losses (Miles and Andersson 2016). The piston speed gets reduced for the same angular speed of the crank. This helps rev up the engine to higher speeds, leading to higher specific power output. B/S is inversely proportional to the k_{factor} of the engine. This indicates that a higher value of B/S results in lower k_{factor} and decreases air utilization. The surface-to-volume ratio near the TDC increases with the B/S. This leads to a higher heat loss during combustion. However, the overall heat transfer may be lower because of the lower surface-to-volume ratio at the BDC (Flowers et al. 2010). There are several advantages and disadvantages of an oversquare engine. The B/S value is close to 1 for most commercial engines. Overall, an oversquare engine should be favored for higher power density, while a near square engine is preferred for higher efficiency.

2.4.3.2 Piston Design Evolution and Current State

Several piston designs have been developed over the years for optimizing the performance and emissions of diesel engines. The most common design for light-duty engines is the 're-entrant piston geometry.' The fundamental idea behind the development of this shape is to re-direct the impinging spray plume towards the piston bowl center for better utilization of air. Large-bore engines use an open bowl design as the increased number of nozzles govern air utilization. The re-entrant bowl forms the basis for design optimization. Several studies investigated the effect of flow on combustion using different bowl geometries (Pei et al. 2019; Leach et al. 2018; Genzale et al. 2009). Stepped lip design is one of the successful development of the base model, which is used by many leading automakers these days (Yoo et al. 2013; Kogo et al. 2016; Crosse 2010). The stepped lip is used to split the impinging plume towards the upper and lower regime of the impinging plane, as shown in Fig. 2.15.

Studies suggest that the stepped lip geometry increases the air utilization in the squish zone without compromising with the k_{factor}. (Yoo et al. 2013; Dahlstrom et al. 2016). The higher air utilization facilitates the use of increased EGR, thus a reduction in NO_x emissions. Experiments with the stepped lip design of Honda (two-stage chamber) confirmed lower soot and NO_x emissions. Ricardo UK patented a piston design (Twin Vortex Combustion Chamber) based on the stepped lip concept, which successfully met Tier-4 emission norms for off-road engines. The name 'twin vortex' came from the characteristic vortex generated by a stepped lip, which is still under investigation (Busch and Perini 2019). The improved performance and emission characteristics of the stepped lip bowl were numerically investigated and

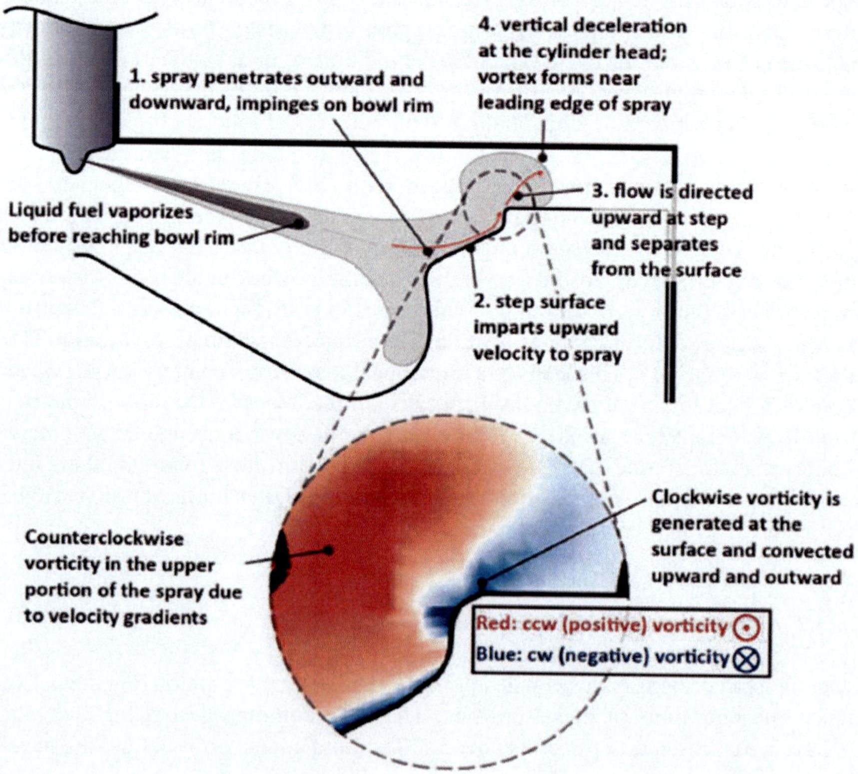

Fig. 2.15 The stepped lip geometry and twin vortex location (Busch and Perini 2019)

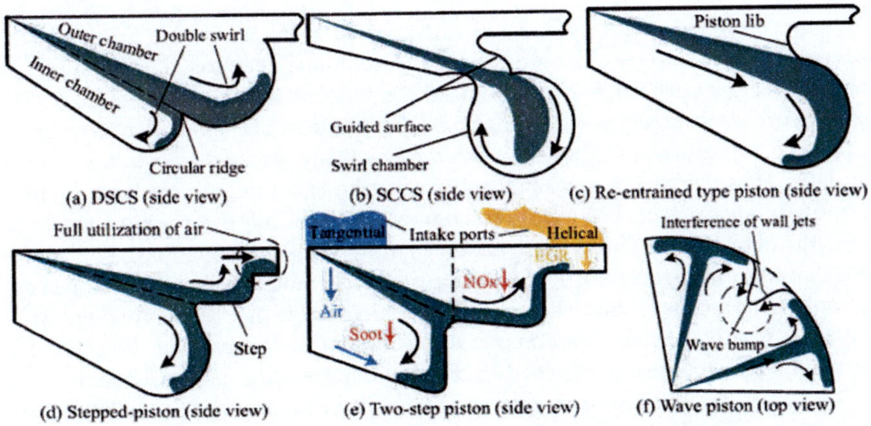

Fig. 2.16 Innovative piston designs for improved air utilization (Chen et al. 2020)

compared with the conventional bowl by many researchers. A higher mixing rate of the stepped bowl geometry is beneficial for soot reduction. Many innovative designs have come up based on the spray splitting concept for increased air utilization. A few of them have been shown in Fig. 2.16.

A state-of-the-art design, which has shown promising results, is the 'wave piston design' (Fig. 2.16f). The wave piston design developed by the Volvo group introduced an additional bump in the bowl periphery of the stepped lip design (Eismark et al. 2019a). Several experimental studies reported significant improvement in emissions with the wave piston (Eismark et al. 2019b; Zhang et al. 2020). Pastor et al. (Pastor et al. 2020) compared the soot oxidation characteristics of conventional design and stepped lip design with the wave piston in a light-duty optical engine. The wave piston showed a faster rate of soot oxidation compared to the other two.

The design evolution of the pistons will continue with the introduction of new fuels and combustion concepts. A completely different piston design is under investigation for RCCI combustion. This piston design has been developed with reduced squish area to mitigate the HC emissions (Hanson et al. 2012). The design has been inspired by heavy-duty engines and scaled down for light-duty applications.

2.5 Future Research Directions

2.5.1 Thermal Aspects

The practical limit analysis of diesel engines indicates the potential areas of improvement for an efficient engine. The optimal placement of heat release rate is the key to achieve higher efficiency and remains an area of focus for the engine developers. However, several other areas are not getting due attention. The optimization of turbocharger performance for advanced combustion strategy is one of them. The change in combustion characteristics from CDC mode to LTC mode also changes the exhaust composition. LTC has been explored by many researchers over the last two decades. However, the changes in the thermodynamics of the exhaust have not been explored to the extent. Electric turbo compounding (e-TC) is a novel concept that ends the turbocharger dependence on the engine speed-load variations. The electric booster and turbo-generator use a separate electric machine for turbocharger and compressor to achieve this. The exhaust heat recovery is independently done while the electric motor drives the compressor according to the transient demand. The concept seems promising and needs more attention from the engine researchers.

Similarly, the variable swirl operation for mode switching engine can be an interesting area of study. Heat loss from the engine is one of the prime causes of exergy destruction of the engine. There is immense potential in the waste heat recovery concept for minimizing exergy destruction. The waste heat recovery can be a retrofitment solution for conventional engines with a mechanical fuel injection system. Furthermore, the emissions from the mechanical engine can be cleaned by utilizing novel alternative fuels such as DME. The engine family also needs due attention considering their large numbers in the market.

2.5.2 *Interdisciplinary Aspects*

The sustainable evolution of engines requires an interdisciplinary approach. Thermal science groups have carried out the bulk of engine research in academia. Thermal engineers have been forced to work under several constraints related to material limits, engine kinematics, tribological aspects, etc. For example, the engine efficiency can be improved thermodynamically by increasing the pressure and temperature in the combustion chamber. However, the thermal loading constraints do not allow to cross a certain limit. The engine friction mitigation is a key to increase the brake thermal efficiency of the engine. The engine tribology needs to be revisited under the changed boundary conditions relevant to advanced combustion technologies. Lubricating oil is identified as a key source of toxic particulate emissions. The noise and vibrations associated with rapid pressure rise are yet another limit. Rapid combustion helps in increasing the degree of constant volume combustion, therefore the efficiency. However, due to the constraints of shock loading and noise, it has to be compromised. Advanced NVH analysis is required to reduce the burden on the combustion optimizers. The material aspects need to be explored to give a free hand to the thermal engineers.

2.6 Summary

The chapter has focused on key areas to provide an overall picture of diesel engine evolution, limitations, potential, and future direction for research and evolution. The efficiency limits discussed subject to various constraints and scope to remove these limitations have been highlighted. Chemical kinetics of soot and hydrocarbons have been discussed for better interpretation of engine-out emissions. The authors hope that this chapter would help access the engine behavior better and plan further optimization by the researchers in the future. Analysis of the state-of-the-art for further evolution of diesel engines has been provided. Finally, the research gap and interdisciplinary areas for further improvement have been identified and highlighted.

References

Akihama K, Kosaka H, Hotta Y, Nishikawa K, Inagaki K, Fuyuto, T, Iwashita Y, Farrell JT, Weissman W (2009) An investigation of high load (compression ignition) operation of the "naphtha engine"—a combustion strategy for low well-to-wheel CO_2 emissions. SAE Int J Fuels Lubr 1(1):920–932. Retrieved 28 June 2021, from http://www.jstor.org/stable/26272061

Attili A, Bisetti F, Mueller ME, Pitsch H (2014) Formation, growth, and transport of soot in a three-dimensional turbulent non-premixed jet flame. Combust Flame 161(7):1849–1865. https://doi.org/10.1016/j.combustflame.2014.01.008

Bauder R, Fröhlich A, Rossi D (2010) The new generation of the Audi 3.0 L V6 TDI engine. MTZ Worldw 71(10):20–27

Bergstrand P, Denbratt I (2001) Diesel combustion with reduced nozzle orifice diameter. SAE technical paper 2001-01-2010. https://doi.org/10.4271/2001-01-2010

Bosch (2014) World map of automotive powertrains: how people drive in Europe, the Americas, and Asia. Bosch Mobility Solutions Press Release, 6 Nov 2014. Accessed through http://www.bosch-presse.de/pressportal/de/en/world-map-of-automotive-powertrains-how-people-d on 21 June 2021

Bryzik W, Kamo R (1983) TACOM/cummins adiabatic engine program. SAE Trans 92:1063–1087. Retrieved 28 June 2021, from http://www.jstor.org/stable/44644435

Busch S, Perini F (2019) Progress toward understanding vortex generation in stepped-lip diesel engine combustion chambers. Res Eng 1:100004. https://doi.org/10.1016/j.rineng.2019.100004

Busch S, Zha K, Miles PC (2014) Investigations of closely coupled pilot and main injections as a means to reduce combustion noise (no. SAND2014-17433C). Sandia National Lab. (SNL-CA), Livermore, CA, USA

Busch S, Zha K, Warey A, Pesce F, Peterson R (2016) On the reduction of combustion noise by a close-coupled pilot injection in a small-bore direct-injection diesel engine. J Eng Gas Turbines Power 138(10). https://doi.org/10.1115/1.4032864

Catania A, D'Ambrosio S, Finesso R, Spessa E, Cipolla G, Vassallo A (2009) Combustion system optimization of a low compression-ratio PCCI diesel engine for light-duty application. SAE Int J Engines 2(1):1314–1326. Retrieved 28 June 2021, from http://www.jstor.org/stable/26308471

Charlton SJ (2005) Developing diesel engines to meet ultra-low emission standards. SAE technical paper no. 2005-01-3628. https://doi.org/10.4271/2005-01-3628

Chen Y, Li X, Li X, Zhao W, Liu F (2020) Verifying the wall-flow-guided assumption of the lateral swirl combustion system in DI diesel engines. Fuel 266:117079. https://doi.org/10.1016/j.fuel.2020.117079

Chi Y, Park S, Lee K, Jo C, Yoo D (2008). New v6 3.0 l diesel engine for Hyundai/Kia's SUVs. MTZ Worldw 69(11):24–30

Cope D, Wright A (2006) Electromagnetic fully flexible valve actuator. SAE technical paper no. 2006-01-0044. https://doi.org/10.4271/2006-01-0044

Crosse J (2010) Going clean-off highway. Ricardo PLC, West Sussex, UK

Curran HJ, Gaffuri P, Pitz WJ, Westbrook CK (1998) A comprehensive modeling study of n-heptane oxidation. Combust Flame 114(1–2):149–177. https://doi.org/10.1016/S0010-2180(97)00282-4

Curran S, Hanson R, Wagner R, Reitz RD (2013) Efficiency and emissions mapping of RCCI in a light-duty diesel engine. SAE technical paper no. 2013-01-0289. https://doi.org/10.4271/2013-01-0289

Cursente V, Pacaud P, Gatellier B (2009) Reduction of the compression ratio on a HSDI diesel engine: combustion design evolution for compliance the future emission standards. SAE Int J Fuels Lubr 1(1):420–439. Retrieved 28 June 2021, from http://www.jstor.org/stable/26272020

Dahlstrom J, Andersson O, Tuner M, Persson H (2016) Experimental comparison of heat losses in stepped-bowl and re-entrant combustion chambers in a light-duty diesel engine. SAE technical paper no. 2016-01-0732. https://doi.org/10.4271/2016-01-0732

Dec J (1997) A conceptual model of DI diesel combustion based on laser-sheet imaging. SAE Trans 106:1319–1348. Retrieved 28 June 2021, from http://www.jstor.org/stable/44730758

Dec JE (2003) Diesel-fueled HCCI engines. In: Zhao F, Asmus T, Assanis D, Dec J, Eng J, Najt P (eds) Homogeneous charge compression ignition (HCCI) engines. Society of Automotive Engineers, Inc., Warrendale, PA

Denny M, Matamis A, Persson H, Richter M, Andersson Ö (2019) Interaction between fuel jets and prevailing combustion during closely-coupled injections in an optical diesel engine (No. 2019-01-0551). SAE technical paper no. 2019-01-0551. https://doi.org/10.4271/2019-01-0551

Denny M, van den Ende L, Persson H, Andersson Ö (2020) Manipulating heat release features to minimize combustion noise. Fuel 263:116613. https://doi.org/10.1016/j.fuel.2019.116613

Deraad S, Fulton B, Gryglak A, Hallgren B, Hudson A, Ives D, Morgan P, Styron J, Waszczenko E, Cattermole I (2010) The new ford 6.7 L V-8 turbocharged diesel engine. SAE technical paper no. 2010-01-1101. https://doi.org/10.4271/2010-01-1101

Diesel Technology Forum (2021) Web source: https://www.dieselforum.org. Accessed on 21 June 2021

Edwards CF, Ramakrishnan S, Svrcek M, Roberts G, Heberle JR, Mobley P, Calbry-Muzyka A, Pass R (2010) Realizing ultra-high-efficiency engines: understanding. In: Advanced energy systems laboratory department of mechanical engineering. Stanford University

Edwards KD, Wagner RM, Briggs TE, Theiss TJ (2011) Defining engine efficiency limits. In: 17th DEER conference. Detroit, MI, pp 3–6

Eidenböck T, Mayr K, Neuhauser W, Staub P (2012) The new BMW six-cylinder diesel engine with three turbochargers. MTZ Worldw 73(10):18–25

Eismark J, Andersson M, Christensen M, Karlsson A, Denbratt I (2019a) Role of piston bowl shape to enhance late-cycle soot oxidation in low-swirl diesel combustion. SAE Int J Engines 12(3):233–250. Retrieved 28 June 2021, from https://www.jstor.org/stable/26840425

Eismark J, Christensen M, Andersson M, Karlsson A, Denbratt I (2019b) Role of fuel properties and piston shape in influencing soot oxidation in heavy-duty low swirl diesel engine combustion. Fuel 254:115568. https://doi.org/10.1016/j.fuel.2019.05.151

Fallahzadeh F, Subrahamanyam J, Sharma V, Gajendra Babu M (2005) Simulation and evaluation of a variable valve timing single-cylinder spark-ignition engine. SAE technical paper no. 2005-01-0765. https://doi.org/10.4271/2005-01-0765

Fasolo B, Doisy AM, Dupont A, Lavoisier F (2005) Combustion system optimization of a new 2-liter diesel engine for EURO IV. SAE technical paper no. 2005-01-0652. https://doi.org/10.4271/2005-01-0652

Flaig U, Polach W, Ziegler G (1999) Common rail system (CR-system) for passenger car DI diesel engines; experiences with applications for series production projects. SAE technical paper no. 1999-01-0191. https://doi.org/10.4271/1999-01-0191

Flowers DL, Martinez-Frias J, Cleeves JM (2010) US patent application no. 12/624,276

Frenklach M, Wang H (1991) Detailed modeling of soot particle nucleation and growth. Symp Int Combust 23(1):1559–1566. https://doi.org/10.1016/S0082-0784(06)80426-1

Ganser M (2000) Common rail injectors for 2000 bar and beyond. SAE Trans 109:827–839. Retrieved 28 June 2021, from http://www.jstor.org/stable/44634268

Genzale C, Reitz R, Musculus M (2009) Effects of piston bowl geometry on mixture development and late-injection low-temperature combustion in a heavy-duty diesel engine. SAE Int J Engines 1(1):913–937. Retrieved 28 June 2021, from http://www.jstor.org/stable/26308329

Greeves G, Khan I, Wang C, Fenne I (1977) Origins of hydrocarbon emissions from diesel engines. SAE Trans 86:1235–1251. Retrieved 28 June 2021, from http://www.jstor.org/stable/44648147

Hanson R, Curran S, Wagner R, Kokjohn S, Splitter D, Reitz R (2012) Piston bowl optimization for RCCI combustion in a light-duty multi-cylinder engine. SAE Int J Engines 5(2):286–299. Retrieved 28 June 2021, from http://www.jstor.org/stable/26278359

Herfatmanesh MR, Zhao H (2013) Experimental investigation of effects of dwell angle on fuel injection and diesel combustion in a high-speed optical CR diesel engine. Proc Inst Mech Eng Part D J Automob Eng 227(2):246–260. https://doi.org/10.1177/0954407012450656

Hotta Y, Nakakita K, Fuyuto T, Inayoshi M, Fujiwara K, Sakata I (2002) Cause of exhaust smoke and its reduction methods in an HSDI diesel engine under high-speed and high-load conditions. SAE Trans 111:2064–2074. Retrieved 28 June 2021, from http://www.jstor.org/stable/44743222

Hotta Y, Inayoshi M, Nakakita K, Fujiwara K, Sakata I (2005) Achieving lower exhaust emissions and better performance in an HSDI diesel engine with multiple injection. SAE technical paper no. 2005-01-0928. https://doi.org/10.4271/2005-01-0928

Imarisio R, Rossi Sebastiano GM (2001) Potential of future common rail DI diesel engines. Ing Automob (Paris) 748:74–78

Jaichandar S, Tamilporai P (2003) Low heat rejection engines—an overview. SAE technical paper no. 2003-01-0405. https://doi.org/10.4271/2003-01-0405.

Jena A (2020) Optical diagnostics of spray development in diesel engines. In: Simulations and optical diagnostics for internal combustion engines, pp 53–68

Jena A, Singh H, Agarwal A (2021) Effect of swirl ratio and piston geometry on the late-compression mean air-flow in a diesel engine, SAE technical paper no. 2021-01-0647. https://doi.org/10.4271/2021-01-0647

Kanda T, Hakozaki T, Uchimoto T, Hatano J, Kitayama N, Sono H (2005) PCCI operation with early injection of conventional diesel fuel. SAE Trans 114:584–593. Retrieved 28 June 2021, from http://www.jstor.org/stable/44722026

Kashdan J, Mendez S, Bruneaux G (2007) On the origin of unburned hydrocarbon emissions in a wall guided, low NO_x diesel combustion system. SAE Trans 116:234–257. Retrieved 28 June 2021, from http://www.jstor.org/stable/44650861

Kogo T, Hamamura Y, Nakatani K, Toda T, Kawaguchi A, Shoji A (2016) High-efficiency diesel engine with low heat loss combustion concept-Toyota's inline 4-cylinder 2.8-liter ESTEC 1GD-FTV engine (no. 2016-01-0658). SAE technical paper no. 2016-01-0658. https://doi.org/10.4271/2016-01-0658

Kosaka H, Wakisaka Y, Nomura Y, Hotta Y, Koike M, Nakakita K, Kawaguchi A (2013) Concept of "temperature swing heat insulation. In: Combustion chamber walls and appropriate thermo-physical properties for heat insulation coat. SAE Int J Engines 6(1):142–149. Retrieved 28 June 2021, from http://www.jstor.org/stable/26277604

Kurtz E, Styron J (2012). An assessment of two piston bowl concepts in a medium-duty diesel engine. SAE Int J Engines, 5(2):344–352. Retrieved 28 June 2021, from http://www.jstor.org/stable/26278364

Lachaux T, Musculus MP (2007) In-cylinder unburned hydrocarbon visualization during low-temperature compression-ignition engine combustion using formaldehyde PLIF. Proc Combust Inst 31(2):2921–2929. https://doi.org/10.1016/j.proci.2006.07.044

Le MK, Zhang Y, Zhang R, Rao L, Kook S, Chan QN, Hawkes ER (2017) Effect of jet–jet interactions on soot formation in a small-bore diesel engine. Proc Combust Inst 36(3):3559–3566. https://doi.org/10.1016/j.proci.2016.07.025

Leach F, Ismail R, Davy M, Weall A, Cooper B (2018) The effect of a stepped lip piston design on performance and emissions from a high-speed diesel engine. Appl Energy 215:679–689. https://doi.org/10.1016/j.apenergy.2018.02.076

Lee E, Kwak S, Kim M, Joo S, Chun J, Pae S, Yu J, Grimm J (2009) The new 2.0 l and 2.2 l four-cylinder diesel engine family of Hyundai-Kia. MTZ Worldw 70(10):14–19

Lequien G, Li Z, Andersson O, Richter M (2015) Lift-off length in an optical heavy-duty diesel engine: effects of swirl and jet-jet interactions. SAE Int J Engines 8(5):2188–2198. Retrieved 28 June 2021, from http://www.jstor.org/stable/26278112

Li X, Zhou H, Zhao LM, Su L, Xu H, Liu F (2016) Effect of split injections coupled with swirl on combustion performance in DI diesel engines. Energy Convers Manage 129:180–188. https://doi.org/10.1016/j.enconman.2016.09.011

Lutsey N (2018) The ever-improving efficiency of the diesel engine. ICCT report

Ma J, Schock H, Carlson U, Hoglund A et al. (2006) Analysis and modeling of an electronically controlled pneumatic, hydraulic valve for an automotive engine. SAE technical paper no. 2006-01-0042. https://doi.org/10.4271/2006-01-0042

Mao Q, Van Duin AC, Luo KH (2017) Formation of incipient soot particles from polycyclic aromatic hydrocarbons: a ReaxFF molecular dynamics study. Carbon 121:380–388. https://doi.org/10.1016/j.carbon.2017.06.009

Markit HIS (2020) US vehicle in operation data (transit buses and school buses, 2010 model or newer) data. Accessed through https://www.dieselforum.org/ on 25 June 2021

Michelsen HA (2017) Probing soot formation, chemical and physical evolution, and oxidation: a review of in situ diagnostic techniques and needs. Proc Combust Inst 36(1):717–735. https://doi.org/10.1016/j.proci.2016.08.027

Miles PC (2009) In-cylinder turbulent flow structure indirect injection, swirl-supported diesel engines. In: Arcoumanis C, Kamimoto T (eds) Flow and combustion in reciprocating engines. Springer, Berlin, Heidelberg (Chapter 4)

Miles PC, Andersson Ö (2016) A review of design considerations for light-duty diesel combustion systems. Int J Engine Res 17(1):6–15. https://doi.org/10.1177/1468087415604754

Minami T, Takeuchi K, Shimazaki N (1995) Reduction of diesel engine NO_x using pilot injection. SAE Trans 104:1104–1111. Retrieved 28 June 2021, from http://www.jstor.org/stable/44633283

Musculus MP, Miles PC, Pickett LM (2013) Conceptual models for partially premixed low-temperature diesel combustion. Prog Energy Combust Sci 39(2–3):246–283. https://doi.org/10.1016/j.pecs.2012.09.001

Nam K, Choi SB (2012) Development of a camless engine valve actuator system for robust engine valve timing control. Int J Veh Syst Model Test 7(4):372–389. https://doi.org/10.1504/IJVSMT.2012.049429

Nehmer D, Reitz R (1994) Measurement of the effect of injection rate and split injections on diesel engine soot and NO_x emissions. SAE Trans 103:1030–1041. Retrieved 28 June 2021, from http://www.jstor.org/stable/44632855

Olmeda P, Martín J, García Martínez A, Villalta-Lara D, Warey A, Doménech Llopis V (2017) A combination of swirl ratio and injection strategy to increase engine efficiency. SAE Int J Engines 10(3):1–13. https://doi.org/10.4271/2017-01

Parrish S, Zhang G, Zink R (2012) Liquid and vapor envelopes of sprays from a multi-hole fuel injector operating under closely-spaced double-injection conditions. SAE Int J Eng 5(2):400–414. Retrieved 28 June 2021, from http://www.jstor.org/stable/26278368

Pastor JV, García A, Micó C, Lewiski F (2020) Soot reduction for cleaner compression ignition engines through innovative bowl templates. Int J Engine Res 1468087420951324. https://doi.org/10.1177/1468087420951324

Pei Y, Pal P, Zhang Y, Traver M, Cleary D, Futterer C, Brenner M, Probst D, Som S (2019) CFD-guided combustion system optimization of a gasoline range fuel in a heavy-duty compression ignition engine using automatic piston geometry generation and a supercomputer. SAE Int J Adv Curr Pract Mobility 1(2019-01-0001):166–179. https://doi.org/10.4271/2019-01-0001

Perini F, Miles PC, Reitz RD (2014) A comprehensive modeling study of in-cylinder fluid flows in a high-swirl, light-duty optical diesel engine. Comput Fluids 105:113–124. https://doi.org/10.1016/j.compfluid.2014.09.011

Pickett LM, Siebers DL (2004a) Soot in diesel fuel jets: effects of ambient temperature, ambient density, and injection pressure. Combust Flame 138(1–2):114–135. https://doi.org/10.1016/j.combustflame.2004.04.006

Pickett LM, Siebers DL (2004b) Non-sooting, low flame temperature mixing-controlled DI diesel combustion. SAE technical paper no. 2004-01-1399. https://doi.org/10.4271/2004-01-1399

Pierpont D, Montgomery D, Reitz R (1995) Reducing particulate and NO_x using multiple injections and EGR in a DI diesel. SAE Trans 104:171–183. Retrieved 28 June 2021, from http://www.jstor.org/stable/44615074

Rakopoulos CD, Giakoumis EG (2006) Second-law analyses applied to internal combustion engines operation. Prog Energy Combust Sci 32(1):2–47. https://doi.org/10.1016/j.pecs.2005.10.001

Reitz RD, Duraisamy G (2015) Review of high efficiency and clean reactivity controlled compression ignition (RCCI) combustion in internal combustion engines. Prog Energy Combust Sci 46:12–71. https://doi.org/10.1016/j.pecs.2014.05.003

Rinolfi R (1995) The potentials of a new common rail diesel fuel injection system for the next generation of DI diesel engines. Fortschr Ber VDI Reihe 12(239):87

Sahoo D, Petersen BR, Miles PC (2012) The impact of swirl ratio and injection pressure on fuel-air mixing in a light-duty diesel engine. In: Internal combustion engine division spring technical conference, vol 44663. American Society of Mechanical Engineers, pp 393–404. https://doi.org/10.1115/ICES2012-81234

Shayler PJ, Ng HK (2004) Simulation studies of the effect of fuel injection pattern on NO and soot formation in diesel engines. SAE technical paper no. 2004-01-0116. https://doi.org/10.4271/2004-01-0116

Siebers D, Higgins B (2001) Flame lift-off on direct-injection diesel sprays under quiescent conditions. SAE Trans 110:400–421. Retrieved 28 June 2021, from http://www.jstor.org/stable/44724317

Sonawane U, Agarwal AK (2021) Engine emission control devices for particulate matter and oxides of nitrogen: challenges and emerging trends. In: Novel internal combustion engine technologies for performance improvement and emission reduction. Springer, Singapore, pp 197–220

Srivastava DK, Das A, Singh NK (2018) Variable valve actuation systems. In: Advances in internal combustion engine research. Springer, Singapore, pp 63–91

Stanton D (2009) Highly efficient and clean diesel engine technologies. In: Directions in engine efficiency conference, Detroit, MI

Stanton DW (2013) Systematic development of highly efficient and clean engines to meet future commercial vehicle greenhouse gas regulations. SAE Int J Engines 6(3):1395–1480. https://doi.org/10.4271/2013-01-2421

Sugimoto C, Sakai H, Umemoto A, Shimizu Y et al. (2004) Study on variable valve timing system using electromagnetic mechanism. SAE technical paper no. 2004-01-1869. https://doi.org/10.4271/2004-01-1869

Terazawa Y, Nakai E, Kataoka M, Sakono T (2011) The new Mazda four-cylinder diesel engine. MTZ Worldw eMag 72(9):26–33

Totton TS, Chakrabarti D, Misquitta AJ, Sander M, Wales DJ, Kraft M (2010) Modelling the internal structure of nascent soot particles. Combust Flame 157(5):909–914. https://doi.org/10.1016/j.combustflame.2009.11.013

Tow T, Pierpont D, Reitz R (1994) Reducing particulate and NO_x emissions by using multiple injections in a heavy-duty DI diesel engine. SAE Trans 103:1403–1417. Retrieved 28 June 2021, from http://www.jstor.org/stable/44632882

Vaglieco BM (2010) Multiple injection diesel combustion process in the high-speed direct injection diesel engine. In: Advanced direct injection combustion engine technologies and development, pp 155–174. https://doi.org/10.1533/9781845697457.1.155

van den Heuvel B, Willems W, Krämer F, Morris T, Karvounis E (2006) Combustion system development for the new diesel engines in light and medium commercial vehicles from Ford and PSA. MTZ Worldw 67(9):2–5

Velásquez M, Mondragón F, Santamaría A (2013) Chemical characterization of soot precursors and soot particles produced in hexane and diesel surrogates using an inverse diffusion flame burner. Fuel 104:681–690. https://doi.org/10.1016/j.fuel.2012.04.033

Web source: https://www.tu-auto.com/diesel-powertrains-will-reduce-co2-quicker-than-bevs-forum-claims/. Accessed on 20 June 2021

Xuan T, Pastor JV, García-Oliver JM, García A, He Z, Wang Q, Reyes M (2019) In-flame soot quantification of diesel sprays under sooting/non-sooting critical conditions in an optical engine. Appl Therm Eng 149:1–10. https://doi.org/10.1016/j.applthermaleng.2018.11.112

Ya B (1946) The oxidation of nitrogen in combustion and explosions. Acta Physicochim URSS 21:557

Yang K, Yamakawa H, Nishida K, Ogata Y, Nishioka Y (2018) Effect of split injection on mixture formation and combustion processes of diesel spray injected into two-dimensional piston cavity. Proc Inst Mech Eng Part D J Automob Eng 232(8):1121–1136. https://doi.org/10.1177/0954407017724246

Yoo D, Kim D, Jung W, Kim N, Lee D (2013) Optimization of diesel combustion system for reducing PM to meet tier4-final emission regulation without diesel particulate filter. SAE technical paper no. 2013-01-2538. https://doi.org/10.4271/2013-01-2538

Zhang S, Zhao Z, Zhao C, Zhang F, Wang S (2016) Development and validation of electro-hydraulic camless free-piston engine. Appl Therm Eng 102:1197–1205. https://doi.org/10.1016/j.applthermaleng.2016.03.093

Zhang Y, Kim D, Rao L, Zhang R, Kook S, Kim KS, Kweon CB (2017) The soot particle formation process inside the piston bowl of a small-bore diesel engine. Combust Flame 185:278–291. https://doi.org/10.1016/j.combustflame.2017.07.028

Zhang Y, Kim D, Rao L, Kook S, Kim KS, Kweon CB (2019) In-flame soot particle structure on the up-and down-swirl side of a wall-interacting jet in a small-bore diesel engine. Proc Combust Inst 37(4):4847–4855. https://doi.org/10.1016/j.proci.2018.07.104

Zhang T, Eismark J, Munch K, Denbratt I (2020) Effects of a wave-shaped piston bowl geometry on the performance of heavy-duty diesel engines fueled with alcohols and biodiesel blends. Renew Energy 148:512–522. https://doi.org/10.1016/j.renene.2019.10.057

Zhang Y, Nishida K (2003) Vapor/liquid behaviors in split-injection D.I. diesel sprays in a 2-D model combustion chamber. SAE Trans 112:1276–1287. Retrieved 28 June 2021, from http://www.jstor.org/stable/44742351

Zhang Y, Ito T, Nishida K (2001) Characterization of mixture formation in split-injection diesel sprays via laser absorption-scattering (LAS) technique. SAE Trans 110:2189–2200. Retrieved 28 June 2021, from http://www.jstor.org/stable/44724476

Zhou X, Song M, Huang H, Yang R, Wang M, Sheng J (2014) Numerical study of the formation of soot precursors during low-temperature combustion of an-butanol–diesel blend. Energy Fuels 28(11):7149–7158. https://doi.org/10.1021/ef501370u

Chapter 3
Multi-mode Low Temperature Combustion (LTC) and Mode Switching Control

Sadaf Batool, Jeffrey D. Naber, and Mahdi Shahbakhti

Abstract Low temperature combustion (LTC) modes offer high thermal efficiency and low engine-out NOx and soot emissions. The common LTC modes include homogeneous charge compression ignition (HCCI), partially premixed charge compression ignition (PPCI), and reactivity-controlled compression ignition (RCCI). To realize these promising LTC modes, optimal combustion control of the engine in each LTC mode is required. This will require precise control of combustion phasing and engine load, while constraining engine variables including peak in-cylinder gas pressure, maximum pressure rise rate, high intensity ringing or knock, and coefficient of variation of indicated mean effective pressure (IMEP) to allow safe and stable combustion. This chapter explains state-of-the-art of LTC engine control, including dynamic modeling, model predictive combustion control, experimentation, and implementation of real-time closed loop combustion controllers. A well-recognized limitation of LTC engines is a constrained optimal operating range. To this end, multi-mode engines are desired. These include (i) multi-mode LTC engines with mode transition among LTC modes, e.g., dual mode HCCI-RCCI engine, or triple mode HCCI-PPCI-RCCI engine, and (ii) multi-mode engines including LTC and conventional combustion modes, e.g., dual mode HCCI-SI (spark ignition) engines or dual mode RCCI-CDC (conventional diesel combustion) engines. These demands for innovative combustion controllers that capture the engine transient dynamics such as in-cylinder air–fuel ratio variations or residual gas temperature variations during mode transitions. This chapter introduces multi-mode LTC engines and explains controller development for these engines.

Keywords Low temperature combustion (LTC) · Combustion control · Multi-mode LTC engine · Mode switching

S. Batool (✉) · J. D. Naber
Michigan Technological University, Houghton, MI 49931, USA
e-mail: mahdi@ualberta.ca

M. Shahbakhti
University of Alberta, Edmonton, AB, Canada

A. K. Agarwal et al. (eds.), *Advanced Combustion for Sustainable Transport*, Energy, Environment, and Sustainability, https://doi.org/10.1007/978-981-16-8418-0_3

3.1 Introduction

Environmental protection agencies around the world, such as Environmental Protection Agency (EPA) and National Highway Traffic Safety Administration (NHTSA) in U.S., European Environmental Agency (EEA), Ministry of Ecology and Environment the People's Republic of China, etc. have set the stringent regulations for greenhouse gas (GHG) emissions and vehicle fuel economy for light and heavy-duty vehicles. These regulations have compelled the automobile manufacturers to explore the advanced combustion strategies. Advanced combustion technologies including low temperature combustion (LTC) are renowned for high thermal efficiency and low engine-out emissions. Light duty vehicles contribute towards the maximum consumption of fuel in transportation sector. The recent projections by the U.S. Energy Information Administration forecast that the new light duty vehicles running solely on IC engines would contribute to 81% of the U.S. market share by 2050 (EIA 2016). Hybrid electric and plug-in hybrid electric vehicles also use ICE. This leads to market share increase to 85% when hybrid electric and plug-in hybrid electric vehicles are also included (EIA 2016). The transportation sector is one of the major contributors towards the GHG emissions. Therefore, increasing fuel conversion efficiency of ICEs by the use of multi-mode LTC engines will allow reduction in GHG emissions. Furthermore, biofuels produce much lower GHG emissions; thus, providing much cleaner combustion as compared to gasoline fueled vehicles. Given the fuel flexibility of multi-mode engines, they can run with different biofuels. Biofuels have a potential of reducing GHG emissions for sustainable transportation (Ternel et al. 2021; Ogunkunle and Ahmed 2021). LTC strategies using bio-fuels offer reduced GHG emissions together with high thermal efficiency (Singh and Agarwal 1778; Weall and Collings 2007; Adams et al. 2013). When compared to the performance of diesel fueled homogeneous compression ignition engine, biodiesel fuel resulted in reduced HC and CO emissions at low load while reduction in smoke emissions was observed at high load operation (Adams et al. 2013). Singh et al. compared the performance of conventional diesel combustion (CDC) and premixed charge compression ignition (PCCI). The study was extended to the performance comparison of diesel fueled CDC and PCCI modes with a blend of biodiesel with mineral diesel (B20) fuel together with the open-loop CDC-PCCI mode switching. The results showed that PCCI led to significant reduction in NOx and particulate emissions when compared to CDC operation using both fuels. However, diesel fueled PCCI mode produced higher concentrations of SO2, HCHO, etc. B20-fueled PCCI operation offered relatively lower concentrations of unregulated emissions (Singh and Agarwal 1778). In addition, non-fossils based fuel can be accommodated with already existing vehicle powertrains without any engine changes (Ternel et al. 2021). Therefore, multi-mode LTC engines contribute to sustainable transportation via three major areas (i) offering higher fuel conversion efficiency compared to conventional SI and CI engines, (ii) broadening engine speed and load range with high fuel conversion efficiency, and (iii) possibility to run with a range of renewable fuels.

The common LTC modes include homogeneous charge compression ignition (HCCI), premixed charged compression ignition (PCCI), partially premixed charged compression ignition (PPCI) and reactivity-controlled compression ignition (RCCI). In HCCI, fuel is injected during intake stroke which provides enough time to mix homogeneously with air. The air–fuel mixtures undergo autoignition at the end of compression stroke. This mode offers high thermal efficiency and ultra-low NOx emissions. However, it is often prone to excessive pressure rise rates (MPRR) which cause engine knocking. The low load of HCCI is limited by misfire or partial burn while the higher load limitation of HCCI is mainly due to the high MPRR and knocking (Kalghatgi and Head 2006). These drawbacks can be abated by introducing different levels of air–fuel premixing. Premixed or partially premixed mixtures are formed by varying the fuel injection timings and using high levels of dilution. This introduces the desired ignition delay and increases in-cylinder mixing (Cho et al. 2009). Lower compression ratio, high injection pressure, exhaust gas recirculation (EGR) and enhanced mixing can be used to achieve the desired ignition delay (Manente et al. 2009). In literature, ignition delay is either defined as the crank angle difference between start of injection and start of combustion (Cho et al. 2009) or the crank angle difference between start of injection and crank angle where 50% of total heat release occurs (Kalghatgi et al. 2006). NOx formation can be avoided by increasing the ignition delay (Kalghatgi et al. 2006). In addition, MPRR can be prevented by using multiple fuel injections and high intake pressure (Kalghatgi et al. 2006). Dual fuel application in LTC has also been focus of research. Reactivity controlled compression ignition (RCCI) is one such combustion mode which uses dual fuels of different reactivity. Usually, a low reactivity fuel is injected via port fuel injector (PFI) and high reactivity fuel is injected via direct injector (DI). Premixed ratio of the fuels and injection timing of high reactivity fuel provides better control on the combustion phasing. A brief overview of the common LTC modes is shown in Fig. 3.1.

High local equivalence ratio and high local temperature in CDC leads to the formation of NOx and soot (Wissink 2015). Unlike CDC, LTC modes generally involve lean premixed mixtures which reduce the local fuel rich zones. Therefore, very high peak in-cylinder gas temperatures are avoided in the LTC modes that help in restricting NOx and particulate matter (PM) formation (Wissink 2015). However, low combustion temperatures in LTC modes limit the HC and CO oxidation; thus, leading to more HC and CO emissions when compared to CDC (Wissink 2015). The comparison of emissions between LTC strategies and CDC can be seen in Fig. 3.2. Homogeneous reactor simulations were carried out for different equivalence ratios and temperatures in Wissink (2015). Figure 3.2 also shows the relationship between NOx, soot, HC and CO emissions for different equivalence ratios and temperatures for the LTC modes.

Closed-loop control of LTC modes is important to ensure stable and safe engine operation while achieving high thermal efficiency and ultra-low engine-out emissions. The variables to be controlled in LTC modes include combustion timing (CA50), indicated mean effective pressure (IMEP), maximum pressure rise rate

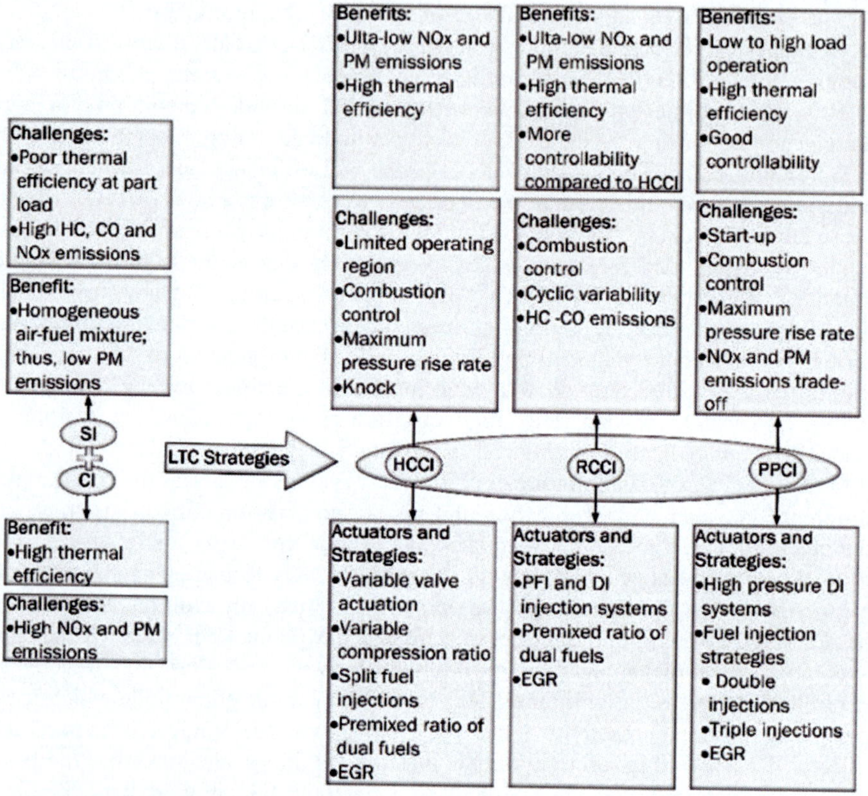

Fig. 3.1 Benefits and limitations of common LTC modes

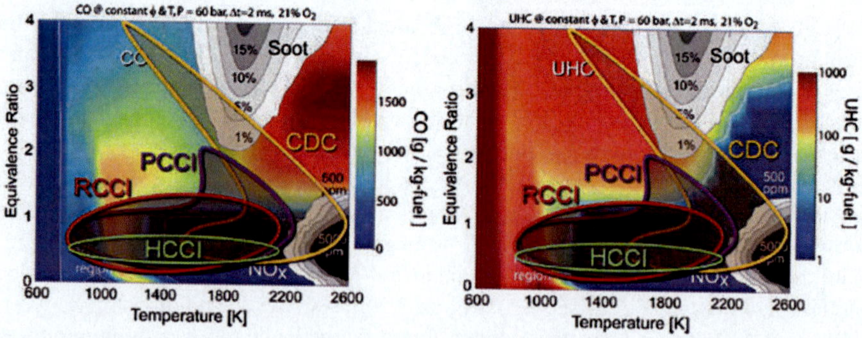

Fig. 3.2 Contours of various combustion strategies and emissions (i.e., NOx, soot, CO, and unburned HC) from homogeneous reactor simulation with n-heptane and air at different equivalence ratios and temperatures. Y-axis shows the local equivalence ratio. The contour plots show the optimal combustion temperature and local equivalence ratio to avoid high unburned HC, CO, NOx and soot emissions [*Sources* Wissink (2015), p. 7; Kim et al. (2009), p. 935]

(MPRR), peak in-cylinder pressure (P_{max}), exhaust gas temperature (T_{exh}) and coefficient of variation of indicated mean effective pressure (COV_{IMEP}). Control of combustion timing is essentially done to achieve maximum work output from the engine and to avoid high pressure rise rates, partial burns, or misfires. An early heat release can result in too high pressure rise rates which may cause engine to knock. Moreover, too late combustion events can lead to incomplete combustion which produces high levels of carbon monoxide (CO) and unburned hydrocarbons (HC). In addition, low temperature combustion event typically results in low exhaust gas temperatures as low as 120 °C which limits the effective performance of the oxidation catalyst in the after treatment system (Williams et al. 2009). Therefore, control of exhaust temperature is important in the LTC modes. The stochastic nature of combustion event can be ascertained by the variations in the engine load output. High cyclic variations in IMEP ($COV_{IMEP} > 5\%$) produce fluctuations in engine torque and speed that affect the drivability, noise, vibration, and harshness (NVH) performance of a vehicle. Thus, COV_{IMEP} is determined over the steady state engine operation and is constrained below 5%. The study reported in Batool et al. (2021) showed 94% of data points have $COV_{IMEP} \leq 5\%$ among 210 different engine operating conditions in HCCI mode. 66% of the data among 300 different RCCI operating conditions showed $COV_{IMEP} \leq 5\%$.

3.1.1 *Limitations of LTC Operation*

LTC modes are prone to limitations in terms of load and speed operating range, engine-out emissions, and thermal efficiency. Despite the high thermal efficiency, ultra-low engine-out emissions, and low cyclic variability, HCCI application is limited owing to the limited operating range. The major factors that limit the operating region of HCCI engine are misfiring due to high dilution and very high pressure rise rates causing high noise levels (Hyvönen et al. 2003). Misfiring limits the low load operation of HCCI. Too much dilution leads to insufficient temperatures required for ignition initiation; thus, resulting in misfire. Similarly, misfire or partial burn occurs at high speed engine operation due to decrease in the available time required for the combustion onset (Hyvönen et al. 2003). High load and high speed HCCI operation faces aspirating limitations. Moreover, lack of dilution in the high load HCCI operation causes very high combustion rates which leads to excessive pressure rise rates bringing forth the engine knocking problem with audible noises and NOx formation (Hyvönen et al. 2005).

Moreover, partially premixed compression ignition mode seems a promising solution to operate a compression ignition engine with gasoline in mid to high load range with high thermal efficiency and low NOx and soot emissions (Kalghatgi et al. 2006). In Kalghatgi et al. (2006), high load up to 15 bar was achieved with single gasoline fuel injection by using exhaust gas recirculation (EGR), boosting and late fuel injections. Multiple fuel injection strategy can be helpful in curbing high pressure rise rates (Ciatti and Subramanian 2011). Gasoline operated PPCI mode is commonly known as gasoline compression ignition (GCI) mode. In Manente et al. (2010), GCI operating range was extended from idle to full load conditions. The performance was characterized based on thermal efficiency and emissions. The study showed 50% brake efficiency at full load condition with low estimated NOx, soot, CO and HC emissions below US 10 and EU VI legislations using low octane number gasoline fuels. However, the poor thermal efficiency of the GCI operation makes it less favorable for idle engine operation despite the low NOx and zero soot emissions (Manente et al. 2010). Therefore, coupling the LTC modes for better thermal efficiency and low engine-out emissions seems an appropriate solution.

3.1.2 Benefits of Multi-mode Operation

The major drawback of LTC modes is their limited operating region. Coupling one or more LTC modes with either SI or CI mode can provide an efficient and clean engine operation. Prior studies (Najt and Foster 1983; Thring 1989) conducted on HCCI mode indicated the possibility of HCCI application for passenger cars by coupling it with the conventional spark ignition gasoline engine, owing to the limited operating region of HCCI. It was anticipated that this would provide diesel engine like fuel economy at idle and light load along with low emissions (Thring 1989). It was observed that LTC modes offer higher thermal efficiency; thus, requiring less amount of fuel for the same desired power output compared to the conventional SI engine. Multi-mode (CSI-HCCI) engine tested on dyno confirmed 10–15% improvements in fuel economy in the drive cycle (Fuerhapter et al. 2004).

HCCI, RCCI and PCCI combustion modes offer less NOx and soot as can be seen in Fig. 3.2. CO and HC emission levels from LTC modes are higher when compared to that of CDC. CDC and SI engines require a 3-way catalytic converter for exhaust after treatment. However, NOx and soot formation can be prevented in LTC modes with proper control. This can be observed in Fig. 3.2, the operating range of the three LTC modes lie outside NOx and soot formation regions which provide advantage over CDC. In Fuerhapter et al. (2004), the comparison of HCCI engine operation with the baseline SI engine has shown 36% reduction in NOx formation. Therefore, a multi-mode engine concept has a potential to meet the stringent emission regulations and the targeted fuel economy in the on-road vehicles. This can be achieved with an optimized mode switching operation with controlled combustion events while minimizing load fluctuations and engine-out emissions. Mode switching operation can be achieved by optimized adjustments of actuators.

3.1.3 Optimal Control of Multi-mode LTC Engine

Combustion events in LTC modes are predominantly governed by the chemical kinetics. The combustion initiation is very sensitive to the temperature and pressure changes at the start of compression stroke (Koopmans et al. 2003). HCCI mode does not offer direct control over the combustion process (Bengtsson et al. 2006). The thermodynamic state of the air–fuel charge in HCCI is dependent on the residual gas fraction trapped in the cylinder. By precise control of diluents and intake air states, combustion timing can be controlled in HCCI (Shaver et al. 2006). In dual fuel HCCI operation, premixed ratio of two fuels can also be used as a control knob (Bidaravatan 2015). In PPCI and RCCI modes, the timing of direct injection of fuel provides better control over combustion timing and rate of heat release. However, LTC modes are susceptible to high pressure rise rates which can cause engine knocking. That is why, closed-loop optimal control is the best choice for LTC engines. Similar challenges are faced during mode switching which depends on combustion process dynamics, cyclic coupling, actuator dynamics and direction of the mode switch. The common challenges encountered during mode transitions are shown in Fig. 3.3. Mode switching challenges can be wielded by precise actuation. High cyclic variations, high MPRR, partial burns or even failed mode transition can occur due to the poor choice of control parameters (Koopmans et al. 2003). Mode switching dynamics vary significantly with the direction of transition. This difference is quite evident in SI-HCCI-SI mode switching. HCCI to SI mode switching is considered simple because the SI mode does not predominantly depend on the previous cycle (Roelle et al. 2004). However, SI to HCCI mode switching is difficult because of the residual gas coupling dynamics (Roelle et al. 2004). The very first cycle after mode transition usually results in complete combustion accompanied by the high rate of heat release and advanced combustion due to the high temperature of residuals trapped from the last SI cycle (Roelle et al. 2004). This advanced combustion resulted in low engine work. It was also observed by Roelle et al. that the cycles

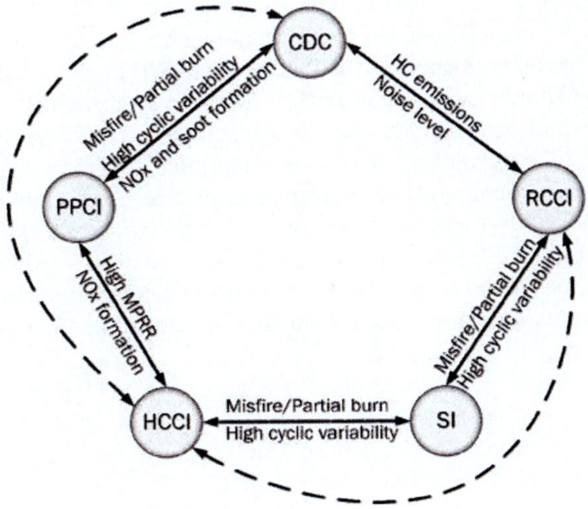

Fig. 3.3 Possible mode switching and complications

after mode transition are prone to misfire (Roelle et al. 2004). This was explained on the basis of heat transfer losses together with less initial chemical energy which led to reduced engine work and unstable HCCI (Roelle et al. 2004). Therefore, residual gas coupling dynamics must be accounted for a successful SI to HCCI mode switching. This requires optimal control over the process to retain appropriate amount of hot residuals to avoid excessive MPRR and to avoid misfire in the subsequent cycles. Although, HCCI to SI mode switching does not involve cyclic coupling but it can still face challenges of unstable combustion. This is mainly because of the actuator dynamics. HCCI is operated under lean conditions with wide open throttle (WOT) while SI requires stoichiometric air–fuel ratio which is commonly achieved with throttle position control for the required amount of intake air. Studies have shown that throttle response is a major factor behind an unsuccessful HCCI to SI mode transition event (Kakuya et al. 2008; Milovanovic et al. 2005). This is because of the slower intake manifold filling dynamics compared to that of valve profile switching. Therefore, it is important to take the actuator dynamics into account. This indicates the necessity of a model-based controlled mode switching event accounting for all the dynamics involved.

A multi-mode engine running in HCCI and PPCI mode using external EGR does not involve throttle dynamics and cyclic coupling. However, HCCI to PPCI mode transition still requires optimal actuation to avoid excessive MPRR and NOx formation. Furthermore, the transition from LTC-CDC results in engine load fluctuations, high MPRR and high engine noise levels (Kim and Bae 2013). High MPRR during LTC-CDC mode transition was mainly due to the reduced amount of EGR which resulted in advanced combustion and consequently led to increased peak cylinder pressure and high MPRR (Kim and Bae 2013). The study suggested that MPRR can be controlled using split fuel injection strategy and by reducing the fuel injection duration (Kim and Bae 2013). The study mentioned that the HC emissions reduced rapidly during LTC to CDC mode switching because of the oxidation of HC at higher combustion temperatures. In Tanabe et al. (2011), potential challenges in CDC-PPCI-CDC mode switching were NOx and soot spikes which were prevented by proper control of fuel injection and air management. Therefore, it is imperative to investigate all the dynamics involved in mode transition. The dynamics can be based on the cyclic coupling, actuator responses or direction of mode transition. A model-based control strategy proves to be the best choice for a stable and smooth transition. Furthermore, mode switching needs finite amount of time to stabilize. Thus, a multi-mode engine should be capable of smooth and fast transitions with minimum load fluctuations. A controlled mode switching assures continuous power delivery while avoiding misfire and safe engine operation. In addition, controlled combustion leads to an optimal combustion phasing which prevents high pressure rise rates; thus, constraining engine noise levels. Appropriate control actions by adjusting the injecting timing, injection duration and split fuel injection help in smooth transition and reduced combustion noise (Kim and Bae 2013).

3.2 Controlled Variables

Controlled variables or parameters are the quantities to be controlled to ensure fuel efficient, stable, and safe engine operation while meeting engine-out emission constraints. The challenges associated with the LTC modes can be addressed by implementing optimal control strategies. Model-based controllers are capable of providing optimized control over the LTC process by controlling combustion phasing, engine load, maximum pressure rise rate, coefficient of variation of indicated mean effective pressure (COV_{IMEP}) and exhaust gas temperature. Optimal control of these parameters helps in expanding the operating range of LTC modes. A brief introduction of the controlled variables is as follows:

3.2.1 Combustion Phasing

Combustion phasing (CA50) can be defined as the crank angle by which 50% of heat is released. Integrated Arrhenius Rate Threshold, Shell auto-ignition and Modified knock integral models can predict the start of combustion in LTC modes. The Shell auto-ignition model is a chemical kinetic based model that includes reaction mechanism from initiation to the termination. The model uses reduced order mechanism and predicts the auto-ignition timing accurately for a variety of temperature and engine speeds. Arrhenius-type reaction rate model can be used to determine the auto-ignition timing in the LTC modes. This is a semi-empirical based model which uses a single global reaction. Arrhenius rate model is used to compute the inverse of ignition delay by explicitly taking into account the fuel and oxygen concentrations. Integration of the inverse of ignition delay with respect to time predicts the onset of combustion when the integral becomes equal to a certain predetermined threshold. Knock-integral model predicts the onset of combustion. This empirical model computes ignition delay using a correlation based on in-cylinder gas mixture pressure, temperature, and engine speed. Integration of the inverse of ignition delay with respect to time predicts the start of combustion (SOC) when the integral becomes equal to 1. The start of combustion is important but not sufficient to explain the combustion process because of varying burn duration of combustion events happening at the same operating conditions. Therefore, CA50 is controlled to enable an optimal combustion event. Wiebe function is commonly used to predict CA50 for model-based control. Feedback from the measured in-cylinder pressure is commonly used to calculate the heat release that provides the combustion phasing. Alternately, in-cylinder ion current sensor can also be used as a feedback sensor, but its performance is limited under lean air–fuel mixtures. Other indirect methods used for the calculation of combustion phasing include measurements from microphones and knock sensors (Souder et al. 2004) and on-board torque sensors (Taraza et al. 1998).

Combustion phasing is one of the most important parameters to be controlled. Variations in CA50 can result in increase or decrease of engine work output, maximum pressure rise rate and emissions. Proper combustion phasing results in maximum brake torque, low maximum pressure rise rates, low NOx, and unburned hydrocarbon emissions. HCCI mode lacks the direct control over the combustion process; therefore, it is indirectly controlled by varying the variable valve actuation to trap the required residual gas fraction or using external exhaust gas recirculation (Asad et al. 2012). Furthermore, intake air temperature can also be used to control combustion phasing. To address the slow response of intake air heating actuation, fast thermal management system has been developed and tested for the closed loop control of combustion phasing (Haraldsson et al. 2004). Combustion phasing can also be controlled with variable compression ratio (Hyvönen et al. 2003; Shaver et al. 2006). In dual fuel application, premixed ratio of the two fuels provides an effective mean to control the combustion phasing (Bidarvatan et al. 2015). In PPCI and RCCI modes, the timing of directly injected fuel is used to control combustion phasing along with EGR (Yang et al. 2020; Raut et al. 2018a).

3.2.2 *Engine Load*

The main purpose of running an engine is to obtain the desired power output. Engine load output is often termed as indicated mean effective pressure (IMEP) or net mean effective pressure (NMEP). It can be easily computed from the in-cylinder pressure trace. Sometimes, peak pressure is considered related to engine load and therefore controlled (Shaver et al. 2005). The total amount of fuel injected in each cycle is used to control IMEP. It is important to control the engine load along with the control of combustion phasing. Partial or incomplete burn in engine cycles result in lower IMEP which is usually followed by high IMEP due to the unburned fuel available from the previous cycle (Milovanovic et al. 2005). This results in high coefficient of variation of IMEP. Simultaneous control of CA50 and IMEP can help in constraining the coefficient of variation of IMEP below 5%.

3.2.3 *Exhaust Gas Temperature*

One of the drawbacks of the LTC modes is the production of carbon monoxide (CO) and unburned hydrocarbons (HC) which can be reduced using three-way oxidation catalysts. In LTC, lean and homogeneous air–fuel mixture leads to lower and uniform temperature distribution throughout the cylinder with lower heat loss resulting in higher thermal efficiency (Krishnamoorthi et al. 2019). Crevice effect results in reduced combustion efficiency and increased HC, CO and NOx emissions. Fuel injection timing, fuel injection pressure, spray wall impingement, increased EGR (mostly beyond ~ 30%) and increased premixed ratio of fuels are the common factors affecting the combustion efficiency (Krishnamoorthi et al. 2019). The crevice effect zone can be avoided by controlling the fuel injection timing (Li et al. 2014). Due to higher thermal efficiency of LTC engines, most of fuel energy is released during expansion stroke; thus, leaving less exhaust gas energy compared to conventional SI and CI engine. In addition, combustion temperature is low in LTC modes due to lean burn or highly diluted engine operation. These lead to low exhaust gas temperature in LTC modes. This low exhaust gas temperature limits the performance of oxidation catalysts. Furthermore, low exhaust gas temperatures offer low exhaust enthalpy; thus, limiting the available work output extracted by the turbocharger. This limits the performance of turbocharger in high load LTC engine operation which requires significant intake boosting (Wissink 2015). In addition, exhaust gas temperatures from LTC modes can be as low as 120 °C, however, catalytic converters typically require a light off temperature of well above 200 °C (Williams et al. 2009). This affects the performance of catalytic converter. Therefore, it is important to control the exhaust gas temperature to meet the light-off requirement of the after treatment system. In Bahri et al. (2013a), artificial neural network modeling was used to predict exhaust gas temperature for ethanol fueled HCCI operation. It was observed that the variations in combustion parameters such as start of combustion (SOC), burn duration (BD) and maximum in-cylinder pressure (P_{max}) were not effectively correlated with variations of exhaust gas temperature. However, exhaust gas temperature was strongly correlated with IMEP and constant-volume adiabatic flame temperature (Bahri et al. 2013a). In Bidarvatan et al. (2015), exhaust gas temperature is controlled along with CA50 and IMEP for HCCI engine using intake manifold pressure, octane rating and mass of fuel as control variables.

There is a trade-off for optimal performance of turbocharger and exhaust after treatment system particularly during the light-off period. Due to high thermal efficiency in LTC engine, the exhaust gas temperature is low that limits the available heating for the exhaust after treatment system components and limits the exhaust exergy available for the turbocharger before entering components of the exhaust after treatment system (e.g., three-way oxidation catalyst). Thus, there exists an optimization problem to provide the desired power output by utilizing the turbocharger when desired, while ensuring the exhaust gas temperature is high enough to heat up the after treatment system quickly to ensure they have reached the light-off temperature (e.g., 300–350 °C).

3.2.4 Maximum Pressure Rise Rate

Maximum pressure rise rate (MPRR) is a measure of engine noise level. In LTC modes, a portion of the air–fuel mixture undergoes autoignition which can result in very high pressure rise rates. Therefore, MPRR is usually constrained below 8 bar/CAD in order to prevent the engine from knocking and controlling engine noise level (Yao et al. 2016). In HCCI mode, the homogeneous air–fuel mixture undergoes autoignition at multiple points inside the cylinder which results in rapid heat release rates causing very high MPRR. That is why, MPRR is an important engine safety parameter to be controlled. Advanced combustion phasing and shorter burn duration in HCCI lead to high MPRR and engine ringing (Bahri et al. 2016). In addition, a strong correlation was found between the engine noise level, in-cylinder peak pressure and in-cylinder pressure variation for HCCI mode (Bahri et al. 2017). In PPCI and RCCI modes, split fuel injections are used to control MPRR along with EGR. By controlling CA50 after TDC, the rate of heat release can be controlled; thus, preventing MPRR. MPRR in RCCI mode can be reduced by increasing the combustion duration which can be achieved by using the fuels with large difference in their reactivity (Ansari et al. 2019).

3.2.5 Engine-Out Emissions

In LTC modes, combustion temperatures are low which helps in preventing the formation of NOx. Diesel fueled PPCI operation requires low compression ratio and excess amount of EGR for low NOx and soot emissions, but it results in low thermal efficiency and high CO and unburned HC emissions (Manente et al. 2010). In Kalghatgi et al. (2006), GCI operation showed very low NOx and soot formation without EGR. In Manente et al. (2009), high octane number fuels result in lower NOx and soot production due to longer ignition delay in PPCI engine operation as compared to diesel. However, long ignition delay results in HC and CO emissions. Retarding the start of injection of fuel increases fuel stratification; thus, increasing the soot level in PPCI (Manente et al. 2009). It was suggested that using high octane number fuels can introduce longer ignition delay which helps in reducing soot formation (Manente et al. 2009). Furthermore, it was argued that improving the injection system can decrease the soot levels by employing higher injection pressures (Manente et al. 2009). Using advanced injection system with common rail pressure up to 2400 bar and EGR in excess of 50%, full load GCI operation was achieved with low NOx, soot and CO emissions simultaneously (Manente et al. 2010). Transient PPCI operation with gasoline fuel can also lead to high MPRR and soot formation (Yang et al. 2017). In order to satisfy low emission levels, double injection strategy has been effective in constraining NOx and soot along with MPRR in GCI operation (Yang et al. 2020). In HCCI mode, HC and CO emissions are strongly correlated with the combustion phasing (Shahbakhti et al. 2010). Bidarvatan et al. developed grey-box predictive model to constrain the CO, HC and NOx emissions in HCCI operation (Bidarvatan et al. 2014; Bidarvatan and Shahbakhti 2014).

Combustion phasing can also be regulated for low engine-out emissions (Kalghatgi et al. 2006). In addition, adding EGR as diluent helps in the prevention of NOx but can lead to soot formation due to reduced oxygen concentration (Tanabe et al. 2011). Moreover, high fuel injection pressure improves the air–fuel mixing quality which prevents soot formation (Manente et al. 2010). Longer ignition delay provides more mixing time for air–fuel mixing which prevents soot formation (Manente et al. 2010). A study showed that proper selection of spray angle, decreasing the injection pressure and premixed ratio while advancing the start of injection timing can significantly reduce HC and CO emissions in RCCI engine operation (Nazemi and Shahbakhti 2016).

3.2.6 COV_{imep}

Cycle to cycle variability in the combustion events is indicated by COV_{IMEP}. COV_{IMEP} is a measure of variability in IMEP for a number of consecutive engine cycles at steady state operating conditions. It is defined as the ratio of standard deviation of IMEP to the mean of IMEP, given by:

$$COV_{IMEP}(\%) = \frac{\sigma_{IMEP}}{\mu_{IMEP}} \times 100$$

where, μ_{IMEP} and σ_{IMEP} are the mean and standard deviation of IMEP, respectively.

High cyclic variability results in engine speed fluctuations and increases engine-out emissions (Mauro et al. 2019). Noise, vibration and harshness (NVH) performance of a vehicle is affected by high COV_{IMEP} (Zhu et al. 2016). LTC modes, operated under ultra-lean conditions, are prone to high COV_{IMEP}. Therefore, combustion cyclic variations need to be controlled to enable stable engine operation while maximizing thermal efficiency and reducing engine-out emissions. Engine operating cycles undergoing partial burn or misfire result in high cyclic variations.

Kalghatgi et al. showed a linear regression model using CA50 and equivalence ratio (φ) to differentiate low and high cyclic variability for HCCI combustion (Kalghatgi and Head 2006). To understand the cyclic variation in HCCI mode, a study by Shahbakhti et al. identified three distinct patterns of cyclic variations in combustion peak pressure, IMEP and ignition timing (Shahbakhti et al. 2007a). These patterns included normal cyclic variations, periodic cyclic variations, and misfired ignition variations. More cyclic variations were observed in the second stage of heat release for different primary referenced fuel (PRF) blends as compared to the first stage of heat release. The study also identified that cyclic variations were highly dependent on the start of combustion timing (Shahbakhti et al. 2007a). Low COV_{IMEP} can be achieved in HCCI mode by using low exhaust gas recirculation (EGR) rate, high equivalence ratio and high intake temperature (Shahbakhti and Koch 2008). Hellstrom et al. regulated CA50 using fuel injection timing as control knob with proportional integral (PI) and linear quadratic Gaussian (LQG) controllers to reduce cyclic variability in HCCI engine operation (Hellstrom et al. 2014). Bahri et al. developed artificial neural network model to detect the misfire cycles for ethanol fueled HCCI operation (Bahri et al. 2013b). In RCCI mode, COV_{IMEP} can be reduced by retarding the injection timing, increasing the injection pressure, and using boosted intake pressure (Jia et al. 2015). CA50 can be adjusted to ensure stable RCCI combustion by limiting COV_{IMEP} for varying load and speed conditions (Raut et al. 2018b). Data-driven classification models for COV_{IMEP} have also been developed for HCCI and RCCI modes. Non-linear model predictive controllers (NMPC) were developed for HCCI and RCCI modes to limit COV_{IMEP} below 3% while regulating CA50 and IMEP (Batool et al. 2021).

3.3 Control Actuators

3.3.1 Variable Valve Actuation

Variable valve actuation (VVA) provides degrees of freedom for cam phasing on intake cam shaft, exhaust cam shaft, cam profile switching or combinations of cam phasing and cam profile switching. VVA can provide variable valve timing and variable valve lift. This helps in achieving the LTC modes by controlling the effective compression ratio and adjusting the amount of residual gas trapped inside the cylinder.

LTC operation is sensitive to charge conditions at intake valve closing (IVC). The charge temperature can be controlled by varying the IVC and exhaust valve closing (EVC) timings. Early or late IVC results in reduced effective compression ratio (Strandh et al. 2005). This can be understood by the fact that compression process begins when the piston starts moving from bottom dead center (BDC) to the top dead center (TDC). Reduction in effective compression ratio as a result of retarding the IVC is intuitive. However, the effective compression ratio in the case of early IVC during the expansion stroke is complicated to comprehend. The in-cylinder charge pressure is essentially equal to the intake pressure at the IVC. In case of early IVC, the cylinder charge undergoes expansion as the piston moves towards the BDC. This reduces the in-cylinder pressure below the intake pressure. When the piston starts moving towards the TDC, the in-cylinder charge undergoes compression in which a portion of compression work is utilized in bringing the pressure of the charge gases equal to the intake pressure; thus, resulting in overall reduced effective compression. Figure 3.4 shows effect of variations in the IVC timing on CA50.

Negative valve overlap (NVO) can be used to trap the required amount of residual gas fraction in the cylinder by altering the EVC. With an early EVC and late intake valve opening (IVO), the residual gases undergo recompression. In-cylinder pressure of HCCI operation with NVO causing an exhaust recompression can be seen in Fig. 3.5. Combustion phasing in HCCI mode can be controlled by varying the EVC to achieve exhaust recompression (Widd et al. 2011). The amount of trapped residual gas affects the bulk gas temperature and dilution rate in the cylinder. Increasing NVO increases the amount of trapped residual gas fraction which results in advanced combustion timing. However, the amount of residual gases are reduced for high load engine operation to reduce the charge temperature to control the combustion timing. This results in reduced dilution which leads to advanced combustion and very high pressure rise rates. Another strategy used to retain the residual gases in the cylinder is exhaust rebreathing which is achieved by re-opening the exhaust valve during the

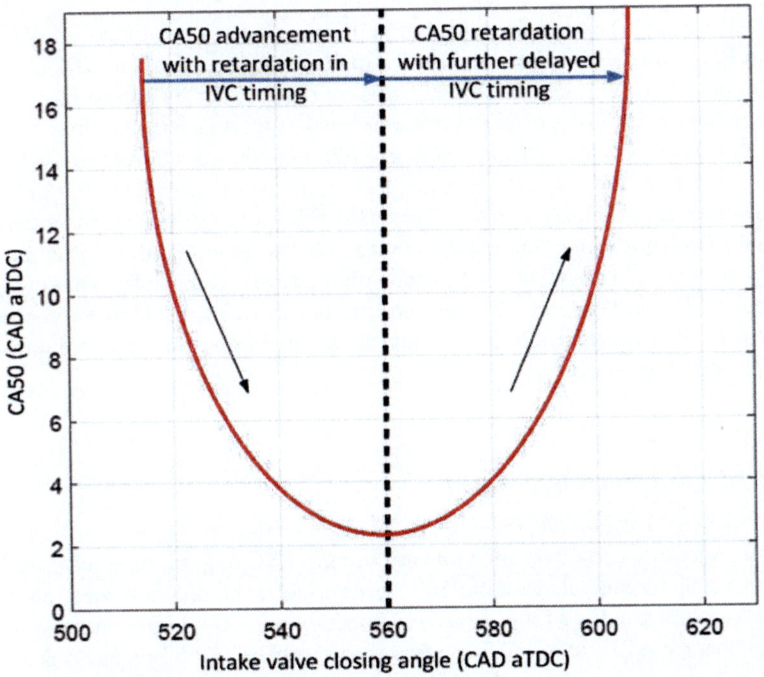

Fig. 3.4 Effect of intake valve closing timing on CA50 (aTDC) in HCCI operating mode based on (Strandh et al. 2005)

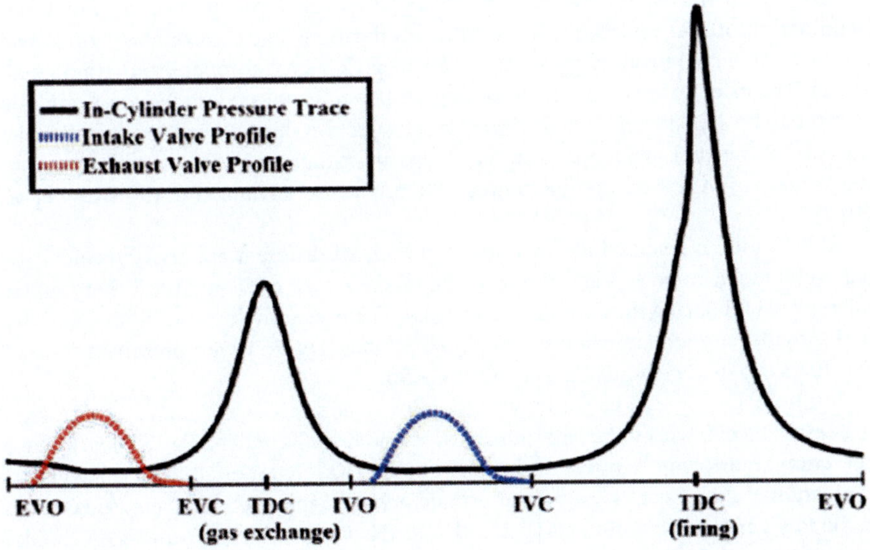

Fig. 3.5 A typical in-cylinder pressure trace of exhaust recompressed HCCI operation due to NVO [*Sources* Fathi et al. (2017), p. 10; Jade et al. (2014), p. 1410]

intake stroke. This helps in the reinduction of hot exhaust gases in the cylinder. Low load in PPCI operation can be achieved using NVO (Borgqvist et al. 2012).

Amount of in-cylinder dilution can be controlled either by controlling NVO by trapping hot gases during exhaust stroke (termed as internal EGR) or using external EGR. NVO can provide better residual gas fraction control on cycle-to-cycle basis; thus, it can be utilized for cycle-by-cycle LTC combustion control in each LTC mode or during mode switching. Changing external EGR will not be as fast as changing internal EGR for engine combustion control. On the other hand, the internal EGR will be hot that will negatively affect the volumetric efficiency of the engine, while external EGR can be cooled before entering the cylinder. In addition, the cooled external EGR is more beneficial in controlling the combustion temperature which reduces NOx formation.

3.3.2 Fuel Injection System

Port fuel injectors and direct injectors are used in LTC engines to adjust fuel injections and fuel–air equivalence ratio. Injection variables include fuel injection timing, fueling amount, number of injections and injection pressure. Different fuel injection strategies are used to realize LTC combustion modes. For homogeneous air–fuel mixture, fuel is injected during the intake stroke for HCCI via PFI or DI. For dual fuel HCCI, usually double PFI systems are used for fuel injection. Premixed ratio of two fuels provides a control knob over the combustion phasing. HCCI mode can use DI system with split fuel injection strategy. Split fuel injection strategy includes main and pilot fuel injections. Pilot injection during exhaust recompression stroke provides better controllability over CA50 in HCCI engine operation (Ravi et al. 2012). The effect of pilot injection timing with a constant main injection timing on CA50 can be seen in Fig. 3.6. An early pilot injection during recompression stroke provides more fuel residence time that helps in initiating the fuel reactions due to moderately high reaction temperatures which leads to advanced CA50 (Ravi et al. 2012).

RCCI mode is realized by injecting two fuels of different reactivity levels. Low reactivity fuel is injected via PFI during the intake stroke while high reactivity fuel is injected via DI during the compression stroke. The injection timing of high reactivity fuel provides a good control over combustion phasing. However, premixed ratio of two fuels can also be adjusted to control CA50.

Multiple fuel injection strategy with pilot injection during early compression stroke results in lower cyclic variations, lower smoke, NOx and maximum heat release rate compared to single injection in GCI engine (Kalghatgi et al. 2007). Kalghatgi et al. studied the effect of single and double gasoline injections on heat release rate in the PPCI engine operation using 25% EGR, 2 bar intake pressure and 40 °C intake temperature (Kalghatgi et al. 2007). The results for a single gasoline injection case and two cases with double fuel injections with constant pilot and main injection timings were discussed. For double injection cases, the pilot injection fueling rate

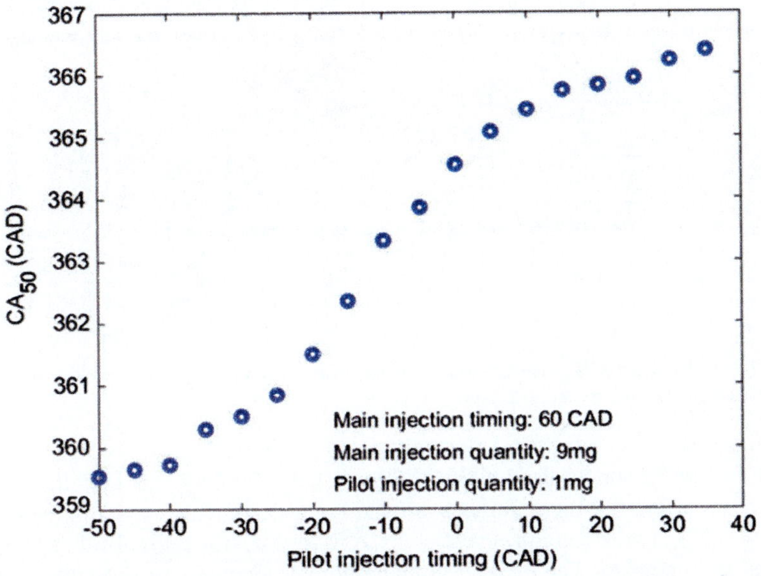

Fig. 3.6 Effect of pilot injection during exhaust recompression on CA50 of an HCCI engine [*Source* Ravi et al. (2012), p. 423]

was varied while keeping the main injection fueling rate constant. Double injection was helpful in attaining higher IMEP, lower NOx, smoke, HC and maximum heat release rate, compared to single injection (Kalghatgi et al. 2007). It was observed that increased pilot fueling rate in double injection led to increased smoke (Kalghatgi et al. 2007). Yang et al. used pilot injection timing to control soot emissions while pilot fueling rate was used to control MPRR for PPCI operation using 80% gasoline and 20% n-heptane (Yang et al. 2017). In addition, high fuel injection pressure can also be helpful in reducing soot emissions (Manente et al. 2009).

3.3.3 Fast Thermal Management (FTM)

Intake air heating is usually employed to realize the LTC modes. Intake air heating affects the combustion in LTC mode but due to the slow response, it cannot be used to control the combustion phasing on cycle-to-cycle basis. Fast thermal management (FTM) can heat the intake air quickly and can be used for control purposes. FTM system consists of a heat exchanger to extract the heat from the hot exhaust gases. The hot air and cold air are then mixed after the heat exchanger to attain the desired temperature of the intake air. The amount of hot and cold air is controlled by the throttle valves. Figure 3.7 shows the schematic of fast thermal management system used for closed-loop control of HCCI mode in reference (Haraldsson et al.

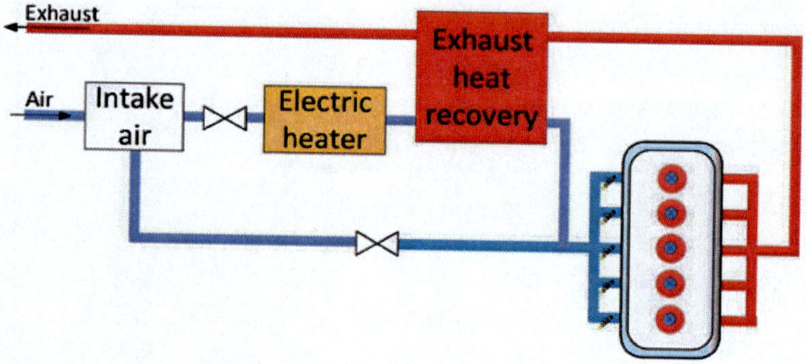

Fig. 3.7 Schematic of a fast thermal management system used for closed-loop control of HCCI mode based on (Haraldsson et al. 2004)

2004). In Haraldsson et al. (2004), CA50 was controlled by changing the inlet air temperature using fast thermal management. For a step change of CA50 from 1 to 9 CAD (aTDC), it took 8 engine cycles for CA50 to reach the desired value using fast thermal management. The inlet air temperature was decreased from 190 to 136 °C to retard CA50 to 9 CAD (aTDC). Moreover, it took 14 engine cycles for CA50 to reach the desired value using variable compression ratio (VCR). FTM proved to be a faster control actuator in controlling CA50 by changing inlet air temperature as compared to VCR (Haraldsson et al. 2004).

LTC is sensitive to air–fuel mixture temperature and cylinder wall temperatures. To this extent, adding a capability of precise control of coolant and oil temperatures should enhance the capability to control LTC combustion. However, the time scale will be in the order of seconds; thus, coolant and oil temperature control cannot be used for cycle-by-cycle combustion control, while they can be used for steady-state engine operation control. These are added to Sect. 3.3 for more explanation.

3.3.4 *Exhaust Gas Recirculation (EGR)*

NOx formation can be prevented by further reducing the combustion temperature in LTC modes. This can be done by either running the engine under lean condition, premixed or with EGR. High levels of cool EGR can delay autoignition; thus, resulting in retarded CA50. Late heat release occurs during expansion stroke with retarded CA50 decreases the overall combustion temperature; therefore, reduces the NOx formation (Kalghatgi et al. 2006). Further improvements can be achieved by optimizing injector design, swirl and multiple fuel injection strategy (Kalghatgi et al. 2006).

3.3.5 Intake Air Pressure Boosting System

Intake air pressure is increased either by using a turbocharger or a supercharger. In LTC modes, intake boosting is usually done to extend the load range (Han et al. 2010). The load range of 5–16 bar was achieved for PPCI operation in a heavy duty engine by using boosting, low pressure EGR and triple fuel injections per cycle (Yin et al. 2020). Variable geometry turbocharger (VGT) was used to maintain the intake manifold pressure at a desired value for the required engine load (Yin et al. 2020). Kalghatgi et al. investigated diesel and gasoline PPCI operation for low NOx and smoke emissions by using EGR at 2 bar intake air pressure using turbocharger (Kalghatgi et al. 2007). Shen et al. investigated the effects of EGR and intake air boosting on PPCI operation using diesel, gasoline and ethanol fuels (Shen et al. 2013). It was observed that increasing EGR led to prolonged ignition delay and reduced MPRR while increasing boost resulted in reduced ignition delay and increased MPRR. Increasing EGR became insignificant at high boost levels. It was suggested that reduced intake pressure or increased EGR could help in keeping low NOx emissions (Shen et al. 2013) However, this could result in high CO, unburned HC and soot emissions (Shen et al. 2013).

3.4 LTC Control

Combustion control in LTC modes is very important. Both open-loop and closed-loop control have been used for the LTC modes. Feedback or closed-loop control refers to the control frameworks that adjust engine control variables based on real-time engine feedback. LTC controllers are grouped into two categories: model-free controllers (e.g., PI, PD, PID, etc.) and model-based controllers (e.g., LQI, MPC, etc.). Model-based controllers are more appropriate for LTC mode switching. This is because they can incorporate understanding of the engine mode transition dynamics and actuator dynamics through control-oriented models. First, a dynamic LTC mode is defined by its mathematical model then control system is designed for the LTC mode. In order to control the combustion event in the LTC engines, control platforms are usually designed for tracking mode in which reference signals are provided and the system has to follow the desired trajectory.

Open-loop control provides the control action without any feedback knowledge of the plant performance because the control framework has no information about the plant outputs. To control the LTC modes, we need to track the desired combustion phasing and load while it is necessary to keep MPRR and peak in-cylinder pressures below a certain limit. This could be possible with a closed-loop control system in which the information about the parameters of interest is acquired from the LTC engine and fed back to the controller for appropriate action. This is critical for multi-mode LTC engines since different modes have different in-cylinder mixture requirements that affect the engine performance. This is particularly important during mode transitions since the engine can easily become unstable (e.g., misfire) if not controlled properly. Moreover, the nonlinear and complex behavior of some of the engine actuators makes it necessary to design a closed-loop LTC control. The controller frameworks employed in LTC mode studies can be divided into two categories namely, model-free and model-based control systems. These systems are discussed in the following sections.

3.4.1 Model-Free Closed-Loop Control Systems

Model-free closed-loop control systems are based on feedback measurement signals and present control laws that do not directly include a model of the system dynamics. The most common of these controllers include proportional control, derivative control, and integral control actions. Derivative control can never be used alone while proportional and integral control actions can used individually or combined. The common combined controllers include proportional-derivative (PD), proportional-integral (PI) or proportional-integral-derivative (PID) controllers. Better performance can be achieved by combining two or three control actions together. Increasing gain of the proportional controller often yields shorter settling time and smaller steady state error while integral action ensures zero steady state error. Derivative action corresponds to the rate of change of the error signal and provides a component proportional to the derivative of error signal. However, derivative action stops once the error becomes constant. Derivative action usually leads to a noisy operation because of the amplification of the signal noise; therefore, a low-pass filter is usually used. The other reason for using a low pass filter is to prevent saturation of the controller by avoiding the integrator windup.

Earlier studies on the control of LTC modes include model-free control framework implementation. These controllers are manually tuned and relatively simple to design. In Haraldsson et al. (2004), two PID controllers were designed to control CA50 and IMEP in HCCI mode by using fast thermal management to heat the intake air and by adjusting fuel quantity, respectively. This study also showed the regulation of CA50 using variable compression ratio. Similarly, Olsson et al. implemented two PID controllers for CA50 and IMEP control. Gain scheduled CA50 controller was designed as a function of sensitivity of CA50 to the octane number (Olsson et al. 2001). CA50 was controlled by adjusting the premixed ratio of two fuels while total fuel quantity injected per cycle was used to control IMEP (Olsson et al. 2001). The study showed the possible control of dual fuel operated HCCI engine (Olsson et al. 2001). CA50 and IMEP are interdependent and therefore, one parameter affects the other. This issue was highlighted in Olsson et al. (2001) while controlling CA50 and IMEP simultaneously. The performance of IMEP controller was compromised by making it slower than CA50 controller to effectively control CA50.

PID controllers are often single-input single-output (SISO) control systems. PID controllers often use a simple transfer function in which gains are tuned manually to achieve the desired performance. Given no model of system dynamics, PID controllers are not robust. In addition, these controllers require substantial calibration efforts for disturbance rejection and cannot provide optimal control action. On the contrary, model-based controllers can offer better performance by taking the interdependence of the engine parameters into account.

3.4.2 Model-Based Closed-Loop Control Systems

Model based controllers use the mathematical model of the LTC mode under consideration. The system dynamics, actuators dynamics and engine outputs can be well explained by the mathematical models that can be integrated into controller design. Model-based controllers can handle multi-input multi-output (MIMO) systems. In reality, mathematical models cannot exactly imitate the real systems. Therefore, robust controllers are designed to handle the inaccuracies or uncertainties in the engine models with minimal performance degradation. Linear quadratic regulator (LQR), linear quadratic Gaussian (LQG), linear quadratic integral (LQI), H_2 optimal control, sliding mode control, adaptive control and model predictive controllers are the commonly used linear optimal control techniques in LTC modes. Controllers, such as model predictive controller, sliding mode control, etc., can also be used for nonlinear control actions. Non-linear model-based controllers have also been developed for LTC modes as will be explained in subsequent sections.

3.4.3 HCCI Control

HCCI mode does not offer direct control over the combustion process. Combustion in HCCI undergoes auto-ignition at multiple points which result in very short burn duration and excessive MPRR. Therefore, it is important to control the combustion process in HCCI mode. In HCCI mode, the charge composition at the IVC is a dominant factor that affects the combustion timing in HCCI mode (Shahbakhti and Koch 2007). The commonly used actuators for the control of combustion timing in HCCI mode include variable valve actuation, variable compression ratio, inlet air heating, fuel injection strategies and premixed ratio of dual fuel. The widely used actuator for HCCI mode realization is the use of VVA. Residual gas fraction trapped in the cylinder and effective compression ratio are varied using VVA to adjust the combustion phasing effectively. Increased negative valve overlap (NVO) traps hot residual gases from the previous cycle which results in advanced combustion phasing. Varying the intake valve closing can change the combustion phasing because of the variation in the effective compression ratio. Shaver et al. designed an H_2 optimal controller based on a two-input and two output state space model to control CA50 and peak pressure by varying IVC and EVC to attain the desired residual gas fraction and effective compression ratio (Shaver et al. 2006). Strandh et al. compared the performance of model-based controllers i.e., LQG and MPC with PID controller for

CA50 control using VVA. The three controllers show good performance with MPC being the best one for CA50 control by adjusting IVC. The study suggested that HCCI control was limited using IVC alone as the control actuator; therefore, inlet air conditioning is required (Strandh et al. 2005). A sensitivity analysis carried out on HCCI identified that start of combustion is more sensitive to the variations in the mixture temperature as compared to the variation in EGR (Shahbakhti et al. 2007b). In Ravi et al. (2012), EVC timing together with pilot fuel injection timing during the exhaust recompression were used to control CA50. Bidarvatan et al. developed discrete sub-optimal sliding mode controller (DSSMC) based on the physics-based control-oriented model of HCCI engine (Bidarvatan et al. 2015). This study (Bidarvatan et al. 2015) controlled exhaust gas temperature along with CA50 and IMEP. An HCCI combustion event can result in very low exhaust gas temperature which can affect the performance of 3-way catalytic converter. Octane number (ON), mass of fuel injected, and intake manifold pressure were adjusted to control CA50, IMEP and exhaust gas temperature as shown in Fig. 3.8. The controller showed good tracking

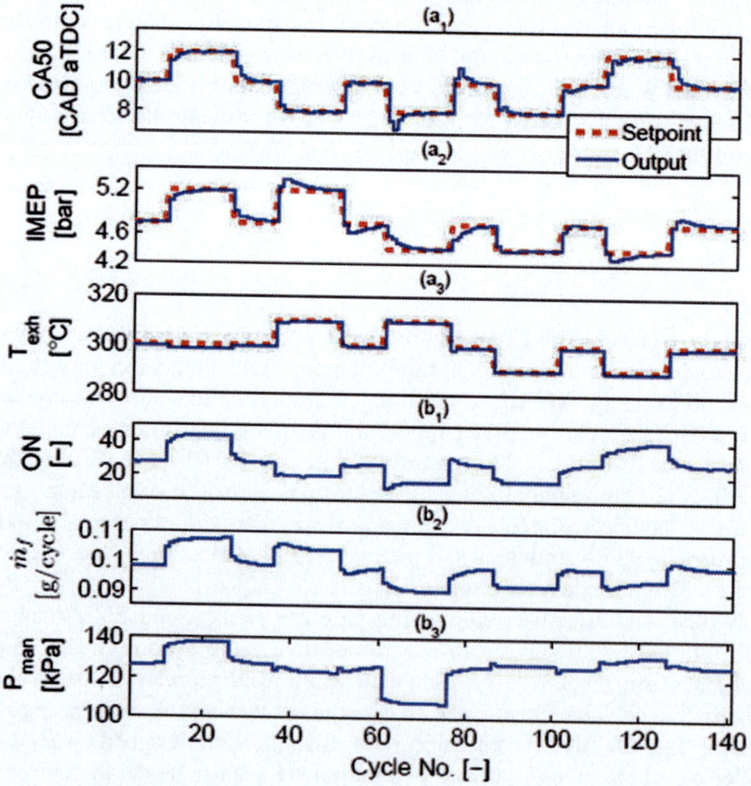

Fig. 3.8 Control of HCCI mode using PID and discrete sliding mode controller. a_x are the plant outputs while b_x are the control inputs [*Source* Bidarvatan et al. (2015), p. 11]

performance while tracking the plant outputs with settling time of 1 engine cycle and zero steady state error. Furthermore, the study in Bidaravatan (2015) compares the performance of DSSMC with the triple PID controllers to control CA50, IMEP and T_{exh}. Step changes for the desired values were provided to validate the controller performance for CA50 and IMEP over a range of engine operation. Comparing the DSSMC and PID controller performances, the output response as a result of PID control action showed fluctuations from the desired values of CA50 and T_{exh} when there were step changes in IMEP (Bidaravatan 2015). This can be explained by the fact that DSSMC is a model-based controller which takes the interdependence of the desired parameters into account which gives DSSMC advantage over PID controller.

Model predictive control (MPC) framework is a control technique which optimizes the control action over a finite time horizon in future. MPC can handle constraints on states, inputs, and outputs. Constraints on MPRR, peak in-cylinder pressure, emissions, and physical limits of engine actuators make MPC a favorable framework to optimally control the LTC modes. The only drawback of MPC is its relatively long computational time. HCCI operation in a heavy duty 6-cylinder Volvo diesel engine using dual fuels (i.e., ethanol and n-heptane) was controlled by an MPC (Bengtsson et al. 2006). A MIMO MPC framework using system identification model was used to control combustion phasing, IMEP and maximum pressure rise rate using IVC timing, amount of fuel injected, intake air temperature and engine speed as inputs (Bengtsson et al. 2006). Combustion phasing and load were simultaneously controlled together with the constraint on cylinder pressure on cycle-to-cycle basis in an HCCI (Bengtsson et al. 2006). The effects of separate constraints on inputs and plant outputs have been studied for an HCCI engine actuated with VVA. A MIMO MPC framework was used to control combustion phasing and IMEP (Ebrahimi and Koch 2015). This study (Ebrahimi and Koch 2015) suggested that the constraints on the outputs were sufficient to avoid ringing and misfire. Nonlinear MPC was developed for dual fuel HCCI operation to control CA50 and IMEP while constraining COV_{IMEP} below 3%. The nonlinear model for COV_{IMEP} of HCCI mode was based on a data-driven classification technique. The controller was capable of regulating CA50 to ensure stable HCCI operation by maintaining COV_{IMEP} below 3% (Batool et al. 2021).

3.4.4 PPCI Control

Unlike HCCI, partially premixed charge compression ignition engine provides a direct control over CA50 by adjusting the injection timing of the fuel. Engine load is controlled by the total amount of fuel injected in the cycle. Model predictive controller has widely been used in closed-loop control of PPCI operation owing to the fact that MPC provides optimal control with constraints handling. Based on the experimental studies, control-oriented models for PPCI are developed for model-based control (Yang et al. 2020; Yin et al. 2020). One such study is conducted by Yang et al. in which a double Wiebe function is used to model the combustion event (Yang et al. 2020). For dual fuel PPCI operation, the developed MPC not only controlled CA50 and IMEP but also constrained NOx, soot and MPRR below 400 ppm, 0.15 mg/m^3 and 8 bar/CAD, respectively. Duration of main fuel injection was used to control IMEP. To counter the effect of increased main injection duration, the timing of main fuel injection manipulated the combustion phasing as shown in Fig. 3.9. The control action to limit NOx emissions can be explained by the fact that when pilot injection advances, it prolongs the mixing period which results in a leaner low stratified mixture. This leads to a more premixed combustion thus limiting the NOx formation. Furthermore, increased amount of pilot injection also results in reduced NOx emission as observed in Kalghatgi et al. (2007). Heat release rate increases due to advanced pilot injection timing and increased pilot fueling rate (Yang et al. 2020). From the combined effect of the control actions, soot was significantly reduced while limiting NOx and MPRR (Yang et al. 2020).

Closed-loop control of PPCI in a heavy-duty engine delivered 5 to 16 bar engine load by utilizing intake air boosting, low pressure EGR and triple fuel injections per cycle (Yin et al. 2020). The desired CA50, intake pressure and EGR flow were determined from the engine load and speed performance. MPC was developed for controlling CA50, intake pressure and EGR flow using main injection timing, variable

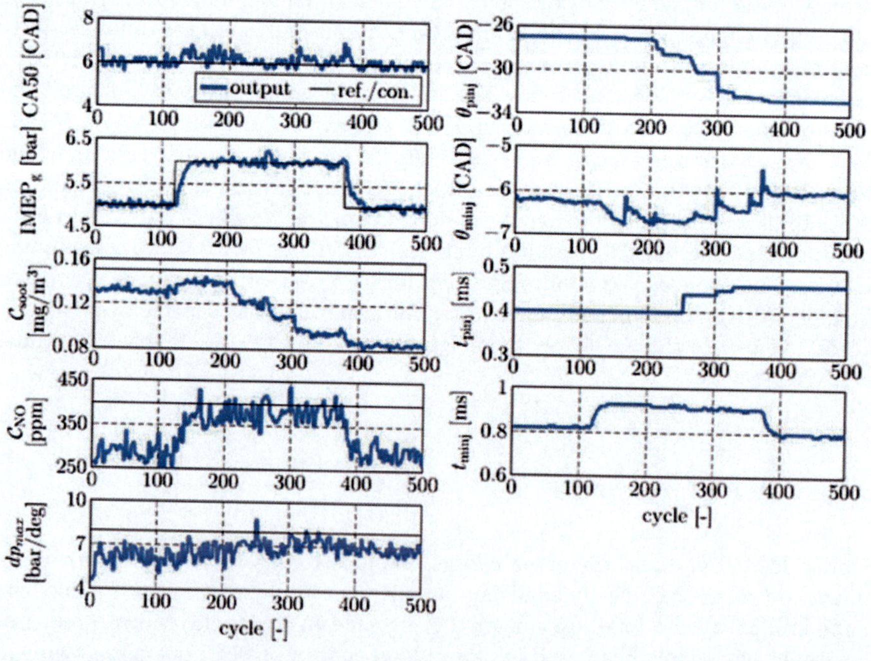

Fig. 3.9 Experimental results for control of PPCI mode in Scania D13 heavy-duty multi-cylinder engine using MPC with constraints on soot, NOx and maximum pressure rise rate. Subplots on the left side show the MPC outputs and state observer outputs. Subplots on right show the corresponding control actions [*Source* Yang et al. (2020), p. 9]

geometry turbocharger valve and EGR flow valve (Yin et al. 2020). However, engine load was controlled by using feedforward and PI controllers using fuel injection duration as manipulated variable (Yin et al. 2020).

3.4.5 RCCI Control

Reactivity control compression ignition mode usually employs port fuel injection system for low reactivity fuel while high reactivity fuel is injected via a DI system. The premixed ratio of two fuels, also known as blend ratio, can be used to control combustion phasing. In addition, injection timing of high reactivity fuel can also effectively control combustion phasing. Gasoline and diesel (Wu et al. 2014), iso-octane and n-heptane (Kondipati et al. 2017), gasoline and ethanol (Splitter et al. 2013), natural gas (NG) and diesel (Indrajuana et al. 2016) are some of the dual fuels used to achieve RCCI. Either physics-based or data driven control-oriented models are used for MIMO control of RCCI engines. In Indrajuana et al. (2016), a robust MIMO feedback controller was developed to control NG-diesel RCCI operation using diesel injection timing and total amount of fuel as manipulated variables to control ignition delay and engine load, respectively. NOx and HC emissions depend on ignition delay and blend ratio. Too long ignition delay and high blend ratio resulted in low combustion efficiency and even misfire. Thus, blend ratio along with ignition delay were controlled to avoid misfire and reduced NOx formation. The controller was based on the multi-zone combustion model (Indrajuana et al. 2016).

Raut et al. developed a physics-based control-oriented model to control RCCI mode (Raut et al. 2018a). RCCI operation was realized with iso-octane and n-heptane fuels. Two control knobs were identified for combustion phasing control based on the sensitivity map shown in Fig. 3.10. Either premixed ratio (PR) of the fuels or start of injection (SOI) of n-heptane was selected as a control knob for combustion phasing based on the sensitivity map. At constant engine speed, intake manifold temperature and pressure, PR and SOI sweeps were conducted to determine the sensitivity of CA50 corresponding to each control knob, see Fig. 3.10. R1 in Fig. 3.10 shows the region in which CA50 is more sensitive to the change in PR while R2 is the region where CA50 is more sensitive to the change in SOI as compared to the change in PR. Using this sensitivity map, the control knob for CA50 was selected based on the desired value. This required gain scheduled MPC capable of regulating different control actions based on the selected control knob for CA50 tracking. Figure 3.11 shows the schematic for a designed MPC controller for RCCI engine based on the sensitivity of CA50 w.r.t PR and SOI. Gain scheduled MIMO MPC controllers were developed to control CA50 by adjusting either PR or SOI timing based on the sensitivity and IMEP by adjusting amount of total fuel injected. In addition, Kalman filter was designed for state estimation. Data-driven RCCI model was developed by Irdmousa et al. using support vector machine (SVM) to obtain linear parameter varying (LPV) models as a function of fuel quantity. MPC was then developed based on the LPV models to control CA50 in RCCI engine operation (Irdmousa et al. 2019).

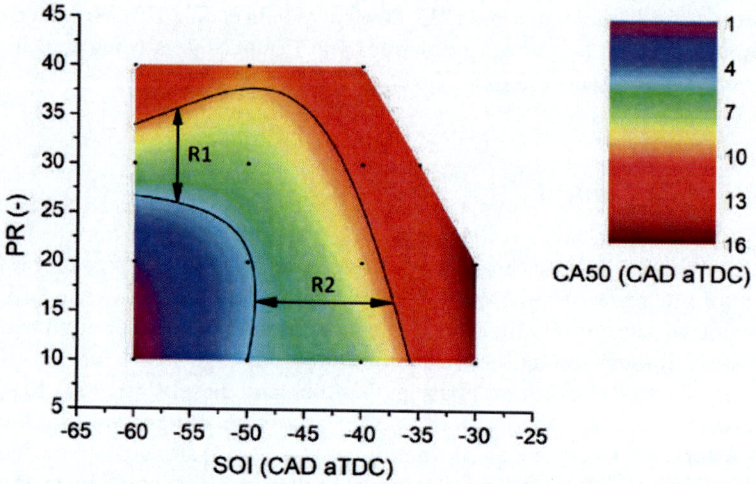

Fig. 3.10 Sensitivity map of CA50 w.r.t premixed ratio of iso-octane and n-heptane and SOI of n-heptane for control of an RCCI engine [*Source* Raut et al. (2018a), p. 140]

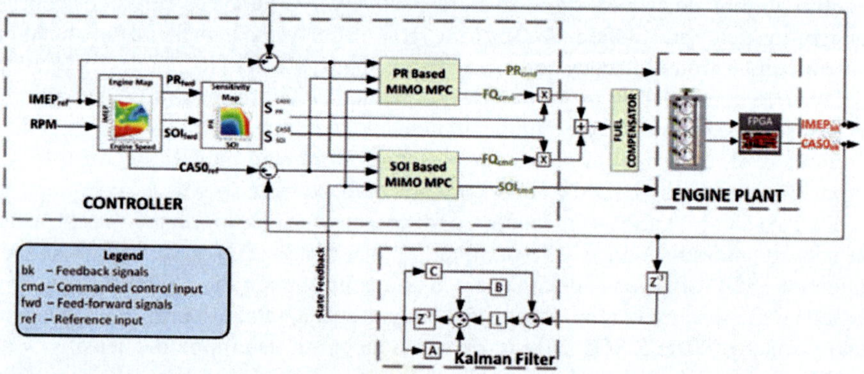

Fig. 3.11 Schematic of a designed MPC for RCCI control [*Source* Raut et al. (2018a), p. 140]

MPRR in RCCI operation can be constrained using MPC as shown in Basina et al. (2020) by constraining MPRR $\leq$ 6 bar/CAD. Based on SVM, an RCCI model was developed using premixed ratio as a scheduling parameter to obtain LPV models (Basina et al. 2020).

For high thermal efficiency and low NOx and soot emissions, Gaussian process regression was used for optimization of RCCI (Xia et al. 2020). Feedforward controller based on data-driven model used nested particle swarm optimization to maximize gross IMEP for fixed amount of fuel injection while satisfying constraints on MPRR, NOx, PM, CO and HC (Xia et al. 2020). It is important to prevent the cyclic variability in RCCI mode below 3% for stable engine operation. In Batool et al. (2021), SVM classification technique was used to develop COV_{IMEP} model for RCCI mode. Based on the model, a nonlinear MPC was developed for RCCI operation. SOI, PR and fuel quantity were used to simultaneously control CA50 and IMEP while constraining COV_{IMEP} below 3%.

3.5 Mode Switching Control

To take the full advantage of LTC modes, multi-mode combustion engine is a viable solution to increase engine speed and load for broad engine applications. This will require developing robust mode switching control strategies. The most common multi-mode engine concept is based on SI and HCCI mode switching. Other multi-mode engines include HCCI-PPCI-CDC, CDC-RCCI, and conventional dual fuel combustion to RCCI mode, as shown in Fig. 3.3. In a multi-mode engine, conventional SI or CI mode is often necessary for engine startup and high load requirements. However, multi-mode engine challenges depend on the modes and type of actuators being used. The common multi-mode engine challenges are summarized in Fig. 3.12.

Ultra-low emissions of LTC modes make a multi-mode engine desirable. Properly controlled mode switching operation can result in minimum load fluctuations and reduced emissions in LTC modes, as observed in Fuerhapter et al. (2004). CO, unburned HC, NOx and soot spikes can occur during mode switching operation. These undesirable emissions can be avoided by a controlled mode switching operation. Increase in dilution helps in avoiding NOx while insufficient in-cylinder oxygen concentration often leads to soot emissions. Therefore, it usually results in a trade-off between NOx and soot emissions. However, soot formation can be prevented by

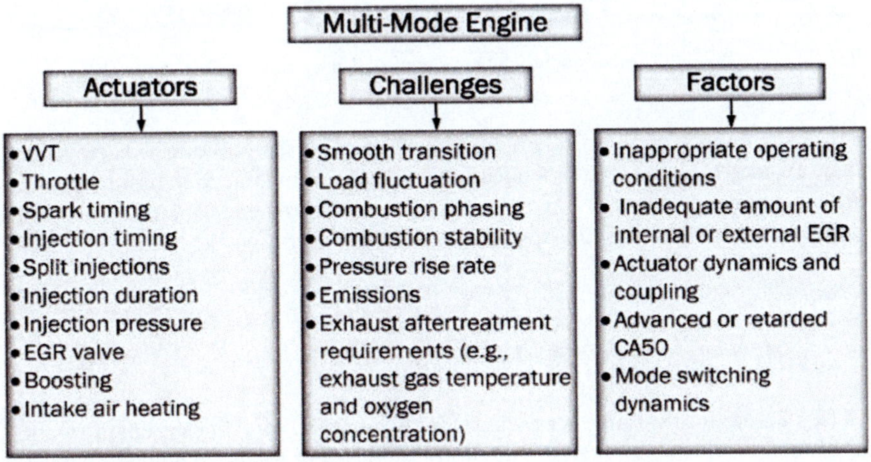

Fig. 3.12 Common multi-mode engine actuators, challenges, and associated factors

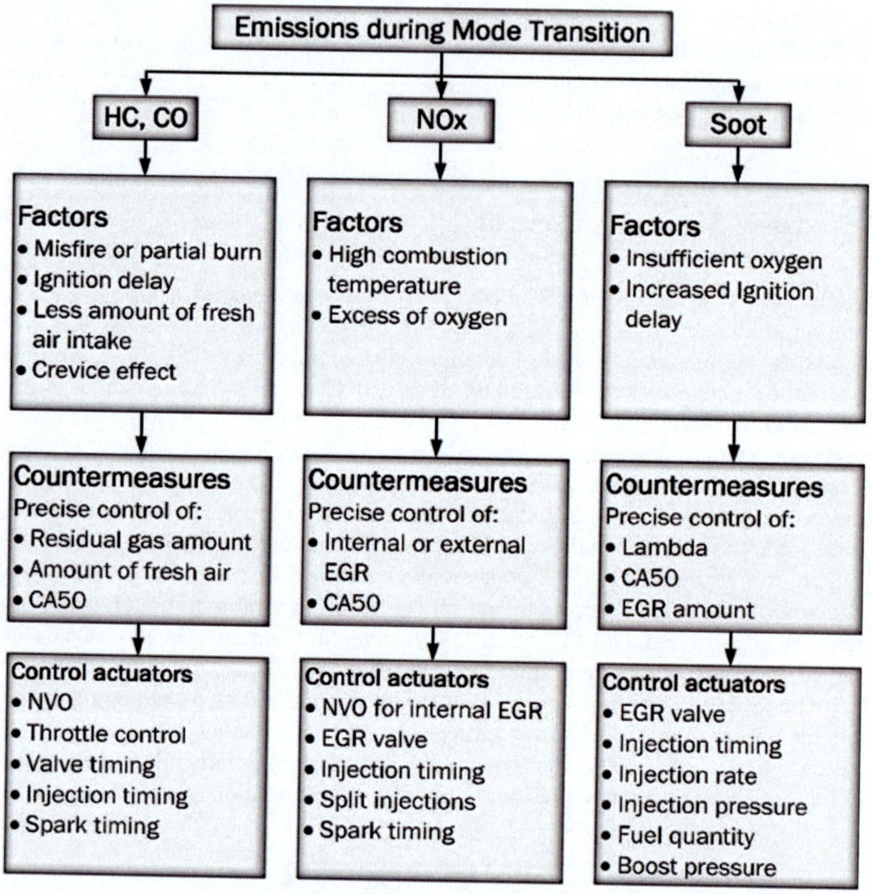

Fig. 3.13 Common multi-mode engine-out emission challenges and possible solutions

increasing injection pressure, injection rate and boost pressure. While NOx emissions can be avoided by using split fuel injections and controlling injection timing together with the amount of dilution. The common mode switching emission challenges and possible countermeasures are listed in Fig. 3.13.

3.5.1 SI-HCCI-SI Mode Switching

In a SI-LTC engine, SI mode is necessary for engine startup. After the engine warm-up, SI mode is switched to HCCI mode for an efficient low load engine operation with low engine-out emissions. Mode switching from HCCI to SI is required to achieve high engine load and speed. HCCI operation at high engine speed is likely to undergo misfire or partial burn due to decrease in the available time required for the combustion onset (Hyvönen et al. 2003). Furthermore, HCCI operation at high load causes very high combustion rates which leads to excessive pressure rise rates due to the lack of dilution (Hyvönen et al. 2005). Moreover, because of low pumping losses and the use of three-way catalyst for the exhaust after treatment, the performance of SI mode is comparable to HCCI mode at high load (Hyvönen et al. 2005). Thus, researchers suggested the viable concept of SI-HCCI-SI dual mode engine for a full speed-load engine operation.

Several rule-based and model-based studies using open-loop and closed-loop control have been conducted on mode switching to develop a multi-mode engine. The challenges in mode switching operation of SI-HCCI multi-mode engine and suggested solutions are shown in Fig. 3.14. To extend the operating range and to avoid frequent mode transition and misfired cycles, assisted spark stratified compression ignition (ASSCI) mode can be effective for successful mode transition between SI and HCCI modes. In ASSCI mode, fuel is directly injected before spark initiation to form a locally rich charge near spark plug. This leads to a two-stage combustion resulting from flame propagation and compression ignition (Li et al. 2014).

Rule-Based Mode Switching. The actuators used to realize SI-HCCI-SI mode switching may include throttle, variable valve timing, cam profile switch (CPS) system and phaser, DI and PFI injectors, spark plug, external EGR and variable compression ratio. The mode switching strategies can be grouped as a single step or two steps based on the actuator dynamics. A single step mode switching strategy is the one in which all the actuators are set to the required set points simultaneously

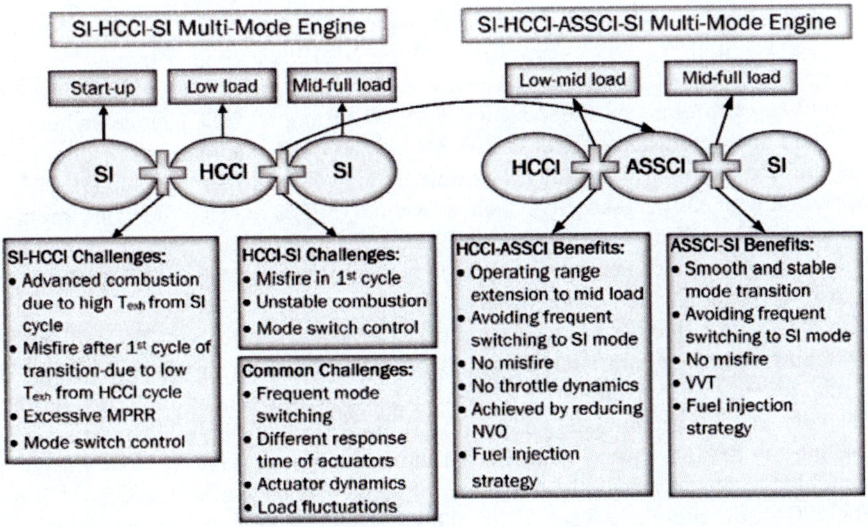

Fig. 3.14 Challenges associated with SI-HCCI-SI mode transition and benefits of introducing ASSCI mode in HCCI-SI mode switching

and mode switching is achieved in a single engine cycle. In comparision, two step strategy is the one in which mode transition takes place in two steps by activating the actuators based on their response time. In Tian et al. (2007), based on the actuator dynamics, cam profile switching, and throttle position were changed in two steps for successful SI-HCCI-SI mode transitions.

HCCI operation can be realized with variable compression ratio (VCR) engine actuated either by varying VVT or tilting the cylinder head (Hyvönen et al. 2005). In Hyvönen et al. (2005), an in-cylinder pressure sensor was used to provide combustion phasing feedback sampled with a fast data acquisition processor board. HCCI to SI mode transition in a multi-cylinder engine with VCR by tilting the mono-head was not successful in the study done by Hyvönen et al. (2005). The possible reason can be the fact that HCCI operates at high compression ratio with air as dilution while SI requires substantially low compression ratio and stoichiometric air–fuel ratio which was not possible to achieve from one engine cycle to the next cycle due to the hardware limitations used in Hyvönen et al. (2005).

The main challenge in mode switching is the precise setting of the control parameters. Poor choice of control parameters can cause high cyclic variability, high MPRR or even failed mode transition (Koopmans et al. 2003). Multi-mode engine realized with VVT can often result in successful mode switching using single step strategy. SI-HCCI-SI mode transition carried out on a Volvo 5-cylinder engine was successful with electronic valve timing control without throttling. For SI to HCCI mode switching, valve timings were changed to achieve NVO to retain hot residual gases from the SI cycle. HCCI is more sensitive to the temperature and pressure at IVC. SI-HCCI was easily realized with NVO resulting in an early combustion timing. Too advanced combustion in first few cycles of SI-HCCI mode transition resulted in reduced load, causing load fluctuations (Koopmans et al. 2003). When HCCI mode was stabilized after a few cycles, the resulting NMEP was higher than that of the SI mode because of higher thermal efficiency (Koopmans et al. 2003). Combustion stability was affected by the variation in NVO during SI-HCCI mode switching. IVC did not significantly affect CA50, but an early IVC had a positive impact on the fuel consumption in a Volvo 5-cylinder engine (Koopmans et al. 2003). HCCI operation with early intake valve closure resulted in 12% reduced fuel consumption as compared to SI operation at the same load. From HCCI to SI mode transition, valve timings were changed from NVO to normal valve overlap. HCCI to SI mode transition has proved to be more robust.

VVT is an effective control actuator to realize HCCI operation and mode switching. However, there still exists probability of misfire during mode switching. It is hard to stabilize the combustion event after a misfire because the unburnt fuel from the previous cycle accumulates and results in advanced combustion and high pressure rise rates in the subsequent cycle. Furthermore, the first cycle of SI to HCCI mode switching results in successful combustion followed by a cycle with misfire (Roelle et al. 2004; Ebrahimi and Koch 2015). This is essentially due to cyclic coupling of the hot residual gases from SI cycle that helps in initiating auto-ignition in the first HCCI cycle. While the residuals trapped from the first transition cycle should be sufficient enough to sustain the HCCI combustion in the subsequent cycles (Santoso et al. 2005). A stable combustion cycle followed by a misfiring event was observed in Santoso et al. (2005) during single step SI-HCCI mode switching. Therefore, two-step strategy was developed for a successful and stable mode switching operation. In Santoso et al. (2005), intake and exhaust valve timings were adjusted to switch from SI to HCCI mode while running both modes under wide open throttle condition to avoid manifold dynamics associated with the throttle control. Kistler 6125 pressure transducer was used for combustion feedback. Spark assistance at TDC was provided during the transition. To avoid misfiring, IVC timing was changed to an intermediate value in the first HCCI cycle followed by the desired IVC timing while EVC timing was changed in one step which resulted in robust transition from SI to HCCI mode. For HCCI to SI, EVC and IVC valve timings were retarded to trap required mass of fresh air. Figure 3.15 shows successful SI-HCCI mode transitions running un-throttled in dual modes using VVT.

For a successful mode switching event, it is very important to understand the effect of different actuators on the coupling of SI and HCCI modes. In addition, SI-HCCI-SI transition dynamics strongly vary depending on the direction of switch and operating conditions. It is important to identify the actuator response times and the coupling dynamics of different actuators. Throttle response has major effect on the mode switching performance even more than valve profile switching (Kakuya et al. 2008; Milovanovic et al. 2005). For instance, a slow or fast throttle response can lead to a misfire (Kakuya et al. 2008). HCCI to SI mode transition is considered relatively easy because of no cyclic coupling through the residual gas (Roelle et al. 2004; Liu et al. 2019). However, it is not always true because different actuators exhibit different dynamics during mode switching and their response times vary. In Milovanovic et al. (2005), SI-HCCI mode transition was successfully accomplished in one engine cycle using throttle control, VVT, cam profile switching and phasing at three different operating conditions. However, mode switch from HCCI to SI was unsuccessful in three different cases with different control strategies (Milovanovic

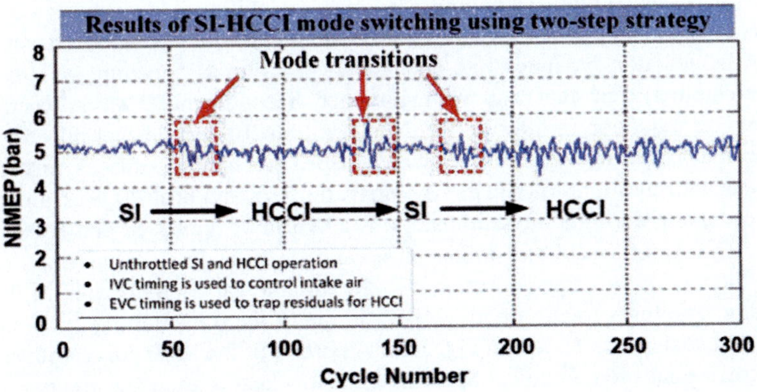

Fig. 3.15 SI-HCCI mode transitions at 1500 rpm and 5 bar NMEP in single cylinder engine with electromagnetic VVT system [*Source* Santoso et al. (2005), p. 10]

et al. 2005). The main reason was slow moving throttle valve which took 5–6 engine cycles to move from WOT to 45% position. Failed mode transition in Kakuya et al. (2008) corroborates the dependence on the direction of mode switching, actuator dynamics, and response times of actuators.

In Kakuya et al. (2008), the actuators' dynamics were investigated during SI-HCCI-SI mode switching. It was observed that the intake manifold pressure response changed at a faster rate as compared to the rate of internally trapped residual gases in SI-HCCI transition. The different response timings of throttle and VVT lead to misfire during mode switching. However, the dynamics were opposite for the HCCI-SI mode switch. Therefore, a two-step mode switching strategy was utilized with an intermediate operating point. In the first step, NVO and throttle opening were slightly increased. NVO was further increased for a stable HCCI operation along with wide open throttle (WOT) in the second step. An accurate NVO period and throttle position are essentially the key factors in mode transition. It was emphasized that a successful combustion event at an intermediate condition could only be possible by retaining precise amount of residuals and fresh air. The intermediate operating point was susceptible to abnormal combustion event because of the following two reasons:

- High internal residuals reduce combustion speed in SI combustion event that leads to prolonged combustion and reduced IMEP. If the internal residual gases are high enough compared with the stable SI operation, then it can cause misfire.
- Low internal residual gases cannot promote auto-ignition.

Optimal spark timing and use of secondary injection can be helpful during mode transition. Advanced spark timing can suppress the effect of hot residual gases on the combustion in SI mode (Kakuya et al. 2008). A second injection during transition creates stratification (Kakuya et al. 2008). In Kakuya et al. (2008), an overlapped timings of second injection and the spark resulted in a combustion event with combined effect of flame propagation and auto-ignition. This strategy resulted in a successful SI-HCCI mode switching.

A two-step strategy for HCCI-SI mode switching resulted in stable transition by taking the actuator dynamics into account. Firstly, throttle opening was reduced, and stoichiometric air–fuel ratio was maintained. Secondly, NVO was reduced along with further reduction in throttle opening. Additionally, second fuel injection with advanced spark timing was employed which helped in achieving a successful combustion event with auto-ignition and flame propagation (Kakuya et al. 2008). Combustion event occurring during a mode transition is a hybrid of typical SI and HCCI operations (Kakuya et al. 2008). Similar observation was made in Zhang et al. (2007) where the mode switching operation utilized variable valve timing and lift without throttling. The mass fraction burn (MFB) profile of the transitioning cycle showed two different slopes as shown in Fig. 3.16. A portion of the MFB curve with smaller slope corresponds to the typical SI event with slow heat release rate and the second portion of the MFB curve with larger slope can be associated with the faster heat release rate which is characteristic of an HCCI event from SI-HCCI mode switch (Zhang et al. 2007). This study used two ECUs, an EFI ECU to control fuel injection,

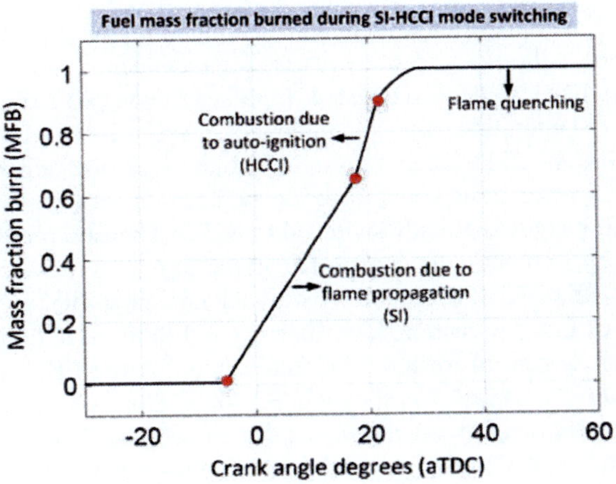

Fig. 3.16 Mass fraction of burn (MFB) fuel based on (Zhang et al. 2007) during SI-HCCI mode transition

ignition, and air–fuel ratio while the other ECU to control intake and exhaust valve timing and lifts (Zhang et al. 2007).

Cam profile switching and phasing have also been used as a control actuator for multi-mode engines realization using VVT and throttle control. Low lift profile with NVO is used for HCCI operation while SI mode is realized with normal overlap and high valve lift. However, the dynamics of actuators, in this case, must be taken into account to ensure stable mode switching. A study on SI-HCCI mode switching was conducted on gasoline direct injection engine equipped with throttle position, variable valve timing and lift profiles (Tian et al. 2007). For SI to HCCI mode transition, cam profile was switched, followed by the opening of throttle to wide open position. For HCCI to SI switch, however, throttle position was closed first followed by the cam profile switching in the subsequent cycle as shown in Fig. 3.17.

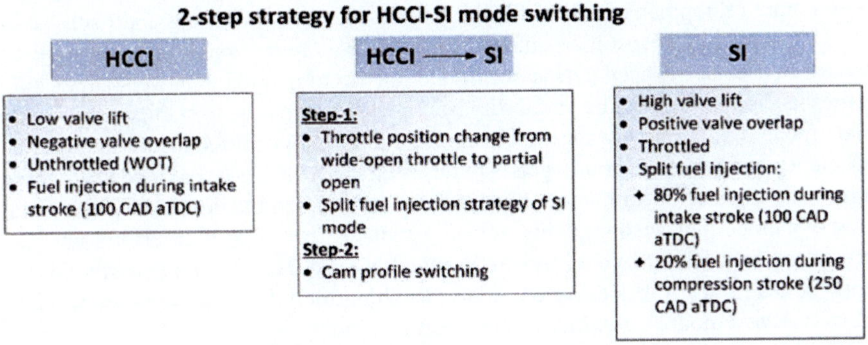

Fig. 3.17 Two-step strategy based on (Tian et al. 2007) for SI-HCCI mode transition

Spark assistance was provided throughout the mode switching process (Tian et al. 2007). In addition to the two-step strategy used for a stable mode switching operation, another strategy has also been suggested. In this strategy, two fuel injections were used for normal SI operation. Main injection took place at 100 CAD aTDC and pilot injection at 250 CAD aTDC in compression stroke to provide rich zone near spark plug. SI to HCCI switch failed because of delayed response of cam switching while fuel injection strategy was already switched to HCCI. The cam profile switch delay was prevented by switching the cam earlier in the last SI cycle to trap enough hot residuals for the first HCCI cycle followed by the change in fuel injection strategy. From SI to HCCI mode transition, low valve lift and short valve timings with NVO could retain less amount of fresh air which resulted in more THC emissions while NOx was reduced dramatically as shown in Fig. 3.18. However, the resulting HCCI cycles showed stable engine operation without knocking. Combustion phasing is usually advanced in HCCI mode because of the hot trapped residual gases from the previous SI cycle which leads to high peak cylinder pressure and MPRR (Tian et al. 2007). For HCCI to SI mode transition, the last four HCCI cycles before the switch were operated under NVO with the SI split fuel injection strategy. This helped in handling the high-lift cam switching. In this way a smooth and stable HCCI to SI mode switch was achieved. Though, there was a slight increase in THC emissions for the engine cycle number 82 and beyond as seen in Fig. 3.18.

Combustion in HCCI is rapid with high rate of heat release because of multiple autoignition points thus leading to high maximum pressure rise rate (Cairns and Blaxill 2007). From SI to HCCI mode transition, temperature of trapped exhaust gases and cylinder walls are high which results in advanced combustion timing of HCCI producing very high MPRR. External EGR can be used to reduce the audible knocking during transition (Cairns and Blaxill 2007). Secondary fuel injection during the compression stroke of SI-HCCI mode transition effectively reduces the rate of heat release and combustion noise without any additional dilution (Nier et al. 2012).

Model-Based Mode Switching. Multi-mode engine performance for high thermal efficiency and low emissions is contingent on mode switching event. A successful mode switching operation takes a few cycles to stabilize and provide a steady state load output (Koopmans et al. 2003). In rule-based mode switching approach, actuators' sequence is determined either through trial-and-error basis or using look-up tables. In case of misfired cycles during mode switching, engine power-output fluctuates leading to high cyclic variability, and it takes more time to stabilize the steady state mode operation with increased engine-out emissions; thus, losing the thermal efficiency and low emission targets. On the contrary, a model-based mode switching approach takes the cyclic coupling dynamics of the combustion modes and actuator dynamics into account. This not only reduces the calibration efforts but also provides a robust control over the mode switching process. The common modelling approaches used for model-based mode switching include semi-empirical models, physics-based models, and finite state machine models.

Furthermore, mode switching is a tricky task which can result in unacceptable engine operation with slight change in the operating conditions. Therefore, either

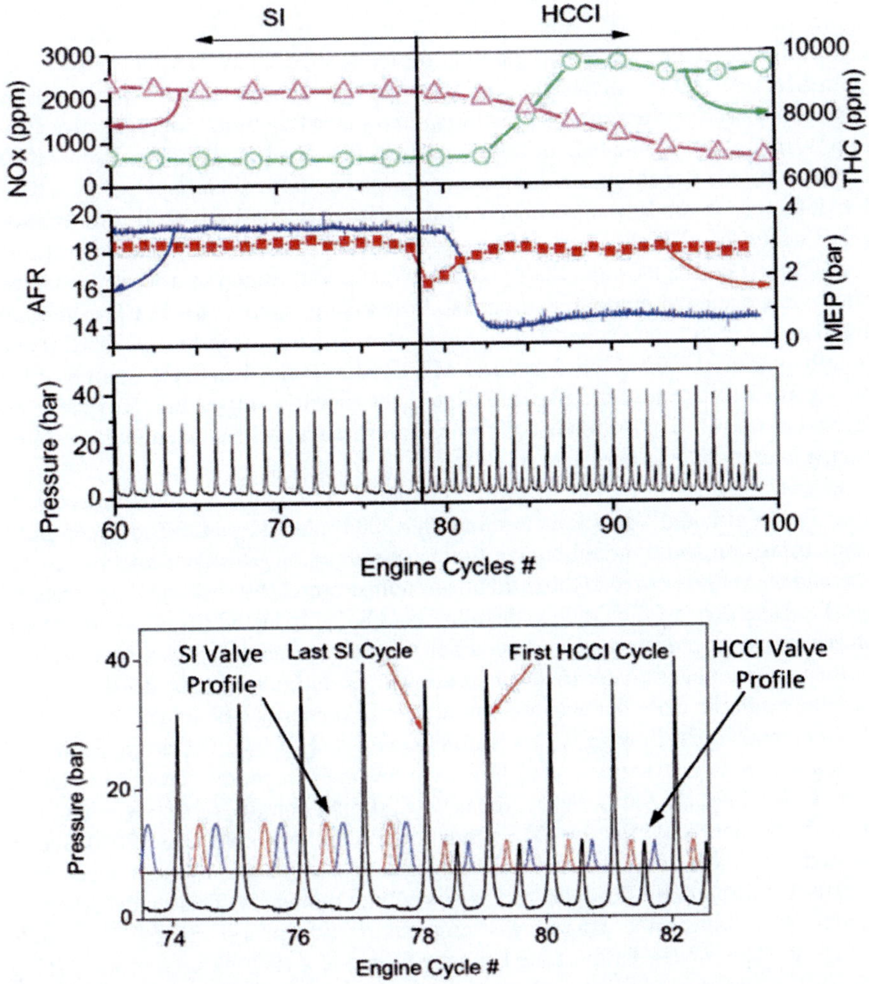

Fig. 3.18 SI-HCCI mode transition in a 2-cylinder gasoline direct injection engine [*Source* Tian et al. (2007), p. 7]

feedforward control or closed-loop transition control must be adopted. One such example of semi-empirical model-based mode transition was implemented for SI-HCCI-SI switching. In Fuerhapter et al. (2004), multi-cylinder gasoline engine was operated in SI and HCCI modes by modifying the intake cam phaser to provide high and low lift profiles and AVL electro-hydraulic valve actuation (EHVA) system was added at exhaust valve of each cylinder. EHVA system ensures fully flexible valve timing, and being able to vary valve opening, duration, and valve lift. Closed-loop combustion control was implemented for both SI and HCCI modes while model-based open-loop SI-HCCI-SI mode transitions were carried out. At low load, SI mode was switched to HCCI and back to SI mode using transition algorithm. This algorithm stored the valve closing timing for the first HCCI cycle after transitioning from SI mode and for steady state HCCI operation. Mode transition function was developed based on the stored valve timings and shape function. Model based corrections were made with the help of semi-empirical function based on load-temperature history of mode transition at different operating conditions (Fuerhapter et al. 2004). For instance, if mode was switched from SI to HCCI at high load, the exhaust gases would be at higher temperature as compared to the mode switching at idle conditions (Fuerhapter et al. 2004). Closed-loop combustion control was activated after a few cycles of mode transition. Mode switching from SI to HCCI resulted in significantly reduced NOx because of lower exhaust gas temperature in HCCI mode (Fuerhapter et al. 2004). Furthermore, MPRR less than 8 bar/CAD was observed during mode switching from SI to HCCI with the transition algorithm. Experimental results showed 36% reduction in NOx when compared with the conventional gasoline engine operation (Fuerhapter et al. 2004).

Physics-based SI-HCCI mode switching model for a light duty engine was developed for an efficient transition (Roelle et al. 2004). Actuators used to accomplish mode transition were variable valve timing and duration. Residual affected HCCI operation is very sensitive to the residual gas temperature. Hot residual gases from SI cycle are less dense which reduce the total mass being inducted (Roelle et al. 2004). Moreover, advanced auto-ignition occurs in the HCCI cycle which can be explained by the fact that heat transfer losses increase with early ignition thus, decreasing not only the available work but also less initial chemical energy which leads to unstable HCCI combustion following the transition (Roelle et al. 2004). Therefore, control of residual gases to be trapped during transition becomes imperative from SI to HCCI switch (Roelle et al. 2004). With proper valve timing control, mode switching can be made efficient and effective. The model developed in Roelle et al. (2004) incorporated the intake and exhaust manifold dynamics, combustion kinetics and heat transfer for SI and HCCI modes and the switching dynamics. This model was also capable of capturing the exhaust gas temperature dynamics of SI to HCCI mode switching. The model demonstrated successful mode switching while maintaining constant combustion phasing and work output. This model provided a pathway for closed-loop mode switching controller for constant combustion phasing and work output (Roelle et al. 2004).

Finite state machine can be used for modelling the SI-HCCI-SI mode switching. It can also be used to develop a discrete-time switching model based on the continuous-time system dynamics. One such study was carried out by Karagiorgis et al. (2007). Complete engine cycle was modelled into five mutually exclusive discrete-time subsystems. Each subsystem was modelled as continuous-time differential equation from intake air-flow dynamics to the exhaust. Look-up tables were provided to determine temperature and concentration dependent thermodynamic properties. State and charge gas concentrations computed by each subsystem were stored to set the initial values for the next cycle. Two strategies for SI-HCCI switching were employed in this study. First strategy used cam profile switching followed by changing throttle position to WOT and valve timing. Fuel injection strategy in HCCI mode used pilot

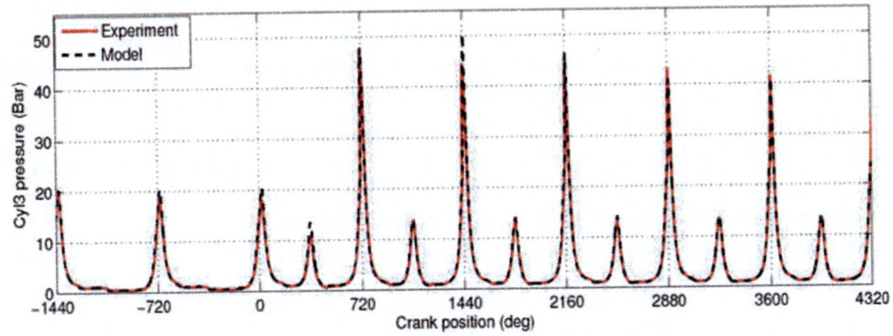

Fig. 3.19 Comparison of experimental SI-HCCI mode transition with the model predictions at 1500 RPM and 2.62 bar IMEP for a 2.0L 4-cylinder GDI Ford engine [*Source* Karagiorgis et al. (2007), p. 67]

injection during NVO and main injection during intake stroke. In this strategy, the last SI cycle ran in lean condition which led to partial burn while the subsequent HCCI cycle showed advanced auto-ignition with rapid combustion. The second strategy was carried out at fixed valve timing and spark timing during mode transition. However, binary valve lift was switched to low lift for HCCI with phased throttle opening. Fuel amount, injection timing and fuel split ratio were adjusted. The developed model for mode switching was able to capture transition dynamics as shown in Fig. 3.19.

Model-based SI-HCCI-SI mode transition for an engine equipped with cam profile switching and phasing system was developed in Gorzelic (2015). Steady state experimental data was used to parameterize the model for a wide range of conditions to include the operating points pertinent to cam switching used for mode transition. First, open-loop mode switching was carried out to study the dynamics during the transition. For SI-HCCI mode switching, throttle was switched to WOT before cam switching while maintaining constant load and AFR. Low lift was achieved by switching the intake and exhaust cams simultaneously during the final SI cycle. Spark timing was fixed at 20 CAD aTDC. EVC timing and start of injection timings were retarded for the first HCCI cycle to retard CA50. However, first HCCI cycle still resulted in advanced combustion timing and high MPRR because of high temperature of residuals from the SI event (Gorzelic 2015). Second cycle of HCCI was retarded because of the low exhaust gas temperature from the previous cycle. The third cycle of HCCI showed significantly enlarged peak pressure after TDC during recompression indicating recompression heat release rate (RCHR) of unburnt fuel from the main combustion event of the previous HCCI cycle (Gorzelic 2015). Combustion timing was advanced in the third HCCI cycle due to RCHR and early fuel injection; thus, stabilizing HCCI combustion. Based on transient data of mode switching, model was augmented with additional parameter to counter the transient effects. Comparing the model performance with the experimental data during transition, a correction factor was introduced for the residual gas temperature. This correction factor was tuned to match the CA50 of only first HCCI cycle after transitioning from SI mode as shown in Fig. 3.20. This correction factor was computed by back-calculation from the data using Newton–Raphson inversion by matching CA50. This can be computationally expensive for real-time application, but this needs to be computed only once for a new mode switch (Gorzelic 2015). Moreover, recursive least squares and simplified projected gradient algorithms along with stochastic counterpart can be used to tune the correction parameter for real-time applications (Gorzelic 2015). Significant improvements in developed model were observed for the conditions where steady state data was not available (Gorzelic 2015).

3.5.2 HCCI-ASSCI-SI Mode Switching

Due to HCCI limited speed-load operation and to avoid frequent mode switching by switching cam profile, VVT and throttle control, a new mode was introduced to bridge the gap between low load HCCI and high load SI operation (Li et al. 2014). This mode is named as assisted spark stratified compression ignition (ASSCI) in which fuel is directly injected before spark initiation to form a locally rich charge near spark plug. Multi-point autoignition is followed by the flame propagation which leads to two staged combustion (Li et al. 2014). ASSCI offers low pressure rise rate because of two-stage combustion resulting from flame propagation and compression ignition (Li et al. 2014). HCCI operation was achieved with NVO and WOT conditions. When the requested load exceeded the HCCI mode limit, NVO was reduced to achieve ASSCI without switching back to SI mode as shown in Fig. 3.21. Furthermore, hysteresis was introduced for a region to avoid frequent mode switching between ASSCI and SI which involved cam profile switch. In Li et al. (2014), split fuel injection in HCCI mode was used with one injection during recompression leading to fuel reforming and the second injection during intake for homogeneous air–fuel mixture. For ASSCI, one injection during intake stroke and the second injection during compression stroke for stratification were used. SI mode was realized with single injection during intake stroke. An open-loop transition was carried out from HCCI-ASSCI-SI modes. Spark assistance was provided at fixed timing (24 CAD bTDC) during mode switching.

ASSCI was achieved at stoichiometric conditions with reduced amount of internal EGR. First step involved closing the throttle from WOT to the ASSCI position. In second step, fuel strategy was changed from HCCI to ASSCI mode requirement and NVO was reduced to reduce the amount of internal EGR. The first cycle resulted in reduced IMEP because of the retarded combustion timing which can be explained by the delayed response of the NVO (Li et al. 2014).

For ASSCI to SI mode switch, fuel injection rate was increased first while spark timing was fixed at 24 CAD bTDC. Fuel injection strategy from ASSCI (50% fuel injection during intake stroke and 50% during compression stroke) was shifted to SI with split injections (60% fuel injection during intake stroke and 40% in compression stroke). Smooth and stable mode switching was completed in 2–3 engine cycles without misfire.

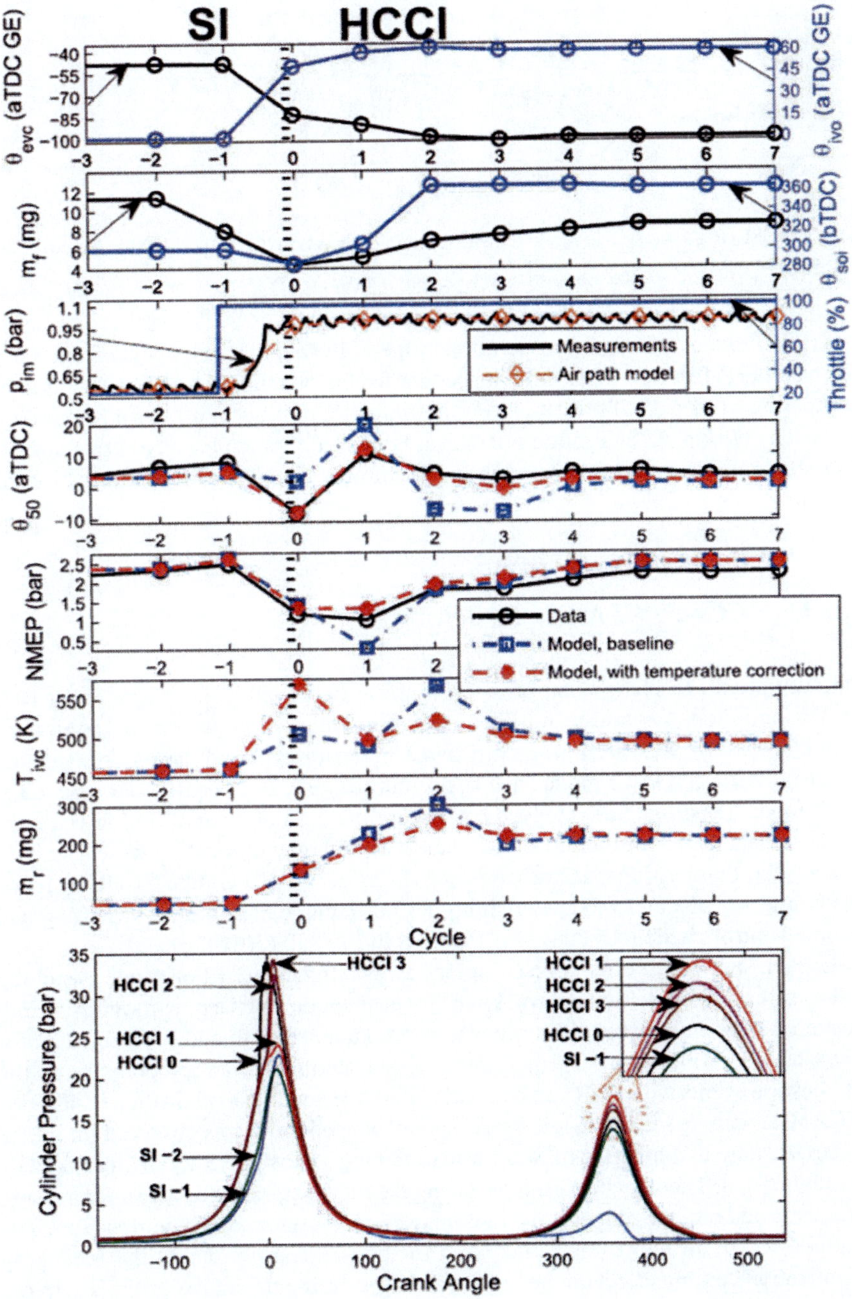

Fig. 3.20 Comparison of experimental open-loop SI-HCCI mode transition with the model prediction on cycle-to-cycle basis for a 2.0L, 4-cylinder engine. SI-1 represents the final SI cycle while HCCI 0 represents the first HCCI cycle [*Source* Gorzelic (2015), p. 50]

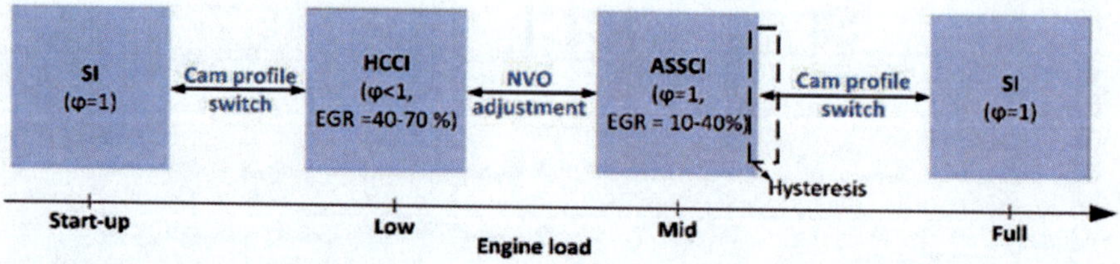

Fig. 3.21 SI-HCCI-ASSCI-SI mode transition strategy based on (Li et al. 2014) for a multi-mode engine operation

ASSCI mode extended the operation range of the engine under NVO from 3.5 to 6.5 bar IMEP. But this study did not account for the benefits of fuel consumption by comparing SI to ASSCI operation. Furthermore, there is no information about NOx emissions during ASSCI mode operation. However, this strategy confirms smooth and stable mode switching without any misfire or unstable combustion (Li et al. 2014).

3.5.3 HCCI-PPCI Mode Switching

Partially premixed charge compression ignition is proved to be a promising LTC combustion mode for mid to full load operation with low NOx and soot emissions (Yin et al. 2020). Research group at Lund University explored the mode transition between HCCI and PPCI modes in a heavy-duty engine. In Li et al. (2016), HCCI to PPCI mode transition was achieved by varying start of injection of fuel from intake stroke to near TDC. Furthermore, external EGR and inlet air heating were also used to keep the combustion phasing constant (Li et al. 2016). Unlike SI-HCCI mode switching, HCCI-PPCI mode switching did not include cam profile switching and throttle control challenges that were reported in Li et al. (2016).

Li et al. studied the effect of SOI, intake air heating and EGR on mode switching and identified the important control knobs for each mode and the transitioning period. Intake air temperature and SOI sweeps were conducted with and without EGR for a constant CA50 (Li et al. 2016). Effects of SOI and intake air temperature on the combustion events with EGR can be divided into five zones to explain the sensitivities to CA50, as can be seen in Fig. 3.22. A 'spoon' shaped trend was observed for intake air temperature as a function of SOI for maintaining a constant CA50 with 55% EGR (Li et al. 2016). The injection timings during the intake stroke and early compression stroke created a homogeneous air–fuel mixture resulting in HCCI combustion. The required intake air temperature increased in zone 1, possibly due to the fuel spray vaporization cooling effect on the inducted charge during the intake stroke. Required intake air temperature in zone 2 is independent of SOI (Li et al. 2016). CA50 was not significantly sensitive to the variations in SOI in zones 1 and 2. However, CA50 was highly sensitive to the variations in SOI and intake air temperature in zones 3 to 5 (Li

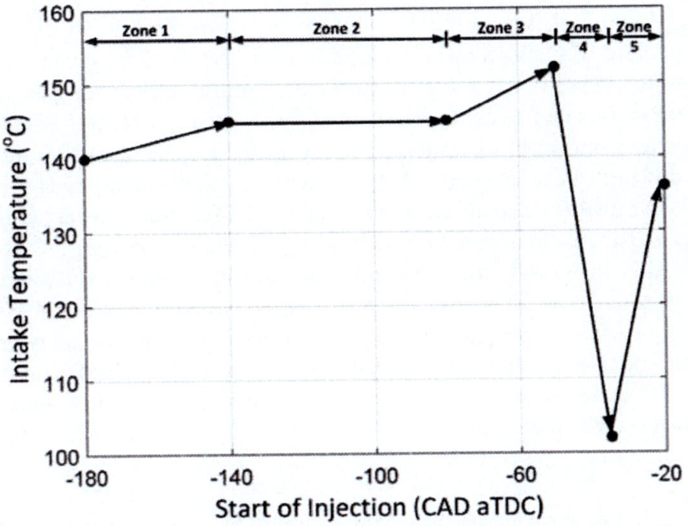

Fig. 3.22 Required intake air temperature to maintain CA50 at 3 CAD (aTDC) as a function of SOI for HCCI-PPCI transition for a single cylinder heavy-duty CI engine based on (Li et al. 2016). Higher T_{in} and lower T_{in} are the perturbations in the intake air temperature for the baseline reference points

et al. 2016). It is common intuition that increase in intake air temperature advances the combustion phasing. A later SOI in zones 3 and 5 retarded CA50; therefore, intake air temperature was increased to maintain constant CA50. SOI around -60 CAD aTDC in zone 3 showed a pit where the required intake air temperature was decreased because the fuel spray was entering into the piston bowl (Li et al. 2016). Fuel injections in zone 4 result in fuel stratification providing a local rich region which required low intake temperature. SOI from − 50 to − 45 CAD aTDC lead to increased HC emissions because a portion of fuel spray hit the top-land. When compared to the case with EGR, similar trends of SOI and intake temperature sensitivities were found without EGR with slight differences. The overall intake air temperature requirement was low without using EGR. In addition, zone 3 (crevice effect zone) and zone 4 (charge stratification) required much lower temperature compared to the case with EGR. The study suggested that SOI was an effective control knob for CA50 for zones 3–5. However, intake air temperature was the only option to control CA50 in zones 1 and 2 (Li et al. 2016). Furthermore, the maximum global temperature stayed between 1500 and 1700 K throughout the SOI sweep using EGR which was beneficial for CO oxidation and suppressing NOx (Li et al. 2016). CO and THC were found sensitive to the crevice effect (Li et al. 2016). Overall lower emissions were produced in the case with EGR (Li et al. 2016).

Shen et al. also studied the transition from HCCI to PPCI mode using gasoline-like fuel (Shen et al. 2015). HCCI to PPCI mode transition was carried out by using single and double fuel injection strategies, start of injection timing, EGR and intake boost pressure. Single fuel injection during intake stroke resulted in HCCI combustion because of homogeneous air–fuel charge. Fuel injection during late compression stroke resulted in premixed charge compression ignition. Transition period between HCCI and PPCI mode was associated with SOI range from − 80 to − 45 CAD aTDC. The challenges during this transition zone included excessive MPRR due to rapid burn rate and high NOx emissions. A double fuel injection strategy can be effective in HCCI-PPCI transition. Second injection with 30% of fuel injected very close to TDC can reduce burn rate and increase thermal efficiency while increasing NOx formation. Using EGR, NOx formation can be reduced. But use of EGR results in a trade-off between NOx and soot formation. Increasing boost pressure has a positive impact on reducing soot and HC emissions because of providing more oxygen concentration input to the cylinders. Therefore, split fuel injection strategy, EGR and high boost levels are effective means of controlling combustion and emissions during HCCI-PPCI transition (Shen et al. 2015).

3.5.4 CDC-PPCI Mode Switching

Mode switching between conventional diesel combustion (CDC) and premixed charge compression ignition is another interesting multi-mode engine concept as shown in Fig. 3.23. The engine runs in CDC mode during start-up and high load requirements. However, PPCI can be employed during low to medium load range for better efficiency and low emissions (Singh and Agarwal 2021). Transition period in CDC-PPCI switching can result in high NOx and soot emissions (Tanabe et al. 2011). Therefore, either rule-based or model-based feedback-controlled mode switching strategy is necessary for smooth transition with low MPRR and emission levels. In Tanabe et al. (2011), switching from CDC to PPCI resulted in NOx spike while smoke spike was observed during PPCI to CDC mode switching. A PID controller was used to regulate CA50 while feedforward controller regulated the mode transition by suppressing NOx and soot. NOx increased with advanced CA50 and increased oxygen concentration (> 17%). NOx becomes insensitive to CA50 if oxygen concentration is below 17% (Tanabe et al. 2011). CA50 was adjusted to avoid NOx spike by feedforward action while transitioning from CDC to PPCI when oxygen concentration decreased due to addition of EGR. Soot can be controlled by adjusting fuel injection and amount of air to maintain air–fuel ratio and by avoiding retarded CA50 (Tanabe et al. 2011).

Response time of actuators can also affect the mode switching process. Therefore, it is important to take the response time of actuators in account. Fuel injection timing, duration and number of injections can be controlled on cyclic basis. EGR delay can be improved by reducing the EGR path (Kim and Bae 2013). Fluctuations in power output and MPRR can be reduced by optimizing fuel injection strategy (Kim and Bae 2013). High fuel injection pressure improves fuel atomization and mixing thus reducing HC emissions (Han et al. 2013).

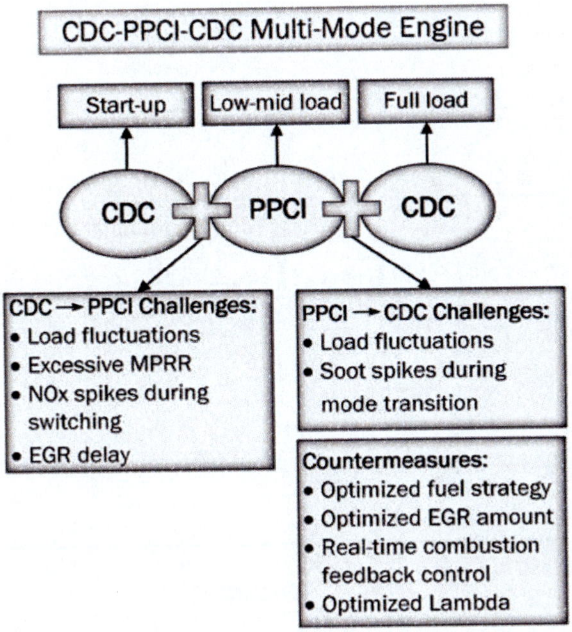

Fig. 3.23 Challenges and possible solutions during CDC-PPCI-CDC mode switching

3.6 CDC-RCCI Mode Switching

CDC to RCCI mode switching has been investigated at low load engine operations (Hanson and Reitz 2015). RCCI mode at low load results in low exhaust gas temperature which is not sufficient for catalytic converter (Hanson and Reitz 2015). In addition, appropriate strategies exist for cold start engine operation which make CDC operation more advantageous for cold start condition (Hanson and Reitz 2015). Therefore, mode switching between CDC and RCCI is required for cold start and low load operation. Moreover, RCCI operation is limited at high load because of excessive pressure rise rates which requires RCCI to CDC mode transition (Hanson and Reitz 2015). Fuels used to realize RCCI mode were non-oxygeneted gasoline (RON96) and ultra-low sulfur diesel (ULSD) while CDC mode used ULSD fuel. Swirl ratio and pedal positions were kept fixed for mode switching operation while two direct injections were used for both CDC and RCCI modes. RCCI mode also used PFI. EGR amount was increased from 28.5% to 50.9% for CDC to RCCI mode transition. Figure 3.24 shows the BMEP during RCCI and CDC mode transitions. RCCI operation showed slight improvement in BMEP as compared to CDC. The mode switching results showed that the combustion phasing and MPRR during mode switching were similar to the baseline RCCI and CDC operations. Furthermore, RCCI operation resulted in lower NOx and PM emissions. Smoke was increased from RCCI to CDC mode switching. HC emissions were observed during first 5 cycles of mode transition

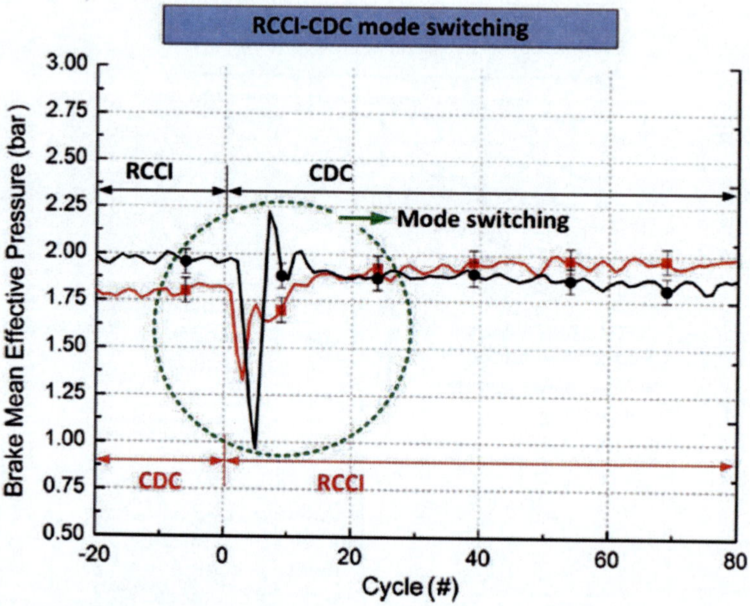

Fig. 3.24 BMEP during CDC-RCCI mode switching in a 1.9L, 4-cylinder Euro IV diesel engine [*Source* Hanson and Reitz (2015), p. 7]

to CDC mode. These emissions were linked to the unburned gasoline fuel which was entering the cylinder from the intake fuel puddle. However, gasoline injections were stopped during RCCI-CDC mode switching (Hanson and Reitz 2015).

3.7 RCCI-CDF Mode Switching

Multi-mode engine concept has also been applied to RCCI and conventional dual fuel (CDF) combustion modes in a heavy-duty engine. Dual fuel compression ignition has been used in marine and automotive applications using diesel with either ethanol or natural gas (Indrajuana et al. 2018). RCCI is a subset of CDF. RCCI utilizes the difference between reactivity of two fuels to control LTC combustion, while this is not the case in all CDF engine. The performance of CDF engine can further be improved in terms of thermal efficiency and reduced engine-out emissions by coupling it with RCCI (Indrajuana et al. 2018). Operating conditions to realize CDF combustion are different than RCCI. That is why, smooth transitions between RCCI and CDF modes are challenging (Indrajuana et al. 2018). Thus, a proper control strategy is essential to ensure smooth RCCI-CDF mode switching and emission control.

In Indrajuana et al. (2018), open-loop and closed-loop CDF and RCCI mode switching has been studied. CDF and RCCI modes were achieved using high pressure

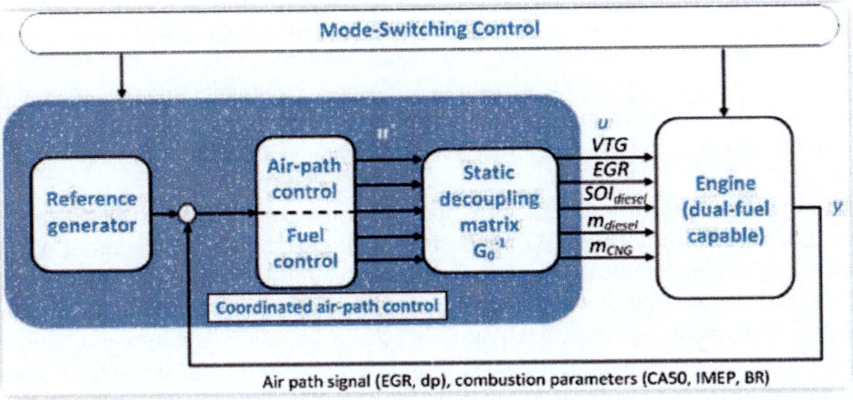

Fig. 3.25 CDF-RCCI mode switching schematic for a heavy-duty engine [*Source* Indrajuana et al. (2018), p. 7]

EGR with natural gas and diesel fuels. Mean-value engine modelling approach was used for air-path dynamic model. Two-zone combustion model for CDF and multi-zone RCCI combustion model were used. Schematic for mode switching control is shown in Fig. 3.25. Static input–output decoupling was used to account for the different dynamics offered by intake air path and fuel path, and to address the inter-dependence of control inputs and engine outputs. By using static decoupling, new control inputs (u*) were obtained as a linear combination of actual control inputs (u). This resulted in five independent single input–single output (SISO) subsystems. EGR valve and variable-turbine geometry (VTG) were the two available control actuators for air-path control.

Two PI controllers and a feedforward controller were developed to control EGR ratio and the pressure difference (dp) across engine. Air–fuel ratio was not closed-loop controlled. Fuel path control actuator actions included diesel injection timing, diesel quantity and natural gas quantity to control CA50, IMEP and blend ratio. High blend ratio results in reduced fuel consumption and low CO_2 and NOx emissions (Indrajuana et al. 2018). RCCI-CDF mode switching was done by controlling the actuators to track CA50, IMEP, blend ratio and pressure rise rate. Closed-loop controller was proved to be a better choice than an open-loop controller.

3.8 Summary

Based on thermal efficiency and engine-out emissions, LTC modes have a potential to meet the stringent fuel economy and emission targets. However, LTC modes are prone to excessive pressure rise rate due to autoignition which must be controlled by adjusting internal or external EGR, fuel injection timings, intake temperature and pressure. High MPRR limits the high load operation in LTC modes. Low load operation in LTC modes is limited due to risk of misfires and partial burns resulting in high engine torque fluctuations, high cyclic variability, HC and CO emissions. Overall, the limited optimal operation of LTC modes compared to conventional SI and CI modes limits the application of these engines. Multi-mode engine is a viable solution to benefit from the high thermal efficiency and ultra-low emissions of LTC modes.

A multi-mode engine can be developed by using conventional SI or CI modes along with one or multiple LTC modes for full load and speed operation. A multi-mode engine can operate in SI mode for start-up and full load operation while LTC mode can operate from low to mid load range. Among LTC modes, PPCI offers the highest load operating range. Each mode of engine operation requires certain inputs to the engine, as shown in Fig. 3.26. The optimal engine mode can be selected based on engine BSFC map and engine-out emission map. The optimization can be done in a high-level combustion optimizer that selects optimal engine mode and determines the desired IMEP and CA50 reference trajectories. Then the feedback combustion controller will adjust engine cycle-by-cycle control inputs, e.g., spark timing, SOI of pilot and main fuel injections, premixed ratio of fuels, valve overlap, etc. For real-time application, an observer-based controller framework can be developed to control CA50 without using an in-cylinder pressure transducer.

Several researchers have explored SI-HCCI, CDC-PPCI, HCCI-PPCI, HCCI-ASSCI-SI and CDF-RCCI mode switching for an optimal multi-mode engine development. These studies include rule-based open-loop and model-based closed-loop control of multi-mode engine with increased efficiency and reduced emissions. The

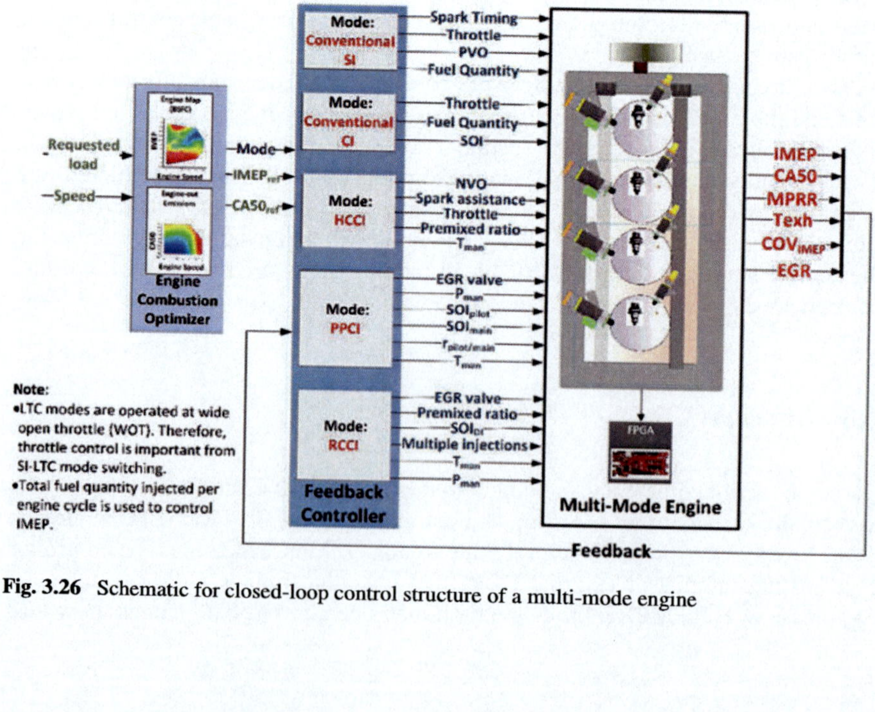

Fig. 3.26 Schematic for closed-loop control structure of a multi-mode engine

common challenges in mode transitioning include misfire, partial burn, advanced combustion causing high pressure rise rates, load fluctuations and engine-out emissions including NOx, soot, HC and CO. A multi-input multi-output (MIMO) model predictive control (MPC) framework is a promising solution to use for mode switching control that incorporates the operating constraints on the actuators (e.g., EGR, VVA, etc.,), state constraints (e.g., MPRR), and it includes mode switching dynamics via a control-oriented plant model. The major challenge is the limited operating range offered by the LTC modes. The mode switching challenges can be successfully addressed by designing optimal mode switching control strategies as discussed in this book chapter. Other emerging studies include electrified LTC powertrains that include mode switching between LTC engine modes and electric/hybrid/electric modes (Solouk et al. 2017).

References

Adams CA, Loeper P, Krieger R, Andrie MJ, Foster DE (2013) Effects of biodiesel-gasoline blends on gasoline direct-injection compression ignition (GCI) combustion. Fuel 784–790

Ansari E, Menucci T, Shahbakhti M, Naber J (2019) Experimental investigation into effects of high reactive fuel on combustion and emission characteristics of the diesel—natural gas reactivity controlled compression ignition engine. App Ener 239:948–956

Asad U, Zheng M, Tjong J, Wang M (2012) A control strategy analysis for clean and efficient combustion in compression ignition engines. In: The 8th international conference on modelling and diagnostics for advanced engine systems (COMODIA 2012)

Bahri B, Aziz AA, Shahbakhti M, Said MFM (2013a) Analysis and modeling of exhaust gas temperature in an ethanol fuelled HCCI engine. J Mech Sci Technol 27:3531–3539

Bahri B, Aziz AA, Shahbakhti M, Muhamad Said MF (2013b) Understanding and detecting misfire in an HCCI engine fuelled with ethanol. App Ener 108:24–33

Bahri B, Shahbakhti M, Kannan K, Aziz AA (2016) Identification of ringing operation for low temperature combustion engines. App Ener 171:142–152

Bahri B, Shahbakhti M, Aziz AA (2017) Real-time modeling of ringing in HCCI engines using artificial neural networks. Energy 125:509–518

Basina LNA, Irdmousa BK, Velni JM, Borhan H, Naber JD, Shahbakhti M (2020) Data-driven modeling and predictive control of maximum pressure rise rate in RCCI engines. In: 2020 IEEE conference on control technology and applications (CCTA), pp 94–99

Batool S, Naber JD, Shahbakhti M (2021) Data-driven modeling and control of cyclic variability of an engine operating in low temperature combustion modes. In: IFAC modeling, estimation and control conference (MECC), Austin, TX, USA

Bengtsson J, Strandh P, Johansson R, Tunestål P, Johansson B (2006) Multi-output control of a heavy duty HCCI engine using variable valve actuation and model predictive control. SAE Tech Pap 2006-01-0873

Bidaravatan M (2015) Physics-based modelling and control of powertrain systems integrated with low temperature combustion engines. Michigan Technological University, Ph.D. thesis

Bidarvatan M, Shahbakhti M (2014) Gray-box modeling for performance control of an HCCI engine with blended fuels. ASME J Eng Gas Turbines Power 136(10):101510

Bidarvatan M, Thakkar V, Shahbakhti M, Bahri B, Aziz AA (2014) Grey-box modeling of HCCI engines. App Therm Engg 70:397–409

Bidarvatan M, Kothari D, Shahbakhti M (2015) Integrated cycle-to-cycle control of exhaust gas temperature, load, and combustion phasing in an HCCI engine. In: 2015 American control conference (ACC), pp 7–12

Borgqvist P, Tuner M, Mello A, Tunestal P, Johansson B (2012) The usefulness of negative valve overlap for gasoline partially premixed combustion, PPC. SAE Int 0148-7191

Cairns A, Blaxill H (2007) The effects of two-stage cam profile switching and external EGR on SI-CAI combustion transitions. SAE Tech Pap 2007-01-0187

Cho K, Han M, Wagner R, Sluder C (2009) Mixed-source EGR for enabling high efficiency clean combustion modes in a light-duty diesel engine. SAE Int J Eng 1:457–465

Ciatti S, Subramanian SN (2011) An experimental investigation of low-octane gasoline in diesel engines. ASME J Eng Gas Turb Pow 133(9):092802

Ebrahimi K, Koch C (2015) Model Predictive control for combustion timing and load control in HCCI engines. In: SAE 2015 World Congress & Exhibition

EIA: International—U.S. Energy Information Administration. EIA

Fathi M, Jahanian O, Shahbakhti M (2017) Modelling and controller design architecture for cycle-by-cycle combustion control of homogeneous charge compression ignition (HCCI) engines—a comprehensive review. Ener Conver Manag 139:1–19

Fuerhapter A, Unger E, Piock W, Fraidl G (2004) The new AVL CSI engine—HCCI operation on a multi cylinder gasoline engine. SAE Tech Pap 2004-01-0551

Gorzelic P (2015) Modeling and model-based control of multi-mode combustion engines for closed-loop SI/HCCI mode transitions with cam switching strategies. University of Michigan, Ph.D. thesis

Han D, Ickes A, Assanis DN, Zhen H, Bohac SV (2010) The attainment and load extension of high-efficiency premixed low-temperature combustion with dieseline in a compression ignition engine. Ener Fuels 24:3517–3525

Han S, Bae C, Choi SB (2013) Effects of operating parameters on mode transition between low-temperature combustion and conventional combustion in a light-duty diesel engine. Int J Eng Res 14:231–246

Hanson R, Reitz R (2015) Investigation of cold starting and combustion mode switching as methods to improve low load RCCI operation. In: Proceeding of the ASME internal combustion engine division 2015 fall technical conference (ICEF 2015), vol 1, Advanced combustion, USA

Haraldsson G, Tunestål P, Johansson B, Hyvönen J (2004) HCCI closed-loop combustion control using fast thermal management. SAE Tech Pap 2004-01-0943

Hellstrom E, Larimore J, Jade S, Stefanopoulou AG, Jiang L (2014) Reducing cyclic variability while regulating combustion phasing in a four-cylinder HCCI engine. IEEE Trans Control Syst Technol 22:1190–1197

Hyvönen J, Haraldsson G, Johansson B (2003) Operating range in a multi cylinder HCCI engine using variable compression ratio. SAE Tech Pap 2003-01-1829

Hyvönen J, Haraldsson G, Johansson B (2005) Operating conditions using spark assisted HCCI combustion during combustion mode transfer to SI in a multi-cylinder VCR-HCCI engine. SAE Tech Pap 2005-01-0109

Indrajuana A, Bekdemir C, Luo X, Willems F (2016) Robust multivariable feedback control of natural gas-diesel RCCI combustion. In: 8th IFAC symposium on advances in automotive control AAC 2016, vol 49, pp 217–222

Indrajuana A, Bekdemir C, Feru E, Willems F (2018) Towards model-based control of RCCI-CDF mode-switching in dual fuel engines. SAE Tech Pap 2018-01-0263

Irdmousa BK, Rizvi SZ, Veini JM, Naber JD, Shahbakhti M (2019) Data-driven modeling and predictive control of combustion phasing for RCCI engines. In: 2019 American Control Conference (ACC), pp 1617–1622

Jade S, Hellström E, Larimore J, Stefanopoulou AG, Jiang L (2014) Reference governor for load control in a multicylinder recompression HCCI engine. IEEE Trans Cont Sys Tech 22(4):1408–1421

Jia M, Dempsey AB, Wang H, Li Y, Reitz RD (2015) Numerical simulation of cyclic variability in reactivity-controlled compression ignition combustion with a focus on the initial temperature at intake valve closing, Int J Eng Res 16:441–460
Kakuya H, Yamaoka S, Kumano K, Sato S (2008) Investigation of a SI-HCCI combustion switching control method in a multi-cylinder gasoline engine. SAE Tech Pap 2008-01-0792
Kalghatgi GT, Head RA (2006) Combustion limits and efficiency in a HCCI engine. Int J Eng Res 7:215–236
Kalghatgi GT, Risberg P, Angstrom HE (2006) Advantages of fuels with high resistance to auto-ignition in late-injection, low-temperature, compression ignition combustion. SAE Trans 115:623–634
Kalghatgi G, Risberg P, Ångström H (2007) Partially pre-mixed auto-ignition of gasoline to attain low smoke and low NO_x at high load in a compression ignition engine and comparison with a diesel fuel. SAE Tech Pap 2007-01-0006
Karagiorgis S, Glover K, Collings N, Petridis A (2007) Hybrid modeling for switching between SI and HCCI combustion modes. In: 2007 European Control Conference (ECC), pp 62–69
Kim SH, Bae C (2013) Strategy for mode transition between low temperature combustion and conventional combustion in a diesel engine. SAE Int J Eng 6:1706–1715
Kim D, Ekoto I, Colban W, Miles P (2009) In-cylinder CO and UHC imaging in a light-duty diesel engine during PPCI low-temperature combustion. SAE Int J Fuels Lubr 1(1):933–956
Kondipati NNT, Arora JK, Bidarvatan M, Shahbakhti M (2017) Modeling, design and implementation of a closed-loop combustion controller for an RCCI engine. In: 2017 American control conference (ACC), pp 4747–4752
Koopmans L, Ström H, Lundgren S, Backlund O, Denbratt I (2003) Demonstrating a SI-HCCI-SI mode change on a Volvo 5-cylinder electronic valve control engine. SAE Tech Pap 2003-01-0753
Krishnamoorthi M, Malayalamurthi R, He Z, Kandasamy S (2019) A review on low temperature combustion engines: Performance, combustion and emission characteristics. Renew Sust Ener Rev 116:109404
Li D, Wang Z, Liu H, Wang J (2014) Combustion mode switch by integrating stoichiometric ASSCI mode in a four-cylinder gasoline SI/HCCI engine. SAE Tech Pap 2014-01-1288
Li C, Yin L, Shamun S, Tuner M, Johansson B, Solsjo R, Bai XS (2016) Transition from HCCI to PPC: the sensitivity of combustion phasing to the intake temperature and the injection timing with and without EGR. SAE Tech Pap 2016-01-0767
Liu Y, Li L, Lu H, Schmitt S, Deng J, Rao L (2019) SI/HCCI mode switching optimization in a gasoline direct injection engine employing dual univalve system. ASME J Eng Gas Turb Pow 141
Manente V, Johansson B, Tunestal P (2009) Partially premixed combustion at high load using gasoline and ethanol, a comparison with diesel. SAE Tech Pap 2009-01-0944
Manente V, Zander C, Johansson B, Tunestal P, Cannella W (2010) An advanced internal combustion engine concept for low emissions and high efficiency from idle to max load using gasoline partially premixed combustion. SAE Tech Pap 2010-01-2198
Mauro AD, Chen H, Sick V (2019) Neural network prediction of cycle-to-cycle power variability in a spark-ignited internal combustion engine. Proc Comb Inst 37:4937–4944
Milovanovic N, Blundell D, Gedge S, Turner J (2005) SI-HCCI-SI mode transition at different engine operating conditions. SAE Tech Pap 2005-01-0156
Najt PM, Foster DE (1983) Compression-ignited homogeneous charge combustion. SAE Tech Pap 830264
Nazemi M, Shahbakhti M (2016) Modeling and analysis of fuel injection parameters for combustion and performance of an RCCI engine. App Ener 165:135–150
Nier T, Kulzer A, Karrelmeyer R (2012) Analysis of the combustion mode switch between SI and gasoline HCCI. SAE Tech Pap 2012-01-1105
Ogunkunle O, Ahmed NA (2021) Overview of biodiesel combustion in mitigating the adverse impacts of engine emissions on the sustainable human–environment scenario. Sustainability 13

Olsson J, Tunestål P, Johansson B (2001) Closed-loop control of an HCCI engine. SAE Tech Pap 2001-01-1031

Raut A, Irdmousa BK, Shahbakhti M (2018a) Dynamic modeling and model predictive control of an RCCI engine. Cont Engg Prac 129–144(81)

Raut A, Bidarvatan M, Borhan H, Shahbakhti M (2018b) Model predictive control of an RCCI engine. In: 2018 Annual American Control Conference (ACC), pp 1604–1609

Ravi N, Liao HH, Jungkunz AF, Widd A, Gerdes JC (2012) Model predictive control of HCCI using variable valve actuation and fuel injection. Cont Engg Prac 20:421–430

Roelle MJ, Shaver GM, Gerdes JC (2004) Tackling the transition: a multi-mode combustion model of SI and HCCI for mode transition control. In: Proceeding series of the ASME 2004 International mechanical engineering congress and exposition. Dynamic systems and control, Parts A and B, pp 329–336

Santoso H, Matthews J, Cheng W (2005) Managing SI/HCCI dual-mode engine operation. SAE Tech Pap 2005-01-0162

Shahbakhti M, Koch CR (2007) Control oriented modeling of combustion phasing for an HCCI Engine. In: 2007 American control conference, pp 3694–3699

Shahbakhti M, Lupul R, Audet A, Koch CR (2007a) Experimental study of HCCI cyclic variations for low-octane PRF fuel blends. In: Proceeding of Combustion Institute/Canadian Section (CI/CS) Spring Technical Conference

Shahbakhti M, Lupul R, Koch CR (2007) Sensitivity analysis and modeling of HCCI auto-ignition timing. In: 5th IFAC symposium on advances in automotive control, vol 40, pp 303–310

Shahbakhti M, Ghazimirsaied A, Koch CR (2010) Experimental study of exhaust temperature variation in a homogeneous charge compression ignition engine. Proc Inst Mech Eng Part D J Auto Engg 224(9):1177–1197

Shahbakhti M, Koch CR (2008) Characterizing the cyclic variability of ignition timing in a homogeneous charge compression ignition engine fueled with n-heptane/iso-octane blend fuels. Int J Eng Res 9:361–397

Shaver GM, Roelle MJ, Caton PA, Kaahaaina NB, Ravi N, Hathout JP, Ahmed J, Kojic A, Park S, Edwards CF, Gerdes JC (2005) A physics-based approach to the control of homogeneous charge compression ignition engines with variable valve actuation. Int J Eng Res 6(4):361–375

Shaver GM, Roelle M, Gerdes JC (2006) A two-input two-output control model of HCCI engines. In: 2006 American control conference, p 6

Shen M, Tuner M, Johansson B, Cannella W (2013) Effects of EGR and intake pressure on PPC of conventional diesel, gasoline and ethanol in a heavy-duty diesel engine. SAE Tech Pap 2013-01-2702

Shen M, Lonn S, Johansson B (2015) Transition from HCCI to PPC combustion by means of start of injection. SAE Tech Pap 2015-01-1790

Singh AP, Agarwal AK (2021) Performance and emission characteristics of conventional diesel combustion/partially premixed charge compression ignition combustion mode switching of biodiesel-fueled engine. Int J Eng Res 22:540–553

Singh AP, Agarwal AK (2021) Performance and emission characteristics of conventional diesel combustion/partially premixed charge compression ignition combustion mode switching of biodiesel-fueled engine. Int J 1778 Engine Res 22:540–553

Solouk A, Shakiba-herfeh M, Shahbakhti M (2017) Analysis and control of a torque blended hybrid electric powertrain with a multi-mode LTC-SI engine. SAE Int J Alter P/t 6:54–67

Souder JS, Mack JH, Hedrick J, Dibble RW (2004) Microphones and knock sensors for feedback control of HCCI engines. In: ASME 2004 internal combustion engine division fall technical conference, California, USA

Splitter D, Wissink M, DelVescovo D, Reitz R (2013) RCCI engine operation towards 60% thermal efficiency. SAE Tech Pap 2013-01-0279

Strandh P, Bengtsson J, Johansson R, Tunestål P, Johansson B (2005) Variable valve actuation for timing control of a homogeneous charge compression ignition engine. SAE Tech Pap 2005-01-0147

Tanabe K, Komatsu F, Nakayama SA (2011) Study on mode transition control between PCI and conventional combustion in a diesel engine. Int J Eng Res 12:69–86

Taraza D, Henein N, Bryzik W (1998) Determination of the gas pressure torque of a multicylinder engine from measurements of the crankshaft's speed variation. SAE Pap 980164

Ternel C, Bouter A, Melgar J (2021) Life cycle assessment of mid-range passenger cars powered by liquid and gaseous biofuels: Comparison with greenhouse gas emissions of electric vehicles and forecast to 2030. Transp Res Part D Transp Environ 97

Thring RH (1989) Homogeneous-charge compression-ignition (HCCI) engines. SAE Tech Pap 892068

Tian G, Wang Z, Ge Q, Wang J, Shuai S (2007) Mode switch of SI-HCCI combustion on a GDI engine. SAE Tech Pap 2007-01-0195

Weall A, Collings N (2007) Highly homogeneous compression ignition in a direct injection diesel engine fuelled with diesel and biodiesel. SAE Tech Pap 2007-01-2020

Widd A, Liao H, Gerdes JC, Tunestål P, Johansson R (2011) Control of exhaust recompression HCCI using hybrid model predictive control. In: Proceedings of the 2011 American control conference, pp 420–425

Williams S, Hu L, Nakazono T, Ohtsubo H, Uchida M (2009) Oxidation catalysts for natural gas engine operating under HCCI or SI conditions. SAE Int J Fuels Lubr 1(1):326–337

Wissink ML (2015) Direct injection for dual fuel stratification DDFS: improving the control of heat release in advanced IC engine combustion strategies. University of Wisconsin, Madison, Ph.D. thesis

Wu Y, Hanson R, Reitz RD (2014) Investigation of combustion phasing control strategy during reactivity controlled compression ignition (RCCI) multicylinder engine load transitions. ASME J Eng Gas Turb Pow 136

Xia L, Jager BD, Donkers T, Willems F (2020) Robust constrained optimization for RCCI engines using nested penalized particle swarm. Cont Engg Prac 99:104411

Yang T, Yin L, Ingesson G, Tunestål P, Johansson R, Long W (2017) Simultaneous control of soot emissions and pressure rise rate in gasoline PPC engine. In: 2017 IEEE Conference on control technology and applications (CCTA), pp 572–577

Yang T, Yin L, Meng X, Tian H, Long W, Tunestål P (2020) Partially premixed combustion optimization using double injection strategy in transient operation. App Ther Eng 169:114963

Yao C, Hu Y, Zhou T, Yang F, Ouyang M, Huang H (2016) Combustion stability control of dieseline PPCI based on in-cylinder pressure signals. In: 8th IFAC symposium on advances in automotive control AAC 2016, pp 333–339

Yin L, Turesson G, Tunestål P, Johansson R (2020) Model predictive control of an advanced multiple cylinder engine with partially premixed combustion concept. IEEE/ASME Trans Mech 25:804–814

Zhang Y, Xie H, Zhou N, Chen T, Zhao H (2007) Study of SI-HCCI-SI transition on a port fuel injection engine equipped with 4VVAS. SAE Tech Pap 2007-01-0199

Zhu Q, Prucka R, Wang S, Prucka M, Dourra H (2016) Control oriented modelling of engine IMEP variation. ASME Int Comb Eng Div Fall Tech Conf

Chapter 4
State of the Art in Low-Temperature Combustion Technologies: HCCI, PCCI, and RCCI

Balendra V. S. Chauhan, Imran Sayyed, Ajitanshu Vedrantam, Akshay Garg, Sawan Bharti, and Mritunjay Shukla

Abstract The need for higher thermal efficiency, lower NOx and Particulate Matter (PM) emissions has attracted researchers to modify the conventional combustion in IC engines. The after-treatment devices are deployed to meet the emission regulations but they have high initial and maintenance costs. The NOx emissions increase at higher temperatures in the conventional combustion in diesel engines, thus, the researchers are inclined towards finding low-temperature combustion techniques (LTCT). The Homogenous Charge Compression Ignition (HCCI), Reactivity Controlled Compression Ignition (RCCI), and Premixed Charge Compression Ignition (PCCI) are some well-known LTCTs to reduce the NOx to nearly 85%, PM emissions to almost 95%, and can reduce fuel consumption to approximately 15%. The chapter thoroughly reviews the emission performance, combustion characteristics, and thermal efficiency of HCCI, RCCI, and PCCI technologies and makes a comparative analysis of their applications for dual fuel and single fuelled engines. The present study focuses on deriving coherence between the aforesaid advanced combustion technologies' emission performance with the Euro VI emission norms.

Keywords LTCT · HCCI · RCCI · PCCI · Heavy-duty · Light-duty · Euro VI

B. V. S. Chauhan
Automotive Fuels and Lubricants Application Division, CSIR-Indian Institute of Petroleum, Dehradun, Uttarakhand, India

I. Sayyed
Panjab University, Chandigarh, India

A. Vedrantam (✉)
Universidad de Alcalá, Alcala de Henares, Spain
e-mail: ajitanshu.m@invertis.org

A. Garg
Department of Mechanical Engineering, College of Engineering Roorkee, Roorkee, India

B. V. S. Chauhan · S. Bharti
Invertis University, Bareilly, India

M. Shukla
Automotive Fuels and Lubricants Application Division, CSIR-Indian Institute of Petroleum, Dehradun, Uttarakhand, India

A. K. Agarwal et al. (eds.), *Advanced Combustion for Sustainable Transport*, Energy, Environment, and Sustainability, https://doi.org/10.1007/978-981-16-8418-0_4

List of Abbreviations

PCCI	Premixed charge compression ignition
RCCI	Reactivity controlled compression ignition
PM	Particulate Matter
NOx	Nitrogen oxides
LTCT	Low-temperature combustion techniques
GHG	Greenhouse gas emissions
UHC	Unburnt hydrocarbons
CO_2	Carbon dioxide
CO	Carbon monoxide
LNT	Lean NOx traps
DPF	Diesel particulate filter
TWC	Three-way catalyst
SCR	Selective catalytic reduction
DOC	Diesel oxidation catalyst
GPF	Gasoline particulate filter
A/F	Air-fuel mixture
CDC	Conventional diesel combustion
LTC	Low Temperature combustion
SCCI	Stratified Charge Compression Ignition
PPLTC	Partially Premixed Low Temperature Combustion
HECC	High-Efficiency Clean Combustion
C	Number of cylinders
S	Number of strokes
BL	Bore length
SL	Stroke length
DV	Volume of displacement
CR	Compression ratio
EGR	Exhaust gas recirculation
RI	Ringing intensity
VCR	Variable compression ratio
VVT	Variable Valve timing
DF	Diesel Fuel
TDC	Top dead centre
BDC	Bottom dead centre
IMEP	Indicated Mean Effective Pressure
ICP	Injection Control Pressure

4.1 Introduction

The IC engines found their application in many ways due to their efficiency and reasonable costs, but the emissions and fuel consumption of engines has always been in research throughout the world (Leach et al. 2020; Shah 2013; Alagumalai 2014; Kalghatgi 2014; Kumar and Vedrtnam 2018). As compared to its counterparts, oil has always been the most important fuel throughout the world, mainly petroleum fuels such as diesel and gasoline cover 65% of the energy needs, whereas coal comes after oil and is mostly used in the Asian continent (Kumar and Vedrtnam 2018; Balat 2007; Shafiee and Topal 2009; Combustion Engine Research Center (CERC) 2016, 2017; None 2017).

The current fuel consumption in the engines has resulted in the depletion of natural fossil fuels and increased their rate of extraction from the earth (Reijnders and Huijbregts 2007; Vickers 2017). Additionally, the increasing number of vehicles has resulted in greenhouse gas emissions (GHG) and higher emissions of harmful gases such as nitrogen oxides (NOx), unburnt hydrocarbons (UHC), carbon dioxide (CO_2), carbon monoxide (CO), particulate matter (PM); also in the NOx environment, the ozone of ground level is formed as a result of the combination of CO and HC, which is again lethal (Paoli et al. 2017, 2018; Levi and Cullen 2018; Tanov 2017; Zhou 2017; Reitz and Duraisamy 2015). Many well established after treatment devices and technologies such as NOx adsorber catalysts (Brogan et al. 1996; Dou and Bailey 1998; Iwachido et al. 2001; Gu and Epling 2019; Wu et al. 2021), lean NOx trap (LNT) (Gregory et al. 1999; West et al. 2004; Kim et al. 2020; Onrubia-Calvo et al. 2020; Rodríguez-Fernández et al. 2021), diesel particulate filter (DPF) (Caliskan and Mori 2017; Yang et al. 2016; Tan 2015; Koltsakis et al. 2013), Three-way catalyst (TWC) (Rood et al. 2020; Wang et al. 2015; Su et al. 2020; Liu et al. 2021), selective catalytic reduction (SCR) (Gholami et al. 2021; Hamada and Haneda 2012; Guan et al. 2014; Fu et al. 2014), diesel oxidation catalyst (DOC) (Beckmann et al. 1992; Nadanakumar et al. 2021; Russell and Epling 2011; Johnson and Kittelson 1996), Particle oxidation catalysts (Lehtoranta et al. 2009, 2007; Liu et al. 2012; Vakkilainen and Lylykangas 2004), and gasoline particulate filter (GPF) (Joshi and Johnson 2018; Belot et al. 2020; Morgan 2015; Minjares and Sanchez 2011; Xia et al. 2017) are used to curb these exhaust emissions.

Researchers around the globe aim to make internal combustion engines fuel economic, highly efficient and with lesser exhaust emissions. So, in the same scenario, the controlled combustion came into the picture and researchers found techniques like low-temperature combustion. The concept of Low-temperature combustion technology (LTCT's) is basically defined as controlled combustion with no flame in the engine cylinder and that too at comparably lower temperatures than in the usual conventional combustion chambers of engines. On the implementation of LTCT, the thermal efficiency of IC engines remains higher with a benefit of reduced particulate matter (PM) and NO_x emissions, additionally, the fuel efficiency also increases (Zheng et al. 2015; Maurya et al. 2018; Singh and Agarwal

2018a; Krishnamoorthi et al. 2019; Li et al. 2020). Though there are some challenges to low-temperature combustions and regulating the release rate of heat and ignition timing as well, apart from this it is observed that the amount of CO and HC emissions remains unaffected or increased (Dev et al. 2017; Imtenan et al. 2014; Agarwal et al. 2017a). Researchers and scientists are working to overcome the challenges against low-temperature technologies. The study of LTCT includes the Reactivity Controlled Compression Ignition (RCCI), Partially Premixed Low-Temperature Combustion (PPLTC), Pre-mixed Charge Compression Ignition (PCCI) and Homogeneous Charge Compression Ignition (HCCI). The aforesaid all are the advanced technologies of combustion, which aims to achieve flameless combustion at low temperature, to achieve good thermal efficiency. This chapter includes the data from a vast variety of studies conducted on engines with data on engine specifications, performance analysis, emission behaviour, combustion characteristics. In our chapter, we have a broader picture of the past work done on combustion technologies which add well with the ongoing research on advanced fuel combustion studies and their future.

4.1.1 Single Fuelled and Dual Fuelled Advance Combustion Technique

The homogeneity of charge and ratio of actual A/F to stoichiometric A/F (also defined as equivalence ratio ϕ) played an important role according to the study conducted by Onishi et al., he observed that pre-injection of a homogeneous mixture of charge could be beneficial for abatement of Particulate matter and NO_x (Onishi et al. 1979; Kamimoto and Bae 1988). After this homogeneity of the mixture, the charge was researched well and other such technologies like PCCI and HCCI were taken into immense consideration to be studied for the advancement of low-temperature combustion technologies, Neely et al., (No 2016; Neely et al. 2005) discussed the ϕ- T (equivalence ratio and temperature) diagram for different low-temperature combustion techniques (for single fuelled engine), this ϕ- T can be easily calculated with the help of kinetics of reaction (Akihama et al. 2001; Kitamura et al. 2002; Golovitchev et al. 2007; Murata et al. 2008), it was introduced for the very first time in 1988 (Kamimoto and Bae 1988) as it proved to be a good analysis tool for combustion analysis (Kook et al. 2005; Li et al. 2009).

In the same scenario Stanglmaier and Roberts, also observed some challenges to the above discussed single-fuelled advance combustions as it is never easy to regulate combustion phenomenon (No 2016; Stanglmaier and Roberts 1999). Yao et al. observed that stability of the phase of combustion is difficult to maintain (Yao et al. 2009), additionally, Gharehghani and An et al., stated that for single fuelled combustion methods it is tough to broaden or increase the range of operating load (Gharehghani 2019; An et al. 2018). Han et al. and Pedrozo et al. studied for single-fuelled advance combustions and observed that incomplete combustion leads to an increase in CO and HC emissions (Han et al. 2012; Pedrozo et al. 2016a). Moreover, researchers also studied the dual-fuelled combustion technologies to better regulate the phase of combustion, as now there will be two distinct fuel reactivity being used. Kokjohn et al. (2011) observed that by the stratification of reactivity of fuel, the phase of combustion can be regulated easily by RCCI, as a result of that mass ratio of fuel with low and high reactivity can be regulated.

4.2 Strategies to Develop Low-Temperature Combustion Technology

The simplest strategy used to achieve LTC in Engines was EGR. However, EGR also has its drawbacks due to particulate emissions. The strategies we will discuss focus a bit more on the composition and injection of charge into the engine. Tan (2015) and O'Connor and Musculus (2013) discussed that depending on the characteristics, low-temperature combustion can be subdivided as the first phase of combustion labelled as 'pre-combustion' based on the flow of charge, the second phase (based on chemical kinetics) of combustion is the main phase in which 'actual combustion' occurs, and lastly, the third phase was stated as 'post combustion' which is based on conditions of turbulent and chemical mixing of charge (Tan 2015; Liu et al. 2019a; O'Connor and Musculus 2013).

The low temperatures in combustion are achieved with certain techniques which help control the Mechanism of fuel consumption. It can be done with the help of different processes such as Stratified Charge Compression Ignition (SCCI), Partially Premixed Low-Temperature Combustion (PPLTC), Premixed charge Compression Ignition (PCCI) combustion, High-Efficiency Clean Combustion (HECC), Homogeneous Charge Compression Ignition (HCCI) combustion and Reactivity Controlled Compression Ignition (RCCI) combustion.(Singh et al. 2020) Fig. 4.1 illustrates the evolution of different LTC strategies adapted from the work of Paykani et al. (2015). The chapter would further go into the details of HCCI, PCCI and RCCI and review their development, benefits and challenges. The basic differences between HCCI, PCCI and RCCI are illustrated in the Table 4.1.

4.2.1 Homogeneous Charge Compression Ignition Combustion (HCCI)

Researchers worldwide found HCCI as a good combustion technique as up to some extent it provides faster combustion, lower emissions and higher thermal efficiency as compared to the ongoing IC engine technology (Mofijur et al. 2019; Raitanapaibule and Aung 2005). HCCI combustion is sometimes termed as "regulated auto-ignition" combustion, as before commencing the combustion, the fuel injected is finely infused

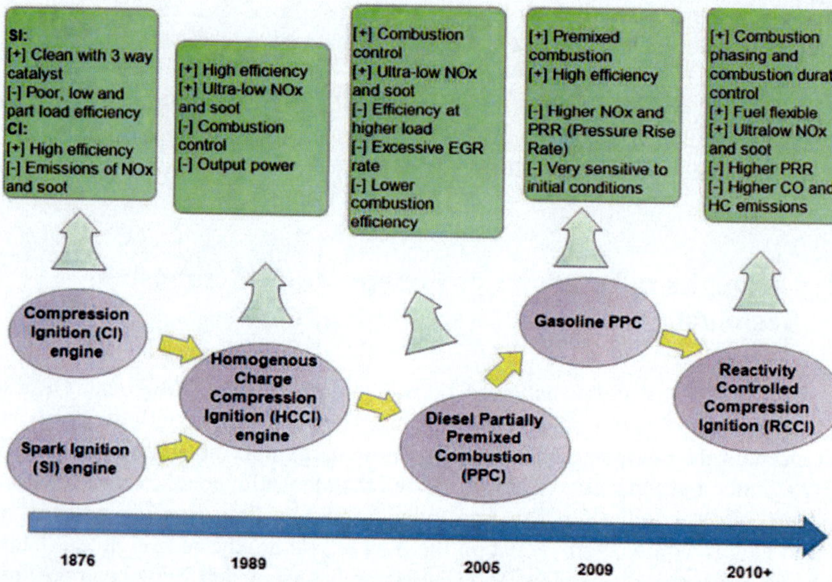

Fig. 4.1 Evolution of LTC Strategies and their advantages [+] and disadvantages [−] (Paykani et al. 2015)

with air. Low temperatures of cylinder gas are attained in HCCI as it works on leaner mixtures, as a result, thermal efficiency is increased by 50% and PM emissions are comparably less from conventional diesel combustion (Komninos and Rakopoulos 2016; Desantes et al. 2016). HCCI is also referred to as a 'flameless and multi-spot technique of combustion', which serves as an advantage (Yap et al. 2004) but multi-spot combustion may contribute to knock (Ganesan 2012; Contino et al. 2013). In biofuels, many fuels such as biodiesel are highly viscous and have very poor atomization characteristics, which result in deposition and also in poor combustion, performance and emissions. Due to enhanced engine out characteristics and charge preparation strategies, HCCI engines are known to run biofuels very well (Saiteja and Ashok 2021). Mai secures lower HC, PM and NOx. Further enhancements in the system such as charge preparation also led to new potential benefits of HCCI engines.

4.2.1.1 Charge Preparation Techniques in HCCI

Charge Preparation techniques in HCCI engines also known as Homogenous charge preparation techniques can be either internal, external type or premixed direct injection types. Internal charge preparations are processed inside the engine combustion chamber (Singh et al. 2014). These techniques include Early direct injection and

Table 4.1 Key differences between HCCI, PCCI and RCCI engines (Duan et al. 2021)

	HCCI	PCCI	RCCI
Fuel	Uses blends of different varieties of liquid and gaseous fuels	Uses blends of different varieties of liquid and gaseous fuels	Uses port fuel injection of high octane and direct injection of high cetane fuel
Power output control	Fuel flow controlled with equivalence ratio ($\phi \leq 1$) or charge dilution	Fuel flow controlled with equivalence ratio ($1 \leq \phi \leq 2$) or high charge dilution	Fuel reactivity stratification and A/F ratio stratification, typically without charge dilution
Fuel consumption control mechanism	Chemical kinetics	Chemical kinetics and injection timing	Chemical kinetics and fuel reactivity
Ignition timing control	Pressure and Temperature	Injection timing	Injection timing
Combustion type	Premixed Volumetric dominated by chemical kinetics	Premixed and diffusion combustion	Premixed and diffusion combustion
Flame front	No flame propagation (homogenous oxidation)	Diffusion flame propagation	Diffusion flame propagation
Emission properties	Higher UHC and CO Lower NOx, CO_2 and PM	Higher UHC and CO Lower NOx, CO_2 and PM	Very high UHC and CO (without after treatment) Ultra-low NOx, CO_2 and PM

ϕ denotes equivalence ratio of the air–fuel mixture

Late direct injection strategies. Early direct injections and late direct injection techniques further use different advanced injection strategies to control the combustion phasing. External homogenous charge preparation on the other hand is processed outside the combustion chamber of the engine. External charge preparation can be applied using the port fuel injection technique or port fuel injection with a vaporizer (Fathi et al. 2011). Premixed direct injection technique involves principles of both internal and external homogenous charge preparation techniques. Premixed direct injection technique helps to control combustion phasing and ignition timing (Yao et al. 2009).

Since HCCI engines lack direct in cycle combustion control, there are certain means to control the combustion in HCCI engines such as compression ratio, variable valve actuation, dual fueling etc. Dual fuelling is one such method that also allows us to use different fuels in HCCI engines. It is observed that HCCI engines perform better in dual fuelling. In comparison to single fuel engines, NOx and PM emissions decrease in dual-fuel engines whereas HC and CO emissions increase (Banapurmath et al. 2014).

To control the combustion in HCCI following methods are used (Fig. 4.2):

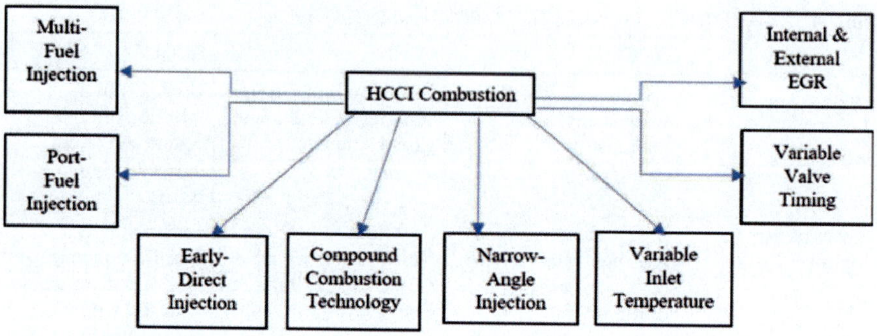

Fig. 4.2 Methods to control the combustion in HCCI (Neely et al. 2005)

For the range of fuels from n-heptane and till iso-octane, the HCCI engines work in the range of compression ratios starting 10:1 and ending to 28:1, however the temperature of inlet changes in the range of 29.85–249.85 °C (Fukushima et al. 2015). HCCI engines are known to operate with biofuels such as methane, ethanol and n-butanol. Additionally, they offer to be operated on low octane number fuels to the large ones (Liu et al. 2008). The oxidation is enhanced by Alcohol fuels, and they offer faster combustion (Yao et al. 2009; Chauhan et al. 2021; Cui et al. 2020). For very little intake charge temperature change, the HCCI's can gain higher temperatures of gas in the cylinder (Hunicz et al. 2022). Table 4.2 provides a thorough review of the behaviour of performance, emission and combustion characteristics observed in the HCCI combustion technologies so far.

Apart from Table 4.2, many other researchers (Yap et al. 2005; Kobayashi et al. 2011; Zhang and Zhao 2014; Aceves et al. 2000; Hyvönen et al. 2005; Lemberger and Floweday 2009; Al-Khairi et al. 2011; Elghawi et al. 2010) worked upon the emission studies on HCCI and found that the CO emissions remained higher as compared to conventional combustion mode. The studies performed (Kobayashi et al. 2011; Silitonga et al. 2019; Bression et al. 2009; Yap et al. 2006; Ganesh et al. 2008) found that incomplete combustion causes the HC emissions also stated that they remained higher as compared to the conventional mode of combustion, but the NOx emissions remain negligible in the HCCI mode (Yap et al. 2005; Polovina et al. 2013; Szybist et al. 2013; Milovanovic et al. 2005).

4.2.1.2 Summary

In addition to the advantages of high efficiency and low emissions of NOx and PM, HCCI Engines prove good at running on almost all fuels such as gasoline, diesel and biofuels like biodiesel, alcohols, alkanes etc. (Agarwal et al. 2017b). While HCCI can operate on different variety of fuels, the combustion phasing may or may not be controlled in all cases of fuels. Different strategies may be implemented to enhance the performance and emission characteristics. Since biofuels like biodiesel, ethers

Table 4.2 Comparison of performance, emission and combustion characteristics of HCCI mode of combustion with conventional combustion

Reference	Engine specifications	Performance analysis		Emissionb Behaviour	Combustion characteristics
		Fuel used	Testing conditions		
Ganesh and Nagarajan (2010)	C = 1, S = 4, BL = 87, SL = 110, CR = 17.5:1. This study was performed on an air-cooled HCCI engine mode	The diesel fuel is used for testing	Operating with and without EGR. Load conditions up to 75% were used	The comparison of CO emissions for conventional combustion was 1.7 g/kWh and for HCCI it was 7 g/kWh. The comparison of HC emissions for conventional combustion was 0.4 g/kWh and for HCCI it was 0.8 g/kWh. For NOx emission conventional combustion shown 9.3 g/kWh and for HCCI NOx was approximately 1.5 g/kWh	The rate of release of heat was 33 J/deg for HCCI and for conventional ones it was 45 J/deg. The pressure in the cylinder of the HCCI engine was 42 bar and for the conventional combustion engine, it was 50 bar

(continued)

Table 4.2 (continued)

Reference	Engine specifications	Performance analysis		Emissionb Behaviour	Combustion characteristics
		Fuel used	Testing conditions		
Lu et al. (2013)	C = 1, S = 4, BL = 98, SL = 105, CR = 18.5:1, DV = 782 cc	The fuel used in testing was n-heptane, Diesel, Ethanol, and Gasoline	The equivalence ratio of 0.26 to 0.4 was used and a premixed ratio of 0.4 to 0.63 is tested	The comparison of CO emissions for conventional combustion was 0.79% and for HCCI it was 0.8%. The comparison of HC emissions for conventional combustion was 270 ppm and for HCCI it was 280 ppm. For NOx emission conventional combustion shown 3 ppm and for HCCI NOx was approximately 20 ppm	The temperature observed in the cylinder of the HCCI engine was 1540 K. The rate of release of heat was 56 J/deg for HCCI

(continued)

Table 4.2 (continued)

Reference	Engine specifications	Performance analysis		Emissionb Behaviour	Combustion characteristics
		Fuel used	Testing conditions		
He et al. (2014)	C = 1, S = 4, Type = VVT, BL = 86, SL = 86, CR = 10.66:1	The fuels used are n-butanol and iso-octane	The comparison data of n-butanol fuel and the Valve lift is presented in this research		The temperature in the cylinder was 1800 K, and the pressure was 36 bar. The rate of release of heat is 70 J/deg
Maurya and Agarwal (2011)	C = 2, S = 4, BL = 87.3, SL = 110, CR = 16.5:1, DV = 1318 cc. The method of cooling was air	The fuel used is diesel	The testing temperature range starts from 120 °C and ends at 150 °C, and the operating pressure range belongs to 2.5 bar to 4.5 bar. The specific fuel consumption was 530 g/kWh	The comparison of CO emissions for conventional combustion was 0.8% and for HCCI it was 0.45%. The comparison of HC emissions for conventional combustion was 225 ppm and for HCCI it was 350 ppm. For NOx emission conventional combustion shown zero emissions and for HCCI NOx was less than 2 ppm	The temperature observed in the cylinder of the HCCI engine was 1350 K, for the conventional one it was 1100 K. The rate of release of heat was 13 J/deg for HCCI and 50 J/deg for conventional one. The pressure in the cylinder of the HCCI engine was 44 bar and for the conventional combustion engine, it was 42 bar. Hence, the observed efficiency of combustion remains to be 90% for HCCI and 92% for conventional engine combustion

(continued)

Table 4.2 (continued)

Reference	Engine specifications	Performance analysis		Emissionb Behaviour	Combustion characteristics
		Fuel used	Testing conditions		
Lawler et al. (2017)	C = 4, S = 4, BL = 86, SL = 86, CR = 13:1, DV = 1998 cc	The diesel fuel is used	For testing purposes, the operating temperature range is from 105 to 119 °C, the operating pressure range starts from 2.3 bar and ends on 8.4 bar. Conditions of operation are IMEP and the amount of injection of water		For HCCI the efficiency of combustion is 98.8% and for conventional combustion, it was 95.7%
Singh and Agarwal (2018b)	C = 1, Type = ERL-3, BL = 95, SL = 110, CR = 16.5:1, DV = 779	The fuels Used were Biodiesel, Kerosene, Diesel, Gasoline	The conditions of operation were AFR. The specific fuel combustion of 280 g/kWh was observed. Additionally, fuel consumption was also seen at the rate of 12 kJ/kWh. The mean effective pressure of 4 bar was seen	The comparison of CO emissions for conventional combustion was 14 g/kWh and for HCCI it was 16 g/kWh. The comparison of HC emissions for conventional combustion was 1.4 g/kWh and for HCCI it was 1.8 g/kWh. For NOx emission conventional combustion shown 6.4 g/kWh and for HCCI NOx was 6 g/kWh	For HCCI the rate of release of heat was 230 J/deg and for conventional combustion, it was 210 J/deg. The pressure inside the cylinder for HCCI was 64 bar and for conventional combustion, it was 63 bar

(continued)

Table 4.2 (continued)

Reference	Engine specifications	Performance analysis		Emissionb Behaviour	Combustion characteristics
		Fuel used	Testing conditions		
Khandal et al. (2019)	C = 1, S = 4, BL = 87.5, SL = 110, CR = 17.5:1	Three fuels namely Diesel, Cottonseed oil and Hydrogen are used. Honge biodiesel fuel was also used in testing	The engine load was varied from 0 to 40%	The comparison of CO emissions for conventional combustion was 0.14% and for HCCI it was 0.07%. The comparison of HC emissions for conventional combustion was 180 ppm and for HCCI it was 590 ppm. For NOx emission conventional combustion shown 22 ppm and for HCCI NOx was approximately 21 ppm. The CO_2 was also observed to be 7.2% for HCCI mode	

(continued)

Table 4.2 (continued)

Reference	Engine specifications	Performance analysis		Emissionb Behaviour	Combustion characteristics
		Fuel used	Testing conditions		
Park et al. (2019)	C = 4, S = 4, Type = Modified engine, BL = 86, SL = 86, CR = 9.2:1	Three fuels were used namely ethanol, diesel, n-butanol	Operating range of pressure: 1–3 bar; Conditions of operation: IMEP	The comparison of CO emissions for conventional combustion was 8 g/kWh and for HCCI it was 90 g/kWh. For NOx emission conventional combustion shown 15 kWh and for HCCI NOx was 10 g/kWh. The observed smoke for conventional combustion was 1.94 FSN and for HCCI it was 1.19 FSN	The temperature in the cylinder for HCCI was 1800 K and for conventional combustion, it was 1860 K. For HCCI the rate of release of heat was 70 J/deg and for conventional combustion, it was 80 J/deg. For HCCI, the delay in the ignition was 6 deg and for conventional combustion, it was 9 deg

C Number of cylinders, *S* Number of Strokes, *BL* Bore length (mm), *SL* Stroke length, *DV* volume of displacement (cc), *CR* Compression Ratio

etc. have great scope to become an alternative to fossil fuels, HCCI combustion is also perfect for future studies of next-generation biofuels in future especially when we face the challenges like fossil fuel resources depletion.

4.2.2 *Premixed Charge Compression Ignition Combustion (PCCI)*

This combustion in PCCI mode is attained by increasing the time gap between the start of injection of charge and the start of combustion, by doing this mixture of fuel and air amplifies before the start of the actual combustion (Zhang and Zhao 2014). Reitz et al. found that at the intake charge temperature of approximately 170 °C, PCCI engine combustion can effectively work in the range of air–fuel ratios of 34:1 to 80:1 which is a much leaner mixture (Reitz and Duraisamy 2015). In the PCCI mode compression ratios remain comparably higher than conventional combustion mode engines, additionally, the initiation of combustion takes place when injection of fuel ends and hence coherence between injection and the combustion can be avoided (Bhiogade and Suryawanshi 2016). In the conventional combustion mode engines, the fuel burning rate is heavily dependent on the time taken for fuel to vaporize and mix with air, unlike this, the PCCI fuel consumption rate is determined by the chemical kinetics of the combustion. Studies investigated 60% of the thermal efficiency of the PCCI, and it was found that lower PM and NOx emissions are observed if dual fuel is used in the PCCI mode of combustion (Parks et al. 2010). Moreover, in the premixed charge combustion mode, the temperature of the gas inside the cylinder and ignition delay is regulated with the help of exhaust gas recirculation (EGR). Premixed Charged combustion mode is preferred over Homogeneous charge combustion, as PCCI offers regulated combustion due to premixed charge. The phase of combustion in the premixed charge compression mode is regulated by parameters such as intake charge temperature, injection timing, injection timing and chemical kinetics. The direct injection premixed charge combustion relies on the quantity of direct injection fuel as it finds the rate of combustion and the time of ignition delay (Parks et al. 2010; Ying et al. 2010). Table 4.3 presents further information about the scientific developments performed on PCCI and gives brief data on performance analysis, emission behaviour and combustion characteristics.

4.2.2.1 Charge Preparation Techniques for PCCI Engines

PCCI uses a charge which is not fully homogenous through an enhanced charge motion which results in the desired ignition delay in combustion (Imtenan et al. 2014). Early injection in PCCI causes high emissions of PM, HC and CO due to injection spray missing the bowl of the piston which can be corrected by implementing the Narrow-Angle Direct Injection (NADI) strategy (Hoang 2020). Generally, there are

Table 4.3 Comparison of performance, emission and combustion characteristics of PCCI mode of combustion with conventional combustion

Reference	Engine specifications	Performance analysis		Emission behaviour	Combustion (characteristics
		Fuel used	Testing conditions		
Lee et al. (2017)	DI EURO IV engine with C = 1, BL = 83, SL = 92, speed = 1500 rpm, CR = 15.5:1, DV = 497 cc	The fuels used were diesel and propane	The variable parameters were premixed ratio 30–70% and indicated mean effective pressure of 0.52–0.55 MPa. 39.49% of thermal efficiency was obtained at operating temperature 85 °C	The comparison of CO emissions for conventional combustion was 2115 ppm and for PCCI it was 1184 ppm. The comparison of HC emissions for conventional combustion was 848 ppm and for PCCI was 5969 ppm. For NOx emission conventional combustion shown 48 ppm and for PCCI NOx was 53 ppm and smoke emission in case of PCCI was 15.74 mg/m^3	In-cylinder pressure of PCCI was 35 bar and for conventional combustion 56 bar. The heat release rate for PCCI was 50 J/deg and for conventional was 45 J/deg

(continued)

Table 4.3 (continued)

Reference	Engine specifications	Performance analysis		Emission behaviour	Combustion (characteristics
		Fuel used	Testing conditions		
Singh and Agarwal (2017)	Model: AVL/5402 CRDI, C = 1, BL = 85, SL = 90, speed = 4200 rpm, CR = 17.5:1, DV = 510.7 cc	The fuels used were diesel and biodiesel	The variable parameters were biodiesel blend 20–40%, injection pressure 400-1000 bar Fuel consumption was 0.35 kg/kWh and thermal efficiency was lower by 26%	The comparison of CO emissions for conventional combustion was 12 g/kWh and for PCCI it was 7 g/kWh. The comparison of HC emissions for conventional combustion was 1 g/kWh and for PCCI was 0.45 g/kWh. For NOx emission conventional combustion shown 4 g/kWh and for PCCI NOx was 5 g/kWh and smoke emission in the case of PCCI was 4.5 g/kWh and for conventional mode was 5 g/kWh	In-cylinder pressure of PCCI was 70 bar and for conventional combustion 71 bar. The heat release rate for PCCI was 250 J/deg and for conventional was 190 J/deg

(continued)

Table 4.3 (continued)

Reference	Engine specifications	Performance analysis		Emission behaviour	Combustion (characteristics
		Fuel used	Testing conditions		
Zhao et al. (2014)	DI Modified engine, S = 4, C = 2, BL = 105, SL = 120, speed = 1700 rpm, CR = 17.01:1, DV = 2078 cc	The fuel used was diesel and dimethyl ether	The variable parameters were premixed ratio of 0–30%, EGR 0–27%, BMEP 2.5–5 bar. SFC was recorded at 0.275 g/kWh and thermal efficiency was 30.2%	The comparison of HC emissions for conventional combustion was 60 ppm and for PCCI was 70 ppm. For NOx emission conventional combustion shown 600 ppm and for PCCI NOx was 400 ppm	In-cylinder pressure of PCCI was 63 bar and for conventional combustion 60 bar. The heat release rate for PCCI was 135 J/deg and for conventional was 125 J/deg. In-cylinder temperature for PCCI was recorded as 1630 K and for conventional medium 1620 K

(continued)

Table 4.3 (continued)

Reference	Engine specifications	Performance analysis		Emission behaviour	Combustion (characteristics
		Fuel used	Testing conditions		
Medhat et al. (2018)	Modified PCCI engine, S = 4, C = 1, BL = 87.5, SL = 110, speed = 1500 rpm, CR = 17.01:1, DV = 825 cc air cooled engine	The fuels used were diesel and ethanol	The variable parameters were equivalence ratio 0–50% and load 0–100%. Specific energy consumption was 12 MJ/kWh and thermal efficiency was 32%	The comparison of CO emissions for conventional combustion was 0.14% and for PCCI it was 0.22%. The comparison of HC emissions for conventional combustion was 225 ppm and for PCCI was 325 ppm. For NOx emission conventional combustion shown 1400 ppm and for PCCI NOx was 670 ppm	In-cylinder pressure of PCCI was 120 bar and for conventional combustion 105 bar. The heat release rate for PCCI was 130 J/deg and for conventional was 126 J/deg. In-cylinder temperature for PCCI was recorded as 2150 K and for conventional medium 2130 K

(continued)

Table 4.3 (continued)

Reference	Engine specifications	Performance analysis		Emission behaviour	Combustion (characteristics
		Fuel used	Testing conditions		
Bhiogade and Suryawanshi (2016)	Model: Kirloskar 2300V3 modified engine. C = 2, BL = 87.5, SL = 110, speed = 1500 rpm, CR = 15.5:1, DV = 1320 cc, water cooled engine	The fuel used was diesel	Variable parameters were EGR 0–30%, load 0–100%. Thermal efficiency was reported at 25%	The comparison of CO emissions for conventional combustion was 1.4 g/kWh and for PCCI it was 5 g/kWh. The comparison of HC emissions for conventional combustion was 0.5 g/kWh and for PCCI was 0.6 g/kWh. For NOx emission conventional combustion shown 8.4 g/kWh and for PCCI NOx was 2 g/kWh. Smoke emission in the case of PCCI was 1.5 BSU and in the conventional medium, it was 3.5 BSU	In-cylinder pressure of PCCI was 28 bar and for conventional combustion 29 bar. The heat release rate for PCCI was 15 J/deg and for conventional was 15 J/deg

(continued)

Table 4.3 (continued)

Reference	Engine specifications	Performance analysis		Emission behaviour	Combustion (characteristics
		Fuel used	Testing conditions		
Park et al. (2019)	CRDI, heavy duty modified engine S = 4, C = 1, BL = 100, SL = 125, speed = 1400 rpm, CR = 17.4:1, DV = 981 cc	The fuels used were diesel and natural gas	The variable parameters were premixed ratio 10–75%, the start of injection 45–20°bTDC and EGR 0–50%. The results reported 37% of thermal efficiency	The comparison of CO emissions for conventional combustion was 8 g/kWh and for PCCI it was 10 g/kWh. The comparison of HC emissions for conventional combustion was 9 g/kWh and for PCCI was 16 g/kWh. For NOx emission conventional combustion shown 7.5 g/kWh and for PCCI NOx was 2.8 g/kWh. PM emission in the case of PCCI was 0.007 g/kWh	The heat release rate for PCCI was 52 J/deg and for conventional was 110 J/deg

(continued)

Table 4.3 (continued)

Reference	Engine specifications	Performance analysis		Emission behaviour	Combustion (characteristics
		Fuel used	Testing conditions		
Smith et al. (2016)	Model: Daimler CRDI turbocharged engine, S = 4, C = 1, BL = 88, SL = 88.34, speed = 1200 rpm, CR = 15.88:1, DV = 537 cc	The fuels used were diesel and dimethyl carbonate	Variable parameters were EGR 0–48%, the start of pilot injection 1 26.4–18.9°bTDC, the start of pilot injection 2 17.2–5.5°bTDC and the start of main injector 7.1–4.4°bTDC Thermal efficiency of 43.5% was reported	The comparison of CO emissions for conventional combustion was 3 g/kWh and for PCCI it was 2.5 g/kWh. The comparison of HC emissions for conventional combustion was 5.8 g/kWh and for PCCI was 4.7 g/kWh. For NOx emission conventional combustion shown 0.34 g/kWh and for PCCI NOx was 0.26 g/kWh. PM emission in the case of PCCI was 0.037 g/kWh	In-cylinder pressure of PCCI was 50 bar, heat release rate for PCCI was 15 J/deg and ignition delay was 11.4 deg

(continued)

Table 4.3 (continued)

Reference	Engine specifications	Performance analysis		Emission behaviour	Combustion (characteristics
		Fuel used	Testing conditions		
Wang et al. (2017)	Model: Ricardo E6/MS, pre-chamber engine S = 4, C = 1, BL = 76.2, SL = 110, speed = 800 rpm, CR = 17.2:1 DV = 507 cc	The fuel used was diesel		The comparison of CO emissions for conventional combustion was 2.84E-07 kg/kg fuel and for PCCI it was 0.177 g/kg fuel. For NOx emission conventional combustion shown 0.00756 g/kg fuel and for PCCI NOx was 4.67 g/kg fuel	In-cylinder pressure of PCCI was 175 bar and for conventional combustion 150 bar. The heat release rate for PCCI was 250 J/deg and for conventional was 130 J/deg
Pandey et al. (2018)	Modified engine S = 4, C = 4, BL = 112, SL = 135, CR = 17.5:1	The fuel used was diesel	Specific fuel consumption was 292 g/kWh at EGR 20%	CO emissions for PCCI was 22 g/kWh, NOx was 138 ppm, HC was 1.8 g/kWh and smoke was 0.23 FSN	In-cylinder pressure of PCCI was 78 bar and for conventional combustion 73 bar. The heat release rate for PCCI was 52 J/deg and for conventional was 75 J/deg

(continued)

Table 4.3 (continued)

Reference	Engine specifications	Performance analysis		Emission behaviour	Combustion (characteristics
		Fuel used	Testing conditions		
d'Ambrosio and Ferrari (2015)	EURO V modified engine, S = 4, C = 4, BL = 83, SL = 90.4, CR = 16.3:1, DV = 1956 cc	The fuel used was diesel	Variable parameters were EGR 0–50%, speed 1500–2750 rpm and BMEP 2–12 bar. Specific fuel consumption was 240 g/kWh	CO emissions for PCCI was 2 g/kWh. HC was 0.3 g/kWh. NOx emission was 1.4 g/kWh and smoke 0.17 g/kWh	In-cylinder pressure of PCCI was 58 bar and for conventional combustion 60 bar. The heat release rate for PCCI was 45 J/deg and for conventional was 47 J/deg. In-cylinder temperature for PCCI was 2600 K and for conventional was 2700 K

C Number of cylinders, *S* Number of Strokes, *BL* Bore length (mm), *SL* Stroke length, *DV* volume of displacement (cc), *CR* Compression Ratio

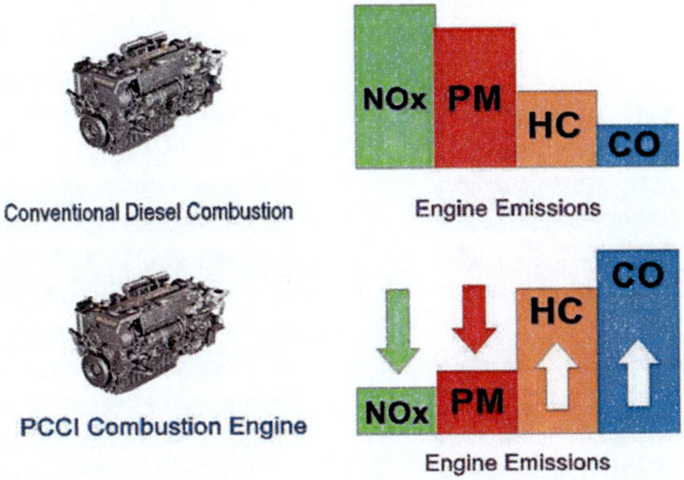

Fig. 4.3 Comparison of emission performance of the diesel-fuelled PCCI with conventional diesel combustion (Benito de la Piedra 2015)

three techniques which are advanced direct injection technique, port fuel injection technique and late direct injection technique. There is also an approach of air–fuel premixed early injection which later is followed by a late fuel pulse injection in compression stroke which controls the ignition (Musculus et al. 2013; Liu et al. 2015).

Apart from the achievements and developments are seen in Table 4.3, Benito de la Piedra (Benito de la Piedra 2015) compared the emission performance of PCCI diesel fuel combustion concerning the conventional diesel combustion, which is as shown in Fig. 4.3.

Moreover, Cao et al., (2020) also discussed the effects of methods of fuel injection on the combustion and emission behaviour of the PCCI combustion engine, he also mentioned that a high-pressure fuel injection technique with several injections per cycle is also shown to effectively reduce NOx and PM emissions. Additionally, Cao et al., stated that the concurrent decrease of NOx, as well as PM emissions from premixed charge combustion mode, can be obtained by integrating a strategic plan of pressurized fuel injection (for efficient atomization as well as the rapid development of a relatively homogenous mixture of air and fuel) along with the suitable ratio of injection (to develop a lean mixture of fuel), leading to a mixture of air and fuel combustible at relatively low temperature. The lowering of brake thermal efficiency and loss of engine power should be pondered as they are crucial factors in the development of engine performance balance.

Shim et al. (2020) analyzed the techniques of combustion namely PCCI, HCCI and diesel-fuelled PCCI, and defined their suitability for practicable use as shown in Fig. 4.4, it depicts the benefits and detriments of dual and single fuel advanced combustion technologies.

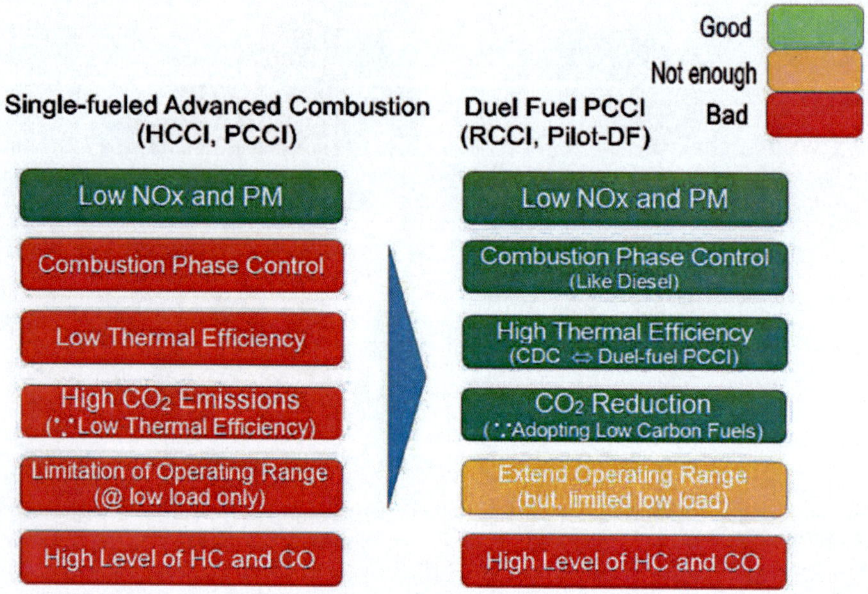

Fig. 4.4 Advantages and disadvantages of dual and single fuel advanced combustion technologies (Shim et al. 2020)

PCCI engines due to their advanced charge preparation techniques offer better control over the combustion of engines for a variety of fuels, unlike HCCI which under certain cases may misfire or undergo knocking due to lack of in-cycle combustion control. The main drawback of PCCI is that it requires higher EGR rates also it is known that due to the high boiling point in PCCI, large fuel injections face problems in the engine (Martin et al. 2009). PCCI engines will certainly do find their applications in better engine control, however, due to sensitivity to operating conditions and limited applications PCCI engines have the least scope of future applications in comparison to HCCI and RCCI engines.

4.2.3 Reactivity Controlled Compression Ignition (RCCI)

As the name reactivity-controlled compression ignition suggests, the fuels of varying reactivity are supplied at predetermined times to maintain the cylinder's anticipated reactivity and achieve the required magnitude and phasing of combustion. The mixing of differently reactive fuels inside the cylinder and employing different patterns of injecting fuel can control the phases of combustion, the combustion magnitude and the duration of combustion. The good thing about the RCCI mode of combustion is that after-treatment devices are not required as they can fulfill the regulations NOx and PM emissions easily and can generate up to 56% of thermal efficiencies (Reitz and

Duraisamy 2015; Benajes et al. 2014). The fuels with lesser reactivity are introduced with the help of port fuel injection to develop a homogeneous mixture of remains of combustion, air, and fuel; however, the highly reactive fuels are introduced directly by just assigning multiple or single time injections, before commencing the PCC reactions (Liu et al. 2019a; Benajes et al. 2014). Reactivity controlled compression ignition model engines can perform on a broader range of load conditions and that too with lesser NOx emissions and with an adequate rise in pressure (Liu et al. 2019b). Pedrozo et al., and Zheng et al., stated the threshold value (RI:5 MW/m^2) of rising in pressure under which the reactivity-controlled compression ignition model engine can work on entire load conditions and ranges of the speed of the engine (Han et al. 2018; Pedrozo et al. 2016b; Zheng et al. 2018). In another observation by Wu et al. and Splitter et al., they showed that at the compression ratio of 14.4:1 the engine load conditions in the reactivity-controlled ignition combustion is narrowed to nearly half, as a result of which the proportionality or agreement between PM emissions and NOx is altered (Han et al. 2020; Wu et al. 2014; Splitter et al. 2013).

The observation from Table 4.4 can be made that CI engines are mostly used for reactivity-controlled compression ignition mode exploration and analysis inclusive of the system of pressurized common rail injection of fuel. To control the combustion phasing the lower exhaust gas recirculation is employed when working at conditions of higher load and speeds (Wu et al. 2017). Zhou et al., and Guardiola et al., introduced the concept of the gradient of reactivity (according to the strategy of fuel injection such as high and lower cetane number and late injection or early injection of fuel) and the global reactivity (depending on type and quantity of fuel injected) (Zhou 2017; Guardiola et al. 2017).

4.2.3.1 Summary

RCCI engines have their advantages over Commercial CI Engines as well as HCCI and PCCI engines. The best advantage of RCCI engines is low emissions of NOx and PM emissions without the use of any exhaust after treatment. RCCI engines and it does so with high engine efficiency which is a drawback of PCCI engines. In RCCI engines there is a little drawback of high production of CO and HC. NOx and PM productions are slightly higher than PCCI but values are in the desired range and there is the compensation of no after treatment. RCCI engines show that in the future it may be possible to obtain the targeted values of engine emissions of NOx and PM without having any engine after-treatment systems. There is also a scope of reduction of HC and CO which are produced in high amounts in RCCI combustion.

There are various organisation working on the projects to develop the HCCI, PCCI, and RCCI technologies, some of them are listed as follows (Table 4.5).

Table 4.4 Comparison of performance, emission and combustion characteristics of RCCI mode of combustion with conventional combustion

Reference	Engine specifications	Performance analysis		Emission behaviour	Combustion characteristics
		Fuel used	Testing conditions		
Kakoee et al. (2018)	Model: Ricardo E6, VCR, C = 1, BL = 76.2, SL = 110, max speed = 3000 rpm, DV = 507	The tested fuels were diesel and natural gas-biodiesel	Variable load conditions. Maximum thermal efficiency was obtained as 41.8% at 1200 rpm and CR = 17:1	Comparison of CO emissions for conventional combustion was 1380 ppm and for RCCI it was 2630 ppm. The comparison of HC emissions for conventional combustion was 40 ppm and for RCCI was 245 ppm. For NOx emission conventional combustion shown 250 ppm and for RCCI NOx was 30 ppm	The rate of release of heat was 50 J/deg for RCCI and for conventional ones it was 29 J/deg. The pressure in the cylinder of the RCCI engine was 45 bar and for the conventional combustion engine, it was 37 bar
Pan et al. (2017)	C = 4, S = 4, BL = 92, SL = 94, speed = max speed = 3600 rpm, CR = 18.5:1, DV = 2498 cc, water cooled engine	The tested fuels were diesel and n-butanol	Engine operating at variable load 25–100% and fuel blend 35–45%. The maximum thermal efficiency obtained was 35% at 2800 rpm	The comparison of CO emissions for conventional combustion was 1400 ppm and for RCCI it was 1100 ppm. The comparison of HC emissions for conventional combustion was 400 ppm and for RCCI was 650 ppm. For NOx emission conventional combustion shown 350 ppm and for RCCI NOx was 620 ppm	The rate of release of heat was 60 J/deg for RCCI and for conventional ones it was 24 J/deg. The pressure in the cylinder of the RCCI engine was 106 bar and for the conventional combustion engine, it was 90 bar. The operating temperature for RCCI was 1350 K and for conventional was 1900 K. Ignition delay for RCCI was 18 deg and that for conventional was 11 deg

(continued)

Table 4.4 (continued)

Reference	Engine specifications	Performance analysis		Emission behaviour	Combustion characteristics
		Fuel used	Testing conditions		
Liu et al. (2019a)	C = 1, S = 4, BL = 105, SL = 125, speed = 1500 rpm, CR = 16:01, DV = 1081 cc, water cooled engine	Fuels used were isooctane and n-heptane	EGR = 0–45% and premixed ratio 1–100%. The increase in thermal efficiency obtained was 39.5%	The comparison of CO emissions for conventional combustion was 63 g/kWh and for RCCI it was 40 g/kWh. The comparison of HC emissions for conventional combustion was 8.2 g/kWh and for RCCI was 5 g/kWh. For NOx emission conventional combustion shown 7.7 g/kWh and for RCCI NOx was 1.5 g/kWh	The rate of release of heat was 130 J/deg for RCCI and for conventional ones it was 230 J/deg. The pressure in the cylinder of the RCCI engine was 105 bar and for a conventional combustion engine, it was 115 bar. Combustion efficiency for RCCI was 91% and for conventional was 88%

(continued)

Table 4.4 (continued)

Reference	Engine specifications	Performance analysis		Emission behaviour	Combustion characteristics
		Fuel used	Testing conditions		
Krishnasamy (2017)	C = 1, S = 4, BL = 87.5, SL = 110, speed = 1500 rpm, CR = 17.01:1, air cooled engine	Fuels used were gasoline and diesel	CR- 15–17.5, fuel ratio 2-20 mg and DI timing 30–70° bTDC. BTE decreased by 30% using gasoline	The comparison of CO emissions for conventional combustion was 5 g/kWh and for RCCI it was 50 g/kWh. The comparison of HC emissions for conventional combustion was 4.8 g/kWh and for RCCI was 8 g/kWh. For NOx emission conventional combustion shown 2.8 g/kWh and for RCCI NOx was 0.2 g/kWh	The rate of release of heat was 25 J/deg for RCCI and for conventional ones it was 15 J/deg. The pressure in the cylinder of the RCCI engine was 55 bar and for a conventional combustion engine, it was 42 bar

(continued)

Table 4.4 (continued)

Reference	Engine specifications	Performance analysis		Emission behaviour	Combustion characteristics
		Fuel used	Testing conditions		
Benajes et al. (2014)	Model: YC6J Guangxi Yuchai, ω chamber, C = 1, BL = 105, SL = 125, Speed = 1500 rpm, CR = 16.01:1, DV = 1080 cc water cooled engine	Fuel used in the study were biodiesel, ethanol, n-butanol/ 2,5-dimethylfuran	EGR varying from 0 to 30%. BTE increased by 47.5%	The comparison of CO emissions for conventional combustion was 35 g/kWh and for RCCI it was 14 g/kWh. The comparison of HC emissions for conventional combustion was 6.8 g/kWh and for RCCI was 7 g/kWh. For NOx emission, conventional combustion was shown 13 g/kWh and for RCCI NOx was 0.5 g/kWh	The rate of release of heat was 220 J/deg for RCCI and for conventional ones it was 140 J/deg. The pressure in the cylinder of the RCCI engine was 120 bar and for a conventional combustion engine, it was 108 bar. Ignition Delay for RCCI was 13.5 deg and for conventional was 10.5 deg

(continued)

Table 4.4 (continued)

Reference	Engine specifications	Performance analysis		Emission behaviour	Combustion characteristics
		Fuel used	Testing conditions		
Liu et al. (2019b)	C = 1, S = 4, supercharging, BL = 105, SL = 125, speed = 2500 rpm, CR = 16.01:1, DV = 1081.8 cc	The fuels used were diesel and ethanol	The variable parameters were the start of injection timing 14–12°bTDC, EGR 60–80%, in charge take temperature 40–100 °C. Thermal efficiency increased by 45.6%	The comparison of CO emissions for conventional combustion was 7.5 g/kWh and for RCCI it was 10.5 g/kWh. The comparison of HC emissions for conventional combustion was 8 g/kWh and for RCCI was 15 g/kWh. For NOx emission conventional combustion shown 10.5 g/kWh and for RCCI NOx was 8 g/kWh	In the case of HCCI rate of release of heat was 170 J/deg, pressure in the cylinder of the engine was 126 bar, the in-cylinder temperature was 1710 K with a combustion efficiency of 95%

(continued)

Table 4.4 (continued)

Reference	Engine specifications	Performance analysis		Emission behaviour	Combustion characteristics
		Fuel used	Testing conditions		
Park and Yoon (2016)	Model: DOHC 4 valves, C = 1, S = 4, BL = 75, SL = 84.5, speed = 1200 rpm, CR = 17.8:1, DV = 373.3 cc	The fuels used were diesel gasoline biogas	The variable parameters were 0.2–0.8 premixed ratio and the 40–5°bTDC start of injection timing	The comparison of CO emissions for conventional combustion was 0.016 g/kWh and for RCCI it was 0.321 g/kWh. The comparison of HC emissions for conventional combustion was 0.097 g/kWh and for RCCI was 1.44 g/kWh. For NOx emission conventional combustion shown 18.42 g/kWh and for RCCI NOx was 10.88 g/kWh. Smoke emissions for RCCI were 0.421 g/kWh and 0.004 g/kWh for conventional combustion	The rate of release of heat was 75 J/deg for RCCI and for conventional ones it was 100 J/deg. The pressure in the cylinder of the RCCI engine was 80 bar and for a conventional combustion engine, it was 78 bar

(continued)

Table 4.4 (continued)

Reference	Engine specifications	Performance analysis		Emission behaviour	Combustion characteristics
		Fuel used	Testing conditions		
Paykani et al. (2015)	C = 1, S = 4, BL = 137.2, SL = 165.1, CR = 16.1:1	Fuel used was diesel and natural gas	Variation in premixed ratio 50–100 and EGR 0–40%. Thermal energy increased by 50.74% at injection pressure 1.75 and injection charge temperature of 60 °C	The comparison of CO emissions for conventional combustion was 31.7 g/kWh and for RCCI it was 1.22 g/kWh. The comparison of HC emissions for conventional combustion was 20 g/kWh and for RCCI was 2.72 g/kWh. For NOx emission conventional combustion shown 0.01 g/kWh and for RCCI NOx was 0.016 g/kWh. Smoke emissions for RCCI were 0.0012 g/kWh	The rate of release of heat was 428 J/deg for RCCI, the pressure in the cylinder was 132 bar and the combustion temperature was 1830 K
Ghazal (2018)	C = 1, S = 4, diesel engine, BL = 130, SL = 160, speed = 1300 rpm	Fuel used was diesel and gasoline	Variable parameter of premixed ratio 50–80%. The comparison point was 60% vs neat diesel. Thermal efficiency increased by 49% at injection charge pressure of 3.1 bar	The comparison of CO emissions for conventional combustion was 40 g/kWh and for RCCI it was 25 g/kWh. The comparison of HC emissions for conventional combustion was 8 g/kWh and for RCCI was 10 g/kWh. For NOx emission conventional combustion shown 0.2 g/kWh and for RCCI NOx was 1.6 g/kWh	The rate of release of heat was 145 J/deg for RCCI and for conventional ones it was 130 J/deg. The pressure in the cylinder of the RCCI engine was 92 bar and for the conventional combustion engine, it was 88 bar Combustion temperature in RCCI was 2200 K whereas in conventional combustion was 2050 K

(continued)

Table 4.4 (continued)

Reference	Engine specifications	Performance analysis		Emission behaviour	Combustion characteristics
		Fuel used	Testing conditions		
Jo et al. (2019)	Model: DOHC, 4 valves, S = 4, C = 1, BL = 83, SL = 92, speed = 1800 rpm, CR = 17.7:1, DV = 498 cc	The fuels used were diesel and ethanol	Injection charge temperature was varied from 20–80 °C and injection timing was varied from 27–6°bTDC. The comparison point was 20 °C and D80/E20. An increase of 4.7% in the mean effective pressure was observed	The comparison of CO emissions for conventional combustion was 15 g/kWh and for RCCI it was 7 g/kWh. The comparison of HC emissions for conventional combustion was 3.5 g/kWh and for RCCI was 1.5 g/kWh. For NOx emission, conventional combustion was shown 5 g/kWh and for RCCI NOx was 6.5 g/kWh	In-cylinder pressure of RCCI was 80 bar and for conventional combustion 50 bar. In-cylinder temperature for RCCI was 1450 K and ignition delay for RCCI was 22 deg and for conventional was 18 deg

C Number of cylinders, *S* Number of Strokes, *BL* Bore length (mm), *SL* Stroke length, *DV* volume of displacement (cc), *CR* Compression Ratio

Table 4.5 Ongoing Projects on HCCI, PCCI, and RCCI

HCCI	Organisation name	Project name
	University of Birmingham	CHASE V6 HCCI Engine, Modelling and Thermal Management
	University of Birmingham	CHASE V6 HCCI Engine & Reforming
	Research Council of Norway and Norwegian University of Science and Technology	Ignition Quality of a Fuel for a HCCI Engine
	Ministry of Higher Education of Malaysia and Universiti Sains Malaysia (USM)	Combustion and Exhaust Emissions in a Partial HCCI-DI Diesel Engine
	Sakarya University of Applied Science	HCCI engine technologies
PCCI	Jiangsu University, Zhenjiang 212,013, China and National Natural Science Foundation of China	Combustion and emissions in a PCCI diesel engine
	China Post doctoral Science Foundation and Science Research Institute, Weichi Power Co. Ltd. China	PCCI engine analysis
	Purdue University, West Lafayette, IN 47,907, United States	PCCI Combustion Timing 2013-SHAV-66351
RCCI	Iran Heavy DieselManufacturing Company (DESA) and Mazandaran University of Science and Technology, Iran	RCCI Heavy Duty engines
	University of Vaasa, Wolffintie 34, 65,200 Vaasa, Finland	Clean Propulsion Technologies project
	Dalian University of Technology, Dalian, Liaoning 116024, P.R. China	RCCI and Reverse RCCI Project
	Key Laboratory of Fluid and Power Machinery, Ministry of Education, Xihua University, Chengdu, 610039, China	RCCI combustion in a heavy-duty diesel engine

4.3 Concluding Remarks

A review of advanced combustion technologies' performance behaviour, emission analysis and combustion characteristics has been done. The chapter thoroughly reviews the emission performance, combustion characteristics, and thermal efficiency of HCCI, RCCI, and PCCI technologies and makes a comparative analysis of their applications for dual fuel and single fuelled modes of operation.

- It is clear from the review that, the HCCI mode of combustion can operate with any type of fuel with a condition that control on combustion becomes challenging. If leaner mixtures of fuel are used in HCCI engines, then there is a possibility of a misfire; and if rich mixtures of fuel are used then knock may take place. To improve the range of operation of the HCCI engine, techniques such as variable valve timing, exhaust gas recirculation, fuel blends, and variable compression ratio etc. are employed. The mission performance of HCCI mode is such that it generates approximately zero PM and NOx emissions; anyhow while the engine is operational, it is very difficult to attain the variable valve timing and variable compression ratios.

- In the PCCI mode of combustion PM and NOx emissions are slightly higher but they remain lesser as compared to conventional combustion mode, however, in the PCCI engines the hydrocarbon and carbon monoxide emissions are observed to be lesser than HCCI engine mode.
- The RCCI technique is the one which tries to overcome the shortcomings of conventional combustion mode (as PM and NOx emissions are higher); premixed charge compression ignition mode (as the outcome of power is lesser in PCCI); homogeneous charge combustion mode (as the range of operation of HCCI mode is lower). As of now, petro-fuels are used in most of the studies on reactivity-controlled compression ignition mode of engines, due to their availability and good heating value; however, it is observed that the operating range of compression ratios for RCCI mode starts from 9.2:1 and ends on 19.9:1, for most of the cases in this review it remains 16:1.

4.4 Declaration of Competing Interest

The authors declare that they have no known competing financial interests or personal relationships that could have appeared to influence the work reported in this paper.

Acknowledgements The research study has been performed at the Institute of Fundamental Technological Research, Polish Academy of Sciences, Poland and Engine After-treatment Area, Automotive Fuels and Lubricants Application Division, CSIR-Indian Institute of Petroleum, Dehradun, U.K, India, in collaboration with Universidad de Alcalá, Spain and Invertis University Bareilly, India.

References

Aceves SM, Flowers DL, Westbrook CK, Smith JR, Pitz W, Dibble R, Christensen M, Johansson B (2000) A multi-zone model for prediction of HCCI combustion and emissions. SAE Trans 431–441

Agarwal AK, Singh AP, Maurya RK (2017a) Evolution, challenges and path forward for low-temperature combustion engines. Prog Energy Combust Sci 61:1–56

Agarwal AK, Agarwal RA, Gupta T, Gurjar BR (eds) (2017b) Biofuels: technology, challenges and prospects. Springer, Berlin

Akihama K, Takatori Y, Inagaki K., Sasaki S, Dean AM (2001) Mechanism of the smokeless rich diesel combustion by reducing temperature. SAE Trans 648–662

Alagumalai A (2014) Internal combustion engines: progress and prospects. Renew Sustain Energy Rev 38:561–571

Al-Khairi NN, Naveenchan P, Aziz AR (2011) Comparison of HCCI and SI characteristics on low load CNG-DI combustion. J Appl Sci 11(10):1827–1832

An Y, Jaasim M, Raman V, Pérez FEH, Sim J, Chang J, Im HG, Johansson B (2018) Homogeneous charge compression ignition (HCCI) and partially premixed combustion (PPC) in compression ignition engine with low octane gasoline. Energy 158:181–191

Balat H (2007) Role of coal in sustainable energy development. Energy Explor Exploit 151–174

Banapurmath NR, Khandal SV, Bagi SM, Kulkarni SR (2014) CNG-HOME operated dual fuel and HCCI engines. In: 2014 1st International conference on non conventional energy (ICONCE 2014), pp 126–130. https://doi.org/10.1109/ICONCE.2014.6808704

Beckmann R, Engeler W, Mueller E, Engler BH, Leyrer J, Lox ES, Ostgathe K (1992) A new generation of diesel oxidation catalysts (No. 922330). SAE Technical Paper

Belot I, Vidal D, Votsmeier M, Hayes RE, Bertrand F (2020). Numerical investigation of the impact of wash coat distribution on the filtration performance of gasoline particulate filters. Chem Eng Sci 221:115656

Benajes J, Molina S, García A, Belarte E, Vanvolsem M (2014) An investigation on RCCI combustion in a heavy duty diesel engine using in-cylinder blending of diesel and gasoline fuels. Appl Therm Eng 63(1):66–76

Benito de la Piedra P (2015) Evaluation of a PCCI combustion in euro VI diesel engines

Bhiogade G, Suryawanshi J (2016) A comparative experimental study on engine operating on PCCI and CI mode. Thermal Sci 87

Bression G, Soleri D, Savy S, Dehoux S, Azoulay D, Hamouda HBH, Doradoux L, Guerrassi N, Lawrence N (2009) A study of methods to lower HC and CO emissions in diesel HCCI. SAE Int J Fuels Lubr 1(1):37–49

Brogan MS, Brisley RJ, Moore JS, Clark AD (1996) Evaluation of NOx adsorber catalysts systems to reduce emissions of lean running gasoline engines. SAE Trans 1916–1930

Caliskan H, Mori K (2017) Environmental, enviro economic and enhanced thermodynamic analyses of a diesel engine with diesel oxidation catalyst (DOC) and diesel particulate filter (DPF) after treatment systems. Energy 128:128–144

Cao DN, Hoang AT, Luu HQ, Bui VG, Tran TTH (2020) Effects of injection pressure on the NOx and PM emission control of diesel engine: a review under the aspect of PCCI combustion condition. Energy Sources Part A Recovery Utilization Environ Effects 1–18

Chauhan BV, Shukla MK, Dhar A (2021) Effect of n-butanol and gasoline blends on si engine performance and emissions. In: Alcohol as an alternative fuel for internal combustion engines. Springer, Singapore, pp 175–190

Combustion Engine Research Center (CERC), Annual report 2016. Chalmers University of Technology, Sweden

Combustion Engine Research Center (CERC), Annual report 2017. Chalmers University of Technology, Sweden

Contino F, Foucher F, Dagaut P, Lucchini T, D'errico G, Mounaïm-Rousselle C (2013) Experimental and numerical analysis of nitric oxide effect on the ignition of iso-octane in a single cylinder HCCI engine. Combust Flame 160(8):1476–1483

Cui Y, Liu H, Geng C, Tang Q, Feng L, Wang Y et al (2020) Optical diagnostics on the effects of fuel properties and coolant temperatures on combustion characteristic and flame development progress from HCCI to CDC via PPC. Fuel 269. https://doi.org/10.1016/j.fuel.2020.117441

d'Ambrosio S, Ferrari A (2015) Effects of exhaust gas recirculation in diesel engines featuring late PCCI type combustion strategies. Energy Convers Manage 105:1269–1280

Desantes JM, García-Oliver JM, Vera-Tudela W, López-Pintor D, Schneider B, Boulouchos K (2016) Study of the auto-ignition phenomenon of PRFs under HCCI conditions in a RCEM by means of spectroscopy. Appl Energy 179:389–400

Dev S, Chaudhari HB, Gothekar S, Juttu S, Walke NH, Marathe NV (2017) Review on advanced low temperature combustion approach for BS VI

Dou D, Bailey OH (1998) Investigation of NOx adsorber catalyst deactivation. SAE Trans 1624–1637

Duan X, Lai M-C, Jansons M, Guo G, Liu J (2021) A review of controlling strategies of the ignition timing and combustion phase in homogeneous charge compression ignition (HCCI) engine. Fuel 285:119142. ISSN 0016-2361

Elghawi UM, Mayouf A, Tsolakis A, Wyszynski ML (2010) Vapour-phase and particulate-bound PAHs profile generated by a (SI/HCCI) engine from a winter grade commercial gasoline fuel. Fuel 89(8):2019–2025

Fathi M, Saray RK, Checkel MD (2011) The influence of exhaust gas recirculation (EGR) on combustion and emissions of n-heptane/natural gas fueled homogeneous charge compression ignition (HCCI) engines. Appl Energy 88(12):4719–4724

Fu M, Li C, Lu P, Qu L, Zhang M, Zhou Y, Yu M, Fang Y (2014) A review on selective catalytic reduction of NO x by supported catalysts at 100–300 C—catalysts, mechanism, kinetics. Catal Sci Technol 4(1):14–25

Fukushima N, Katayama M, Naka Y, Oobayashi T, Shimura M, Nada Y, Tanahashi M, Miyauchi T (2015) Combustion regime classification of HCCI/PCCI combustion using Lagrangian fluid particle tracking. Proc Combust Inst 35(3):3009–3017

Ganesan S (2012) Performance and emission analysis on mixed-mode homogeneous charge compression ignition (HCCI) combustion of biodiesel fuel with external mixture formation (No. 2011–01–2450). SAE Technical Paper

Ganesh D, Nagarajan G (2010) Homogeneous charge compression ignition (HCCI) combustion of diesel fuel with external mixture formation. Energy 35(1):148–157

Ganesh D, Nagarajan G, Ibrahim MM (2008) Study of performance, combustion and emission characteristics of diesel homogeneous charge compression ignition (HCCI) combustion with external mixture formation. Fuel 87(17–18):3497–3503

Gharehghani A (2019) Load limits of an HCCI engine fueled with natural gas, ethanol, and methanol. Fuel 239:1001–1014

Ghazal OH (2018) Reducing diesel engine emission using reactivity controlled approach. J Ecol Eng 19(1)

Gholami Z, Luo G, Gholami F, Yang F (2021) Recent advances in selective catalytic reduction of NOx by carbon monoxide for flue gas cleaning process: a review. Catal Rev 63(1):68–119

Golovitchev VI, Calik AT, Montorsi L (2007) Analysis of combustion regimes in compression ignited engines using parametric φ-T dynamic maps (No. 2007–01–1838). SAE Technical Paper

Gregory D, Marshall RA, Eves B, Dearth M, Hepburn J, Brogan M, Swallow D (1999) Evolution of Lean-NOx traps on PFI and DISI lean burn vehicles. SAE Trans 1457–1469

Gu Y, Epling WS (2019) Passive NOx adsorber: an overview of catalyst performance and reaction chemistry. Appl Catal A 570:1–14

Guan B, Zhan R, Lin H, Huang Z (2014) Review of state of the art technologies of selective catalytic reduction of NOx from diesel engine exhaust. Appl Therm Eng 66(1–2):395–414

Guardiola C, Pla B, García A, Boronat V (2017) Optimal heat release shaping in a reactivity controlled compression ignition (RCCI) engine. Control Theory Technol 15(2):117–128

Hamada H, Haneda M (2012) A review of selective catalytic reduction of nitrogen oxides with hydrogen and carbon monoxide. Appl Catal A 421:1–13

Han D, Ickes AM, Bohac SV, Huang Z, Assanis DN (2012) HC and CO emissions of premixed low-temperature combustion fueled by blends of diesel and gasoline. Fuel 99:13–19

Han W, Li B, Pan S, Lu Y, Li X (2018) Combined effect of inlet pressure, total cycle energy, and start of injection on low load reactivity controlled compression ignition combustion and emission characteristics in a multi-cylinder heavy-duty engine fuelled with gasoline/diesel. Energy 165. https://doi.org/10.1016/j.energy.2018.10.029.

Han W, Lu Y, Jin C, Tian X, Peng Y, Pan S et al (2020) Study on influencing factors of particle emissions from a RCCI engine with variation of premixing ratio and total cycle energy. Energy 202. https://doi.org/10.1016/j.energy.2020.117707

He BQ, Yuan J, Liu MB, Zhao H (2014) Low-temperature combustion characteristics of an-butanol/isooctane HCCI engine. Energy Fuels 28(6):4183–4192

Hoang AT (2020) Critical review on the characteristics of performance, combustion and emissions of PCCI engine controlled by early injection strategy based on narrow-angle direct injection (NADI). Energy Sources Part A Recovery Utilization Environ Effects. https://doi.org/10.1080/15567036.2020.1805048

Hunicz J, Mikulski M, Shukla PC, Gęca MC (2022) Partially premixed combustion of hydrotreated vegetable oil in a diesel engine: Sensitivity to boost and exhaust gas recirculation. Fuel 307:121910. ISSN 0016-2361. https://doi.org/10.1016/j.fuel.2021.121910

Hyvönen J, Haraldsson G, Johansson B (2005) Operating conditions using spark assisted HCCI combustion during combustion mode transfer to SI in a multi-cylinder VCR-HCCI engine (No. 2005-01-0109). SAE Technical Paper

Imtenan S, Varman M, Masjuki HH, Kalam MA, Sajjad H, Arbab MI, Fattah IR (2014) Impact of low temperature combustion attaining strategies on diesel engine emissions for diesel and biodiesels: a review. Energy Convers Manage 80:329–356

Iwachido K, Tanada H, Watanabe T, Yamada N, Nakayama O, Ando H, Hori M, Taniguchi S, Noda N, Abe F (2001) Development of the NOx adsorber catalyst for use with high-temperature condition. SAE Trans 953–964

Jo S, Park S, Kim HJ, Lee JT (2019) Combustion improvement and emission reduction through control of ethanol ratio and intake air temperature in reactivity controlled compression ignition combustion engine. Appl Energy 250:1418–1431

Johnson JE, Kittelson DB (1996) Deposition, diffusion and adsorption in the diesel oxidation catalyst. Appl Catal B 10(1–3):117–137

Joshi A, Johnson TV (2018) Gasoline particulate filters—a review. Emission Control Sci Technol 4(4):219–239

Kakoee A, Bakhshan Y, Aval SM, Gharehghani A (2018) An improvement of a lean burning condition of natural gas/diesel RCCI engine with a pre-chamber by using hydrogen. Energy Convers Manage 166:489–499

Kalghatgi GT (2014) The outlook for fuels for internal combustion engines. Int J Engine Res 15(4):383–398

Kamimoto T, Bae MH (1988) High combustion temperature for the reduction of particulate in diesel engines. SAE Trans 692–701

Khandal SV, Banapurmath NR, Gaitonde VN (2019) Performance studies on homogeneous charge compression ignition (HCCI) engine powered with alternative fuels. Renew Energy 132:683–693

Kim BS, Jeong H, Bae J, Kim PS, Kim CH, Lee H (2020) Lean NOx trap catalysts with high low-temperature activity and hydrothermal stability. Appl Catal B: Environ 270:118871

Kitamura T, Ito T, Senda J, Fujimoto H (2002) Mechanism of smokeless diesel combustion with oxygenated fuels based on the dependence of the equivalence ration and temperature on soot particle formation. Int J Engine Res 3(4):223–248

Kobayashi K, Sako T, Sakaguchi Y, Morimoto S, Kanematsu S, Suzuki K, Nakazono T, Ohtsubo H (2011) Development of HCCI natural gas engines. J Nat Gas Sci Eng 3(5):651–656

Kokjohn SL, Hanson RM, Splitter DA, Reitz RD (2011) Fuel reactivity controlled compression ignition (RCCI): a pathway to controlled high-efficiency clean combustion. Int J Engine Res 12(3):209–226

Koltsakis G, Haralampous O, Depcik C, Ragone JC (2013) Catalyzed diesel particulate filter modeling. Rev Chem Eng 29(1):1–61

Komninos NP, Rakopoulos CD (2016) Heat transfer in hcci phenomenological simulation models: a review. Appl Energy 181:179–209

Kook S, Bae C, Miles PC, Choi D, Pickett LM (2005) The influence of charge dilution and injection timing on low-temperature diesel combustion and emissions. SAE Trans 1575–1595

Krishnamoorthi M, Malayalamurthi R, He Z, Kandasamy S (2019) A review on low temperature combustion engines: performance, combustion and emission characteristics. Renew Sustain Energy Rev 116:109404

Krishnasamy A (2017) A comparison of conventional and reactivity controlled compression ignition (RCCI) combustion modes in a small single cylinder air-cooled diesel engine (No. 2017–01–2365). SAE Technical Paper

Kumar A, Vedrtnam A (2018) Energy-exergy analysis of biodiesel fuels produced from waste cooking oil and mustard oil. Fuel 214:386–408

Lawler B, Splitter D, Szybist J, Kaul B (2017) Thermally stratified compression ignition: a new advanced low temperature combustion mode with load flexibility. Appl Energy 189:122–132

Leach F, Kalghatgi G, Stone R, Miles P (2020) The scope for improving the efficiency and environmental impact of internal combustion engines. Transport Eng 1:100005

Lee J, Chu S, Kang J, Min K, Jung H, Kim H, Chi Y (2017) Operating strategy for gasoline/diesel dual-fuel premixed compression ignition in a light-duty diesel engine. Int J Automot Technol 18(6):943–950

Lehtoranta K, Matilainen P, Åsenbrygg JM, Lievonen A, Kinnunen TJ, Keskinen J, Solla A (2007) Particle oxidation catalyst in light duty and heavy duty diesel applications (No. 2007–24–0093). SAE Technical Paper

Lehtoranta K, Matilainen P, Kinnunen TJ, Heikkilä J, Rönkkö T, Keskinen J, Murtonen T (2009) Diesel particle emission reduction by a particle oxidation catalyst (No. 2009–01–2705). SAE Technical Paper

Lemberger I, Floweday G (2009) 25cc HCCI engine fuelled with DEE. SAE Int J Engines 2(1):1559–1573

Levi PG, Cullen JM (2018) Mapping global flows of chemicals: from fossil fuel feedstocks to chemical products. Environ Sci Technol 52(4):1725–1734

Li T, Suzuki M, Ogawa H (2009) Characteristics of smokeless low temperature diesel combustion in various fuel-air mixing and expansion of operating load range (No. 2009–01–1449). SAE Technical Paper

Li S, Liu J, Li Y, Wei M, Xiao H, Yang S (2020) Effects of fuel properties on combustion and pollutant emissions of a low temperature combustion mode diesel engine. Fuel 267:117123

Liu H, Yao M, Zhang B, Zheng Z (2008) Effects of inlet pressure and octane numbers on combustion and emissions of a homogeneous charge compression ignition (HCCI) engine. Energy Fuels 22:2207–2215. https://doi.org/10.1021/ef800197b

Liu Z, Ge Y, Tan J, He C, Shah AN, Ding Y, Yu L, Zhao W (2012) Impacts of continuously regenerating trap and particle oxidation catalyst on the NO2 and particulate matter emissions emitted from diesel engine. J Environ Sci 24(4):624–631

Liu H, Ma S, Zhang Z, Zheng Z, Yao M (2015) Study of the control strategies on soot reduction under early-injection conditions on a diesel engine. Fuel 139:472–481. https://doi.org/10.1016/j.fuel.2014.09.011

Liu H, Tang Q, Ran X, Fang X, Yao M (2019a) Optical diagnostics on the reactivity controlled compression ignition (RCCI) with micro direct-injection strategy. Proc Combust Inst 37:4767–4775. https://doi.org/10.1016/j.proci.2018.06.180

Liu H, Tang Q, Yang Z, Ran X, Geng C, Chen B et al (2019b) A comparative study on partially premixed combustion (PPC) and reactivity controlled compression ignition (RCCI) in an optical engine. Proc Combust Inst 37:4759–4766. https://doi.org/10.1016/j.proci.2018.06.004

Liu H, Li Z, Zhang M, Xu H, Ma X, Shuai S (2021) Exhaust non-volatile particle filtration characteristics of three-way catalyst and influencing factors in a gasoline direct injection engine compared to gasoline particulate filter. Fuel 290:120065

Lu X, Zhou X, Ji L, Yang Z, Han D, Huang C, Huang Z (2013) Experimental studies on the dual-fuel sequential combustion and emission simulation. Energy 51:358–373

Martin GC, Mueller CJ, Milam DM, Radovanovic MS, Gehrke CR (2009) Early direct-injection, low temperature combustion of diesel fuel in an optical engine utilizing a 15-hole, dual-row, narrow-included-angle nozzle. SAE Int J Eng 1(1):1057–1082. https://doi.org/10.4271/2008-01-2400

Maurya RK, Agarwal AK (2011) Experimental study of combustion and emission characteristics of ethanol fuelled port injected homogeneous charge compression ignition (HCCI) combustion engine. Appl Energy 88(4):1169–1180

Maurya RK, Maurya RK, Luby (2018) Characteristics and control of low temperature combustion engines. Springer, Cham

Medhat E, Elzahaby AM, Bastawissi HA, Mostafa SE, Naceb AM (2018) Kinetic modeling and experimental study on the combustion, performance and emission characteristics of a PCCI engine fueled with ethanol-diesel blends. Egypt J Petrol 27(4):927–37. https://doi.org/10.1016/j.ejpe.2018.02.003

Milovanovic N, Blundell D, Pearson RJ, Turner JWG, Chen R (2005) Enlarging the operational range of a gasoline HCCI engine by controlling the coolant temperature

Minjares RJ, Sanchez FP (2011) Estimated cost of gasoline particulate filters. The International Council on Clean Transportation–Working Paper, 8

Mofijur M, Hasan MM, Mahlia TMI, Rahman SM, Silitonga AS, Ong HC (2019) Performance and emission parameters of homogeneous charge compression ignition (HCCI) engine: a review. Energies 12(18):3557

Morgan C (2015) Platinum group metal and washcoat chemistry effects on coated gasoline particulate filter design. Johnson Matthey Technol Rev 59(3):188–192

Murata Y, Nishio Y, Kusaka J, Daisho Y, Kawano D, Suzuki H, Ishii H, Goto Y (2008) HC4–2: numerical analysis of miller-PCCI combustion on a dynamic φ-T Map (HC: HCCI Combustion, General Session Papers). In: The Proceedings of the international symposium on diagnostics and modeling of combustion in internal combustion engines. The Japan Society of Mechanical Engineers, pp 343–350

Musculus MPB, Miles PC, Pickett LM (2013) Conceptual models for partially premixed low-temperature diesel combustion. Prog Energy Combust Sci 39(2–3):246–283

Nadanakumar V, Muthiya SJ, Prudhvi T, Induja S, Sathyamurthy R, Dharmaraj V (2021) Experimental investigation to control HC, CO & NOx emissions from diesel engines using diesel oxidation catalyst. Mater Today Proc 43:434–440

Neely GD, Sasaki S, Huang Y, Leet JA, Stewart DW (2005) New diesel emission control strategy to meet US Tier 2 emissions regulations. SAE Trans 512–524

No SY (2016) Application of biobutanol in advanced CI engines—a review. Fuel 183:641–658

None N (2017) FY2016 Advanced Combustion Engine Annual Progress Report (No. DOE/EE-1506). EERE Publication and Product Library

O'Connor J, Musculus M (2013) Post injections for soot reduction in diesel engines: a review of current understanding. SAE Int J Engines 6(1):400–421

Onishi S, Jo SH, Shoda K, Jo PD, Kato S (1979) Active thermo-atmosphere combustion (ATAC)—a new combustion process for internal combustion engines. SAE Trans 1851–1860

Onrubia-Calvo JA, Pereda-Ayo B, Caravaca A, De-La-Torre U, Vernoux P, González-Velasco JR (2020) Tailoring perovskite surface composition to design efficient lean NOx trap Pd–La1-xAxCoO3/Al2O3-type catalysts (with A= Sr or Ba). Appl Catal B Environ 266:118628

Pan S, Li X, Han W, Huang Y (2017) An experimental investigation on multi-cylinder RCCI engine fueled with 2-butanol/diesel. Energy Convers Manage 154:92–101

Pandey SK, Akella SS, Ravikrishna RV (2018) Novel fuel injection strategies for PCCI operation of a heavy-duty turbocharged diesel engine. Appl Therm Eng 143:883–898

Paoli L, Lupton RC, Cullen JM (2017) Probabilistic model allocating primary energy to end-use devices. Energy Procedia 142:2441–2447

Paoli L, Lupton RC, Cullen JM (2018) Useful energy balance for the UK: an uncertainty analysis. Appl Energy 228:176–188

Park SH, Yoon SH (2016) Effect of dual-fuel combustion strategies on combustion and emission characteristics in reactivity controlled compression ignition (RCCI) engine. Fuel 181:310–318

Park H, Shim E, Bae C (2019) Improvement of combustion and emissions with exhaust gas recirculation in a natural gas-diesel dual-fuel premixed charge compression ignition engine at low load operations. Fuel 235:763–774

Parks JE II, Prikhodko V, Storey JM, Barone TL, Lewis SA Sr, Kass MD, Huff SP (2010) Emissions from premixed charge compression ignition (PCCI) combustion and affect on emission control devices. Catal Today 151(3–4):278–284

Paykani A, Kakaee A, Rahnama P, Reitz R (2015) Progress and recent trends in reactivity-controlled compression ignition engines. Int J Engine Res. https://doi.org/10.1177/1468087415593013

Pedrozo VB, May I, Dalla Nora M, Cairns A, Zhao H (2016a) Experimental analysis of ethanol dual-fuel combustion in a heavy-duty diesel engine: An optimisation at low load. Appl Energy 165:166–182

Pedrozo VB, May I, Lanzanova TD, Zhao H (2016b) Potential of internal EGR and throttled operation for low load extension of ethanol–diesel dual-fuel reactivity controlled compression ignition combustion on a heavy-duty engine. Fuel 179:391–405

Polovina D, McKenna D, Wheeler J, Sterniak J, Miersch-Wiemers O, Mond A, Yilmaz H (2013) Steady-state combustion development of a downsized multi-cylinder engine with range extended HCCI/SACI capability. SAE Int J Engines 6(1):504–519

Raitanapaibule K, Aung K (2005) Performance predictions of a hydrogen-enhanced natural gas HCCI engine. In: ASME International Mechanical Engineering Congress and Exposition, vol 42185, pp 289–294

Reijnders L, Huijbregts MA (2007) Life cycle greenhouse gas emissions, fossil fuel demand and solar energy conversion efficiency in European bioethanol production for automotive purposes. J Clean Prod 15(18):1806–1812

Reitz RD, Duraisamy G (2015) Review of high efficiency and clean reactivity controlled compression ignition (RCCI) combustion in internal combustion engines. Prog Energy Combust Sci 46:12–71

Rodríguez-Fernández J, Hernández JJ, Ramos Á, Calle-Asensio A (2021) Fuel economy, NOx emissions and lean NOx trap efficiency: lessons from current driving cycles. Int J Eng Res 14680874211005080

Rood S, Eslava S, Manigrasso A, Bannister C (2020) Recent advances in gasoline three-way catalyst formulation: a review. Proc Inst Mech Eng Part D J Automob Eng 234(4):936–949

Russell A, Epling WS (2011) Diesel oxidation catalysts. Catal Rev 53(4):337–423

Saiteja P, Ashok B (2021) A critical insight review on homogeneous charge compression ignition engine characteristics powered by biofuels. Fuel 285:119202. https://doi.org/10.1016/j.fuel.2020.119202

Shafiee S, Topal E (2009) When will fossil fuel reserves be diminished? Energy Policy 37(1):181–189

Shah R (2013) Automotive air pollution and its control by catalytic converters. University of Illinois at Urbana-Champaign

Shim E, Park H, Bae C (2020) Comparisons of advanced combustion technologies (HCCI, PCCI, and dual-fuel PCCI) on engine performance and emission characteristics in a heavy-duty diesel engine. Fuel 262:116436

Silitonga AS, Mahlia TMI, Kusumo F, Dharma S, Sebayang AH, Sembiring RW, Shamsuddin AH (2019) Intensification of Reutealis trisperma biodiesel production using infrared radiation: Simulation, optimisation and validation. Renew Energy 133:520–527

Singh AP, Agarwal AK (2017) CI/PCCI combustion mode switching of diesohol fuelled production engine (No. 2017–01–0738). SAE Technical Paper

Singh AP, Agarwal AK (2018a) Experimental evaluation of sensitivity of low-temperature combustion to intake charge temperature and fuel properties. Int J Engine Res 19(7):732–757

Singh AP, Agarwal AK (2018b) Low-temperature combustion: an advanced technology for internal combustion engines. In: Advances in internal combustion engine research, pp 9–41

Singh G, Singh AP, Agarwal AK (2014) Experimental investigations of combustion, performance and emission characterization of biodiesel fuelled HCCI engine using external mixture formation technique. Sustain Energy Technol Assess 6:116–128

Singh AP, Sharma N, Agarwal R, Agarwal AK (eds) (2020) Advanced combustion techniques and engine technologies for the automotive sector, vol 6. Springer, Singapore, pp. 111–147

Smith J, Hinterberger M, Schneider C, Koehler J (2016) Energy savings and increased electric vehicle range through improved battery thermal management. Appl Therm Eng 101:647–656

Splitter D, Wissink M, DelVescovo D, Reitz R (2013) RCCI engine operation towards 60% thermal efficiency. SAE International, pp 01-0279

Stanglmaier RH, Roberts CE (1999) Homogeneous charge compression ignition (HCCI): benefits, compromises, and future engine applications. SAE Trans 2138–2145

Su S, Lai Y, Hao C, Hou P, Lv T, Ge Y (2020) Review of rapid ageing testing methods of three-way catalyst for gasoline engine. Int J Vehicle Perform 6(3):277–293

Szybist JP, Edwards KD, Foster M, Confer K, Moore W (2013) Characterization of engine control authority on HCCI combustion as the high load limit is approached. SAE Int J Engines 6(1):553–568

Tan C (2015) Model based control for a modern automotive diesel engine. Doctoral dissertation, University of Birmingham

Tanov S (2017) Investigation of partially premixed combustion in an optical engine: In: Cylinder flow and combustion characterization. Lund University.

Vakkilainen A, Lylykangas R (2004) Particle oxidation catalyst (POC) for diesel vehicles (No. 2004–28–0047). SAE Technical Paper

Vickers NJ (2017) Animal communication: when i'm calling you, will you answer too? Curr Biol 27(14):R713–R715

Wang J, Chen H, Hu Z, Yao M, Li Y (2015) A review on the Pd-based three-way catalyst. Catal Rev 57(1):79–144

Wang P, Li JB, Bai FW, Liu DY, Xu C, Zhao L, Wang ZF (2017) Experimental and theoretical evaluation on the thermal performance of a windowed volumetric solar receiver. Energy 119:652–661

West B, Huff S, Parks J, Lewis S, Choi JS, Partridge W, Storey J (2004) Assessing reductant chemistry during in-cylinder regeneration of diesel lean NOx traps. SAE Trans 1975–1985

Wu Y, Hanson R, Reitz RD (2014) Investigation of combustion phasing control strategy during reactivity controlled compression ignition (RCCI) multicylinder engine load transitions. J En Gas Turbines Power 136(9)

Wu B, Zhan Q, Yu X, Lv G, Nie X, Liu S (2017) Effects of Miller cycle and variable geometry turbocharger on combustion and emissions in steady and transient cold process. Appl Therm Eng 118:621–629

Wu G, Chen B, Bai Z, Zhao Q, Wang Z, Song C, Guo X, Shi C (2021) Cobalt oxide with flake-like morphology as efficient passive NOx adsorber. Catal Commun 149:106203

Xia W, Zheng Y, He X, Yang D, Shao H, Remias J, Roos J, Wang Y (2017) Catalyzed gasoline particulate filter (GPF) performance: effect of driving cycle, fuel, catalyst coating (No. 2017–01–2366). SAE Technical Paper

Yang S, Deng C, Gao Y, He Y (2016) Diesel particulate filter design simulation: a review. Adv Mech Eng 8(3):1687814016637328

Yao M, Zheng Z, Liu H (2009) Progress and recent trends in homogeneous charge compression ignition (HCCI) engines. Prog Energy Combust Sci 35(5):398–437

Yap D, Megaritis A, Wyszynski ML (2004) An investigation into bioethanol homogeneous charge compression ignition (HCCI) engine operation with residual gas trapping. Energy Fuels 18(5):1315–1323

Yap D, Karlovsky J, Megaritis A, Wyszynski ML, Xu H (2005) An investigation into propane homogeneous charge compression ignition (HCCI) engine operation with residual gas trapping. Fuel 84(18):2372–2379

Yap D, Peucheret SM, Megaritis A, Wyszynski ML, Xu H (2006) Natural gas HCCI engine operation with exhaust gas fuel reforming. Int J Hydrogen Energy 31(5):587–595

Ying W, Longbao Z, Wei L (2010) Effects of DME pilot quantity on the performance of a DME PCCI-DI engine. Energy Convers Manage 51(4):648–654

Zhang Y, Zhao H (2014) Investigation of combustion, performance and emission characteristics of 2-stroke and 4-stroke spark ignition and CAI/HCCI operations in a DI gasoline. Appl Energy 130:244–255

Zhao Y, Wang Y, Li D, Lei X, Liu S (2014) Combustion and emission characteristics of a DME (dimethyl ether)-diesel dual fuel premixed charge compression ignition engine with EGR (exhaust gas recirculation). Energy 72:608–617

Zheng Z, Yue L, Liu H, Zhu Y, Zhong X, Yao M (2015) Effect of two-stage injection on combustion and emissions under high EGR rate on a diesel engine by fueling blends of diesel/gasoline, diesel/n-butanol, diesel/gasoline/n-butanol and pure diesel. Energy Convers Manag 90:1–11. https://doi.org/10.1016/j.enconman.2014.11.011

Zheng Z, Xia M, Liu H, Shang R, Ma G, Yao M (2018) Experimental study on combustion and emissions of n-butanol/biodiesel under both blended fuel mode and dual fuel RCCI mode. Fuel 226:240–251. https://doi.org/10.1016/j.fuel.2018.03.151

Zhou D (2017) Numerical investigation of combustion process in reactivity controlled compression ignition (RCCI) engine. Doctoral dissertation, National University of Singapore, Singapore

Chapter 5
Combustion in Diesel Fuelled Partially Premixed Compression Ignition Engines

Sam Joe Chintagunti and Avinash Kumar Agarwal

Abstract Global concerns about depleting fuel reserves, environment degradation, and sustainability have motivated researchers to explore advanced combustion strategies. In this regard, low-temperature combustion (LTC) has emerged as a promising combustion strategy that can simultaneously reduce soot and NOx emissions with the additional benefit of superior fuel economy. However, researchers must address its primary challenges, w.r.t. controlling ignition timing, combustion phasing, and high heat release rate (HRR), to practically implement it on a large scale. Partially premixed compression ignition (PPCI) is a type of LTC that can address these drawbacks. In PPCI, the injection is closer to ignition when compared with homogeneous charge compression ignition (HCCI), which provides superior control over the ignition. It achieves better mixing and low in-cylinder temperatures using medium to high exhaust gas recirculation (EGR) levels. Optical diagnostics are powerful tools used by researchers to understand the physical behaviour of these complex processes in engines or engine-like conditions. The main attempt of this chapter is to summarize the understandings obtained by researchers about the two-stage combustion occurring in PPCI conditions using optical diagnostics. In the build-up, a brief introduction is given to the conventional diesel combustion (CDC) model and the chemical kinetics involved in the LTC to provide an overall understanding.

Keywords Advanced combustion strategies · Partially premixed charge compression ignition · Chemical kinetics · Low temperature combustion

5.1 Introduction

In recent years, the emission norms have become increasingly stringent, and it is done for the right reasons. For many years, the environment has been affected by the pollution from vehicles, and diesel engines are regarded as the major contributors. Diesel engines have higher efficiency and torque than their gasoline counterparts

S. J. Chintagunti · A. K. Agarwal (✉)
Engine Research Laboratory, Department of Mechanical Engineering, Indian Institute of Technology Kanpur, Kanpur 208016, India
e-mail: akag@iitk.ac.in

A. K. Agarwal et al. (eds.), *Advanced Combustion for Sustainable Transport*, Energy, Environment, and Sustainability, https://doi.org/10.1007/978-981-16-8418-0_5

(Heywood 1988). Also, the government has subsidized diesel (Clarke 2015), making diesel engines a preferred choice for commercial applications (Mollenhauer and Tschöke 2010). However, over the past years, high soot and NO_x emissions from diesel engines have been discussed and worried about. Though engine after-treatment solutions have shown great potential to tackle diesel emissions, these devices are expensive and under development for large-scale on-road applications (Shukla et al. 2018). Therefore, innovative technologies must be developed to reduce pollutants in the engine cylinder to comply with future emission norms. LTC is an advanced combustion strategy, different from conventional diesel combustion (CDC) and Spark Ignition (SI) (Agarwal et al. 2017). It provides the benefits of simultaneous reduction of soot and NOx, which has been the prime concern for engine developers for years. It also offers advantages in terms of improved fuel economy (Lu et al. 2011; Stanglmaier and Roberts 1999); however, it has several drawbacks, mainly involving ignition and combustion phasing (Haraldsson et al. 2003). Enormous research has been done to understand the complex combustion processes under LTC conditions and the chemical kinetics. However, it is still a grey area on which researchers continue to work and explore.

PPCI is an LTC strategy where the fuel is injected into the engine cylinder such that a moderate premixing of the fuel and air is achieved. In HCCI, the entire mixture is homogeneous with no fuel-rich regions before the start of combustion (SoC). On the other hand, PPCI has fuel-lean and fuel-rich zones in the combustion chamber (Singh and Agarwal 2021). EGR is essential for PPCI to achieve 'positive ignition dwell' and the LTC. Ignition dwell is the time between the end of injection and the SoC (Idicheria and Pickett 2011). In conventional compression ignition (CI) engines, combustion and injection events overlap, leading to compromised fuel–air mixing and soot emissions. This is termed as 'negative ignition dwell.' However, in PPCI, positive ignition dwell is achieved by injecting away from the SoC, delaying the chemical kinetics of combustion by EGR, and injecting less fuel to reduce the injection duration (Cho et al. 2018). PPCI combustion strategies can be classified into early injection and late injection. In the early injection PPCI, the fuel is injected in the middle or towards the end of the compression stroke but before the injection timing, usually found in conventional CI engine (few crank angle degrees (CAD) before top dead centre (TDC)) (Kanda et al. 2005; Klingbeil et al. 2003). In late injection PPCI, fuel is injected in an early expansion stroke or very close to the TDC (Kimura et al. 1999) but later than the injection timing employed in conventional CI engines (Colban et al. 2007; Cho et al. 2009). Nevertheless, the main objective is to inject the fuel into a low temperature and less dense medium to increase the ignition delay for superior mixing.

Figure 5.1 shows the typical needle lift and apparent heat release rate (AHRR) for an early injection PPCI in a light-duty engine (a) (Singh et al. 2007) and its comparison with a conventional diesel engine (b) (Dec 1997). For both the cases, the engine speed was constant at 1200 rpm, and IMEP was 3.9 bar and 4.1 bar for PPCI and conventional engine, respectively. In PPCI, the injection started at ~ 22° CA before TDC (bTDC) and continued for 7° CA. The AHRR initially showed a slight negative bump due to the evaporation of fuel, which is common in both cases. With the onset of the 1st stage ignition (cool flames) in the PPCI, the heat release resulted in

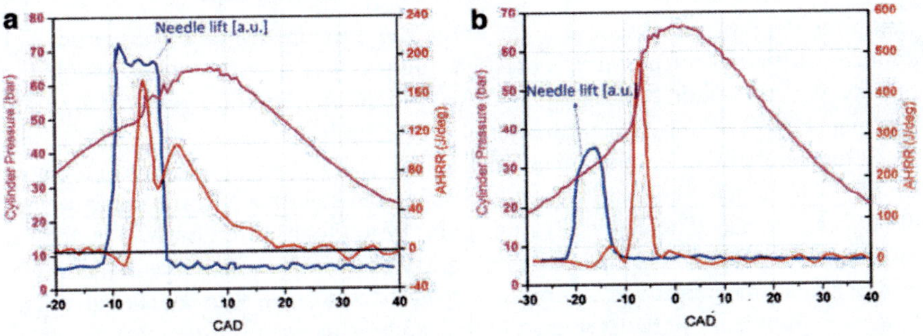

Fig. 5.1 Needle lift, cylinder pressure, and AHRR in **a** conventional diesel engine [adapted from Dec (1997)] and **b** early injection PPCI [adapted from Musculus et al. (2013)]

a small positive peak at around 13° CA bTDC. This was followed by a significant heat release from the main premixed combustion (2nd stage ignition). The main premixed combustion was observed to start after the end of injection (EoI), the trademark of PPCI combustion. There is generally a negligible heat release in the mixing-controlled phase. In the conventional combustion, the negative bump was followed by a premixed heat release and a very prominent diffusion phase heat release. The maximum AHRR in PPCI was close to 500 J/CAD compared to 180 J/CAD for CDC. This renders PCCI combustion engines very noisy.

The conventional understanding of direct injection diesel spray combustion, mainly based on the work documented by Dec (1997), is presented in the next section. This provides a basis to interpret the subsequent studies. Concepts related to the combustion regimes of the PPCI are discussed in the subsequent sections.

5.2 Conventional Diesel Jet Combustion Model

When diesel is injected into the engine cylinder, it quickly penetrates the combustion chamber, entraining ambient air, which gives it a conical shape. The fuel extracts energy from the entrained hot ambient air and evaporates. The length of the fuel jet from the nozzle tip to the point where the entire liquid fuel gets vaporized is called liquid penetration length (LPL). The region before the LPL contains fuel droplets, ligaments, and a liquid core. The vaporized fuel keeps penetrating the chamber due to its momentum and forms a region downstream with a rich yet combustible mixture having an equivalence ratio in the range 2–4. The boundary between the rich fuel jet interior and surrounding air has a small region with a stoichiometric fuel–air mixture. Generally, the start of ignition in diesel combustion is defined as the point where the heat release curve becomes positive after a small negative bump. However, the first chemiluminescence was observed before the positive HRR in the vapour fuel region (Dec 1997). This chemiluminescence is weak and accompanies the formation of HCHO (discussed later in detail). After a very short time (100–200 ms), chemiluminescence is observed in the entire rich mixture region downstream. This coincides with the high premixed HRR. The high temperatures result in the formation of polycyclic aromatic hydrocarbons (PAHs) in the rich mixture zones (called soot precursors), followed by the formation of soot particles.

At the peak of the premixed heat release, a diffusion flame is formed at the outer envelope of the jet. The flames travel upstream and stabilize at some distance from the injector nozzle. The distance of this point from the injector nozzle is called lift-off-length (LOL). The LOL plays a vital role in combustion efficiency as the mixing of fuel and air happens upstream of the lift-off location. The equivalence ratio at the LOL is a significant parameter used by researchers to quantify the mixing process (Singh and Agarwal 2021; Idicheria and Pickett 2011). Dec hypothesized a standing premixed reaction zone between the liquid penetration length and the lift-off length after the onset of diffusion flame. This hypothesis was later confirmed in another study (Bruneaux et al. 2004) that detected HCHO in this region during mixing controlled combustion. Hence, after mixing with air and evaporation, the fuel vapours pass through the premixed reaction zone, where they get partly decomposed and produce HCHO, PAHs, and soot. These species enter the fuel-rich core of the jet and later get oxidized in the diffusion flames at the jet periphery. The shift from premixed to mixing-controlled combustion is not instantaneous but a transition phenomenon.

During the initial premixed combustion, the soot increases in size downstream and is the largest in the head vortex. The diffusion combustion region where stoichiometric combustion of fuel–air mixture occurs is a pool of OH-radicals and excess oxygen. These factors are the primary driving force for soot oxidation; hence soot oxidation occurs in the diffusion region. Also, high temperature and stoichiometric mixture in the diffusion combustion region in the jet periphery are best suited for NOx formation, governed by the Zeldovich mechanism (Dec 1997). After the initial stage (where the jet continues penetrating) and the initial ignition, there is a steady phase where the fuel combusts in a mixing-controlled regime while it is still being injected. This is called the 'quasi-steady' period, and the jet morphology is temporally stable during this period (Fig. 5.2) (Musculus et al. 2013). The jet morphology

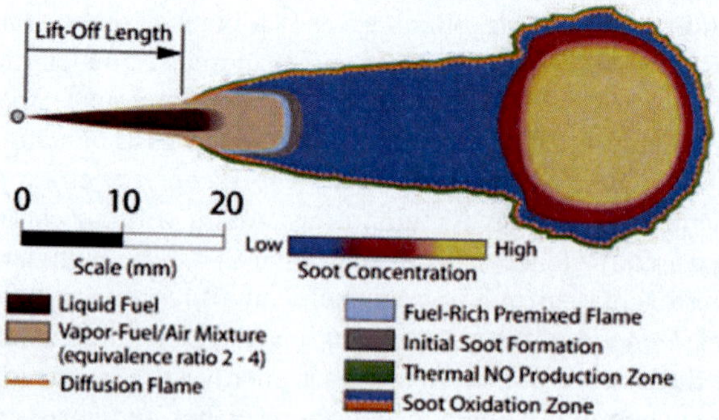

Fig. 5.2 Conceptual diesel jet combustion model (Musculus et al. 2013)

during the quasi-steady period has a liquid region (as the injection is still happening), a vapour region (because of the evaporation due to the high-temperature ambient gases), a premixed reaction zone (just before the LOL), soot concentrated fuel-rich jet interior and a diffusion flame with stoichiometric fuel–air mixture burning at high temperatures.

5.3 Chemical Kinetics

The major difficulties with HCCI/PPCI are the high HRR at high loads and the high cyclic variations at low loads, mainly due to a lack of adequate control over the ignition and subsequent combustion (Haraldsson et al. 2003; Kuwahara et al. 2013). The chemical kinetics dominates the combustion control of HCCI and PCCI mode. Therefore, understanding the chemical kinetics of combustion is essential. Conventional fuels are a mixture of several straight-chain and aromatic hydrocarbons with a wide boiling range. Developing insight into the combustion process of such multi-component fuels is difficult. Fortunately, the ignition behaviour of conventional fuels is similar to some simple single-component fuels such as heptane, dodecane, iso-octane, etc. Comprehensive reaction mechanisms of n-heptane and iso-octane are provided in Curran et al. (1998) and Curran et al. (2002). Several researchers have adapted these reaction mechanisms to understand the combustion behaviour of conventional fuels using simulations for different conditions. The outcomes are then validated against the experimental results obtained in the constant pressure combustion, rapid compression machines (RCM), or shock tubes (Vanhove et al. 2006; Alviso et al. 2020; Ando et al. 2009; Westbrook 2000). In the study by Musculus et al. (2013), simulations for the stoichiometric mixture in a homogeneous constant pressure reactor were performed at 787 K and 50 bar for n-heptane (Fig. 5.3). The results are generally not quantitatively accurate since the charge in the PPCI combustion is not completely premixed. The pressure and temperature stratifications arising during combustion and compression were also ignored. Nevertheless, they provided a qualitative understanding of premixing on the combustion and the changes in molar concentrations of different species. The conclusions provided valuable feedback for developing the optical diagnostics to observe and study various phases.

The temperature plot in Fig. 5.3 followed the trends of HRR in Fig. 5.1b (separate 1st and 2nd stage heat release), showing the evidence of two-stage ignition.

The 1st stage of ignition results from the low-temperature hydrocarbon oxidation, and understanding this complex phenomenon is vital. It was seen that the 1st stage of ignition was mainly dominated by the formation and decomposition of keto-hydroperoxide (KHP). The low-temperature oxidation starts with adding O_2 to an alkyl group forming an RO_2 species (R1) (Westbrook 2000). Higher temperature favours this reaction; hence more RO_2 develops as the temperature increases. The RO_2 molecule isomerizes to produce a QOOH radical species (R2). In isomerization, the hydrogen from the alkyl end shifts to the O_2 end. This is governed by factors such as the size of the alkyl group and the type of CH bond (primary, secondary or

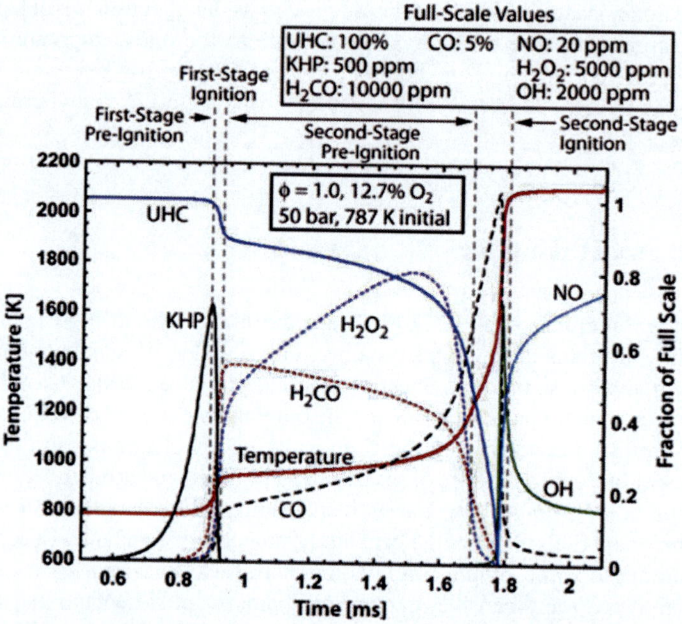

Fig. 5.3 Simulation of vapour phase n-heptane ignition in a stoichiometric mixture (Musculus et al. 2013)

tertiary) being broken. The QOOH formed are seen to take different alternate paths, all chain propagating steps, i.e., one radical reacts and produces one radical. O_2 can add again to the QOOH, creating O_2QOOH radical (R3), which again isomerizes. This isomerized product decomposes into KHP, which is relatively stable below its decomposition temperature. Its decomposition temperature is ~ 800 K, and when the reaction reaches this temperature, the first chain branching reaction (decomposition of KHP) occurs, giving rise to two or more radicals. This marks the 1st stage of ignition. The chain branching step was observed to stop at ~ 850 K because of the shift in the equilibrium in the oxygen addition reaction to the alkyl group and QOOH group (R1 and R3). This temperature is called ceiling temperature. The decomposition of KHP produces OH and other radicals, which are necessary for oxidation and heat release. This acts as a precursor for the 1st stage heat release.

$$R + O_2 + M \rightarrow RO_2 + M \tag{R1}$$

$$RO_2 \rightarrow QOOH \tag{R2}$$

$$QOOH + O_2 \rightarrow O_2QOOH \tag{R3}$$

Apart from the heat release, this decomposition is coupled with the formation of formaldehyde (HCHO) and H_2O_2 (Hydrogen peroxide) (Musculus et al. 2013). The OH and other radicals formed in the 1st stage of ignition oxidize the hydrocarbons to release heat. HCHO is a result of this hydrocarbon oxidation. HCHO is consumed when it reacts with radicals such as OH, but the HCHO concentration follows the UHC concentration closely. HCHO is an important species as it can easily be detected using Laser-Induced Fluorescence (LIF), and its concentration follows the trends of Unburned Hydrocarbons (UHC) (Musculus et al. 2013). Hence detection of HCHO can facilitate the understanding of UHC in the 2nd stage pre-ignition. The other critical species that play a vital role in the 2nd stage of ignition is H_2O_2.

H_2O_2 plays a similar role to KHP (in the 1st stage) for the second stage ignition. Its decomposition, which is a chain branching step, marks the second stage of ignition. H_2O_2 is formed in the 1st stage of ignition mainly by the combination of two HO_2 radicals. The HO_2 radicals are the products of various oxidation reactions of hydrocarbons or decomposed hydrocarbon radicals such as HCHO (R4–R6). H_2O_2 undergoes a regeneration loop, where it undergoes a series of reactions, producing heat energy (R7–R10). This energy is responsible for the increase in temperature that leads to its decomposition temperature (Kuwahara et al. 2013). Apart from the regeneration loop, H_2O_2 also undergoes decomposition, producing OH radicals during the 2nd stage pre-ignition. Still, the formation reaction is quicker than the decomposition reaction giving rise to a large pool of H_2O_2 radicals. At temperatures between 900 and 1000 K (decomposition temperature), the chain branching step producing multiple OH radicals, which dominate the reactions. These OH rapidly consume the unburned fuel molecules and CO (formed by the oxidation of different fuel species and HCHO), releasing a tremendous amount of heat and increasing the temperature. This sudden heat release accompanied by a sudden rise in temperature marks the start of ignition (main premixed combustion) (Kuwahara et al. 2012).

$$2\mathrm{HCHO} + \mathrm{O2} \rightarrow 2\mathrm{H_2O} + 2\mathrm{CO} + 448\,\mathrm{kJ/mol} \tag{R4}$$

$$\mathrm{H} + \mathrm{O_2} + \mathrm{M} \rightarrow \mathrm{HO_2} + \mathrm{M} + 204\,\mathrm{kJ/mol} \tag{R5}$$

$$\mathrm{H_2O_2} + \mathrm{OH} \rightarrow \mathrm{HO_2} + \mathrm{H_2O} + 129\,\mathrm{kJ/mol} \tag{R6}$$

$$\mathrm{H_2O_2} + \mathrm{M} \rightarrow \mathrm{OH} + \mathrm{OH} + \mathrm{M} - 210\,\mathrm{kJ/mol} \tag{R7}$$

$$\mathrm{OH} + \mathrm{HCHO} \rightarrow \mathrm{H_2O} + \mathrm{HCO} + 128\,\mathrm{kJ/mol} \tag{R8}$$

$$\mathrm{HCO} + \mathrm{O_2} \rightarrow \mathrm{CO} + \mathrm{HO_2} + 139\,\mathrm{kJ/mol} \tag{R9}$$

$$\mathrm{HO_2} + \mathrm{HO_2} \rightarrow \mathrm{H_2O_2} + \mathrm{O_2} + 164\,\mathrm{kJ/mol} \tag{R10}$$

Anything before the start of ignition (2nd stage) that facilitates achieving the decomposition temperature expedites the ignition and vice versa. It was seen that increasing the pressure or maintaining the pressure while reducing others parameters decreased the ignition delay. These two factors played an important role in controlling the start of ignition (Westbrook 2000). The bright chemiluminescence of OH radicals facilitates the study of 2nd stage ignition. The decomposition of the pool of OH radicals marks the end of combustion. When the mixture is rich or lean, significant changes were observed in the dwell between the two ignition stages (Musculus et al. 2013). In rich mixtures, 1st stage ignition resulted in higher temperatures than in the stoichiometric mixtures, higher than lean mixtures. These high temperatures mean that the decomposition of H_2O_2 happens earlier, decreasing the dwell between the two ignition stages.

Numerical simulations and constant pressure reactors have given substantial insights into the combustion characteristics of hydrocarbons. However, there are several simplified boundary conditions, unlike the transient condition in engine applications. Therefore, optical experiments are essential for developing predictive models for engine simulations. The laser-induced fluorescence (LIF) technique has been an excellent tool in this regard. In the next section, a brief overview of the LIF measurement of species concentration has been provided.

5.4 Planar Laser-Induced Florescence (PLIF)

PLIF is an optical measurement technique widely used in many applications to study species concentration, temperature, and velocity (Tropea et al. 2007). In IC engines, it is used primarily to detect the presence, spatial distribution, and concentration of a particular species of interest. It is based on fluorescence emitted by a chemical species excited by light (Tagliante et al. 2018). When light (usually laser) with a wavelength in the excitation wavelength range of a particular chemical species is incident, it absorbs a photon from the incident laser light and gets excited. The absorbed energy will be partly or totally dissipated in the vibrational and collisional transfer with the other molecules (Seitzman and Hanson 1993). The remaining energy will be re-emitted in the form of a photon of higher wavelength (and lower energy) with respect to the incident one within a few nanoseconds or microseconds. This de-excitation process emits light energy in a specific wavelength (called fluorescence), characteristic of that particular chemical species. The basic experimental setup of PLIF is shown in Fig. 5.4.

A chemical species that is present in an adequate concentration in the region of interest must be selected. The fluorescence wavelength of the selected species must be different from other species' excitation and fluorescence. Then, laser light in the absorption spectra is passed through appropriate sheet optics to make it planar. This planar light is then adjusted to make it pass through the test section at the required location. Now, the selected chemical species molecules absorb the incident laser light and emit the respective fluorescence. After passing through filters, this fluorescence is

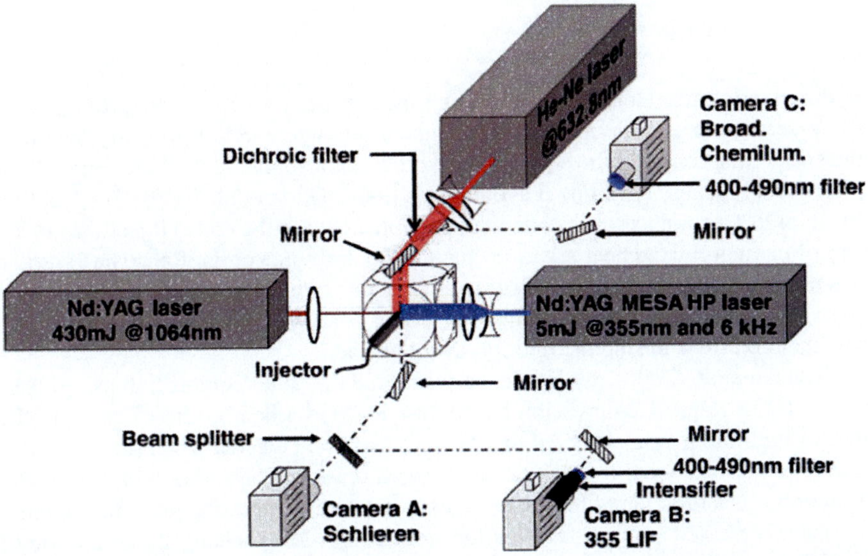

Fig. 5.4 Schematic representation of basic 355 PLIF experimental setup (Tagliante et al. 2018)

collected by a camera (generally an intensified CCD camera suitable for combustion applications) to eliminate the unwanted noise. This signal can be used to understand the presence of the specific chemical species in the plane of interest. The intensity of the acquired fluorescence by the camera can be related to the concentration of the species. However, it may be inaccurate, especially in complex phenomena like combustion in IC engines where many unknown factors are hard to determine (Zhao and Ladommatos 1998). One such factor is quenching, an electronic energy transfer from the excited atoms due to inelastic collisions. It is predominant in the presence of atmospheric oxygen.

The advantage of PLIF is that it can provide both spatially and temporally resolved information. It gives information in the plane of interest, unlike other methods like schlieren, which provide only integrated information along the line of sight. Compared to Raman and Rayleigh scattering techniques (also used to study the concentration of a species), its signal is stronger (depending on the selected species and its concentration), and hence this technique is widely used in fundamental combustion studies (Crimaldi 2008). Researchers have used PLIF to study the fuel–air mixture concentration, cool flames, flame propagation in SI engines, diffusion flames in DI engines (Soid and Zainal 2011; Dec and Coy 1996). HCHO, OH, CH, and fuel dopants are the most widely used chemical species in combustion studies. Recently, a burst mode laser was used for conducting simultaneous PLIF of HCHO, OH, and fuel tracer (Wang et al. 2019). The high temporal ratio of 5 frames/ CAD allowed the researchers to understand the temporal evolution of the fuel–air mixing, ignition, and subsequent combustion without being affected by the cyclic variations.

5.5 First Stage Ignition

As discussed earlier, HCHO is an important species formed in the 1st stage of ignition. The ease of its detection optically has enabled many researchers to study the onset and spatial dispersion of the cool flames, which is the source of the low-temperature oxidation (1st stage ignition) (Fayoux et al. 2004; Onda et al. 2019). The ignition dwell for PPCI combustion is defined as the time between the end of injection and the start of main premixed heat release, which is the 2nd stage of ignition. The 1st stage of ignition is generally initiated during the injection, which may continue for a small interval after the completion of injection (Yang et al. 2016). Let us first understand how the mixture is during the ignition dwell period.

Koshaka et al. (2000) studied the conventional diesel jet combustion using PLIF of HCHO in Rapid Compression Machines (RCM) fuelled by diesel to study the start of ignition. It was observed that HCHO was first detected when the integrated heat released was equal to the heat lost for evaporation. The position of the HCHO detection was a function of the ambient temperature and independent of the ambient pressure. Increased ambient temperature from 580 to 790 K changed the location of HCHO detection from the jet periphery to the central region of the spray head. Also, the increase in ambient temperature resulted in an early HCHO consumption due to an earlier occurrence of the high temperatures, which were responsible for HCHO decomposition. In another study by Musculus and Lachaux (2007), higher ignition delay further delayed the consumption of HCHO and the appearance of OH radicals. This is a clear indication of the presence of LTC even in the CDC, resembling the 1st stage of ignition as seen in the PPC combustion. Still, the high ambient temperatures make the entire phenomenon short-lived, thus projecting an image of a single peak of sudden heat release. Bruneaux et al. (2004) also investigated the conventional diesel jet combustion using HCHO and OH radicals. Excitation at 355 nm, termed as 355 LIF in the experiment, was used to detect HCHO. During the autoignition phase, 355 LIF detected HCHO downstream of the liquid jet. Also, during the mixing-controlled combustion phase, HCHO was detected at the point where the flames were lifted between the liquid core and the soot and fuel-rich zone of the jet interior. It proved the existence of a cool flame where the fuel gets decomposed before the LOL. This LTC zone continues to exist during the entire mixing-controlled combustion. These zones were seen to be similar even when the jet was impinging on a wall (Bruneaux 2008).

Musculus et al. (2007) carried out planar laser-Rayleigh scattering in a CVCC and PLIF in a heavy-duty optical engine with a late injection to study the fuel concentration distribution under the LTC conditions. From both the CVCC and engine studies, it was observed that the spray followed the basic structure and properties of a conventional diesel jet before the end of injection. The portions near the nozzle were fuel-rich, and the equivalence ratio decreased at regions farther away from the nozzle. This was because of air entrainment as the jet propagated into the chamber. But as the fuel injection rate starts to decline at the end of injection (EoI), and the entrainment wave increases at twice the penetration rate, causing a rapid reduction in the spray penetration length. It was seen that the regions near the nozzle became fuel-lean ($\varphi = 0.5$), and away are comparatively fuel-rich (φ near unity) mixtures. With further delayed combustion, the regions near the fuel injector became leaner. In a study conducted on the light-duty engines with early injection, Petersen et al. (2012) reported a similar phenomenon. It was observed that lean mixtures formed in the upper portion of the cylinder near the injector nozzle with equivalence ratios between 0.1 and 0.5. Overly fuel-rich mixtures were formed in the squish region, and the squish flow was insufficient to distribute this fuel-rich mixture to the other leaner central parts of the cylinder.

Ganzale et al. (2009) investigated the ignition in late injection heavy-duty PPC using HCHO PLIF. HCHO fluorescence was first captured at the start of the 1st stage ignition, just 1 CAD before the EoI. HCHO was detected almost simultaneously in all the jet regions, and by the EoI, HCHO was detected in the core (fuel-rich, $\varphi \approx 3.5$) and the periphery (fuel-lean, $\varphi \leq 0.2$). This is interesting as HCHO formation is usually delayed in cooler, fuel-rich regions according to homogeneous, fully mixed chemical kinetics (Musculus et al. 2013). This might be due to the transient dynamics of real spray combustion. These regions do not have specific boundaries, and radicals can freely move between the boundaries. The constant mixing between the regions and the temperature and pressure gradients in these regions cause these inconsistencies. After the EoI, HCHO was detected in the entire jet. This enabled researchers to study the jet structure during the initial stages of combustion using HCHO PLIF (Särner et al. 2005; Kashdan et al. 2004). It was observed that after the EoI, the jet experienced a wavy structure because of the turbulence generated by the jet-air interaction. Though these interactions are common for any fuel jet in the engine conditions, the reduced momentum because of the end of injection is the main reason for these observations (Singh et al. 2009). Also, it was reported that until the start of the main premixed heat release, the spatial distribution of HCHO increased, and the HCHO fluorescence filled the jet without significant changes in the intensity. And once the second stage of ignition started, the HCHO fluorescence decreased rapidly. The model to conceptually portray the above phenomenon is given (Musculus et al. 2013) in Fig. 5.5. The model was given for a late injection heavy-duty LTC combustion. The injection starts at the TDC and lasts for 7° CA. The model incorporated all the phenomena such as (a) conventional type fuel jet up to the EoI, (b) reduction in penetration after the EoI, (c) simultaneous HCHO formation in the vapour fuel–air mixture irrespective of the equivalence ratio, (d) HCHO detection in the entire jet, (e) wavy structure of the jet, and (f) continued HCHO detection up to the second stage of ignition. Since this model was developed for a heavy-duty engine, the spray impinged the wall much later than in the case of a light-duty engine as the cylinder bore was smaller, and the ambient density was lesser for light-duty engines (Petersen et al. 2012).

In another attempt to study the HCHO formation, Hildingsson et al. (2004) compared the HCHO formation and concentrations for HCCI, CDC, and a dual injection mode using simultaneous PLIF for OH and HCHO in an optically accessible diesel engine. Similar simultaneous detection of HCHO across all jet regions during the start of low-temperature oxidation was observed. It was also noted that

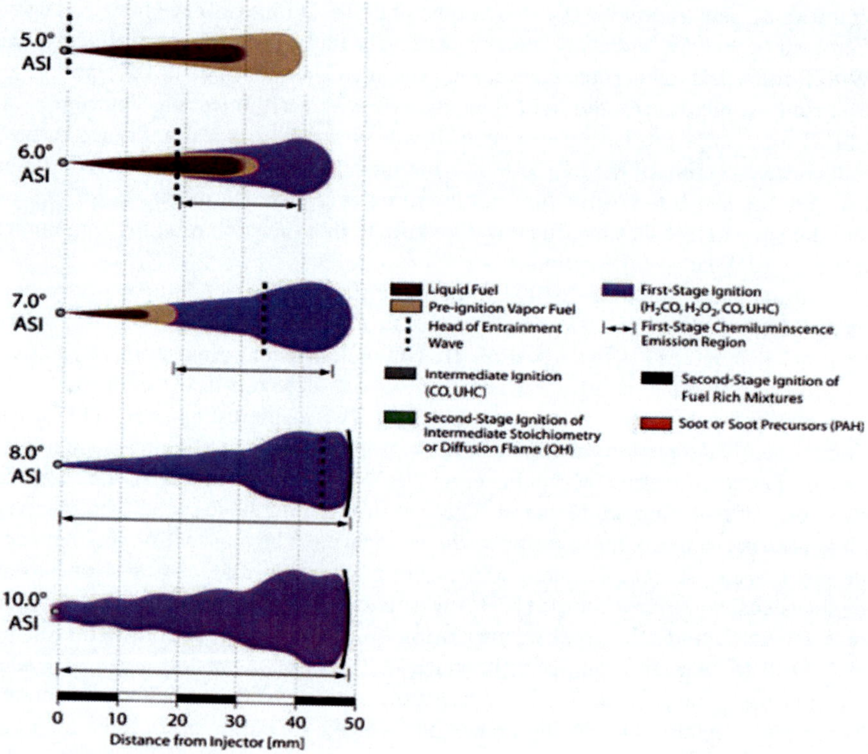

Fig. 5.5 Conceptual model of 1st stage ignition of diesel jet combustion in heavy-duty late injection PPCI (Musculus et al. 2013)

the duration for the detection of HCHO was shorter for the HCCI than the CDC mode. In the traditional diesel injection mode, HCHO was detected up to 40 CAD after TDC (aTDC), whereas in HCCI, the HCHO fluorescence became weak by 25° CA aTDC.

In the light-duty early injection PPCI combustion also, similar observations were made (Petersen et al. 2010). In this study, the fuel (mixture of iso-octane and n-heptane) was port-injected with similar combustion phasing as an early injection PPCI engine. HCHO PLIF studies were conducted, and it was reported that the appearance of HCHO was rapid and simultaneous. The regions with a lean fuel–air mixture showed a low intensity of HCHO fluorescence. However, the entire appearance and disappearance of HCHO followed a similar sequence as in a late injection heavy-duty engine. In other studies (Johansson 2005a, b), chemiluminescence imaging and OH PLIF were used to study the ignition in early injection low-temperature PPCI combustion in an optically accessible heavy-duty engine. As discussed earlier, when the fuel was injected earlier into low ambient density and temperature medium, the spray penetration and ignition delay were higher, resulting in superior fuel–air mixing. This gave rise to leaner regions with an almost stoichiometric mixture that otherwise would have been rich.

Weak chemiluminescence was observed at the tip of the liquid jet (sometimes even overlapping the liquid jet) when the AHRR was maximum for the 1st stage heat release. This chemiluminescence was from the cool flames resembling the 1st stage ignition. Quickly, this chemiluminescence increased by a few orders of magnitudes due to the soot incandescence. This indicated the formation of soot at the end of 1st stage ignition. Also, a sudden increase in fuel evaporation was observed after the significant cool flame heat release. This meant that the heat from the 1st stage combustion played an important role in the vaporization and the fuel–air mixture formation for the subsequent combustion. Petersen et al. (2010) stated that the UHC emissions were mainly dependent on the time taken to release 50% of the fuel energy after the SoC. The shorter this time was, the higher were the cylinder temperatures. At high temperatures, zones of lean fuel–air mixtures were a source of the UHC formation, which underwent oxidation and reduced the UHC concentration. Hence, the heat release phenomenon observed in conventional diesel jets was ideal for lowering the UHC emissions. Therefore, a compromise must be found between a delayed heat release that helps in mixture formation (Jain et al. 2017) and an advanced heat release that reduces the UHC and CO emissions.

5.6 Second Stage Ignition

As seen in the previous section, HCHO is formed during the 1st stage of ignition, which is subsequently consumed in the 2nd stage of ignition. From chemical kinetics, one understands that OH radicals are formed in the 2nd stage of ignition. Hence 2nd stage of ignition is marked by the consumption of HCHO and the appearance of OH radicals in significant concentration. Another important point is that 355 nm PLIF used to detect HCHO fluorescence, is also sensitive to PAH fluorescence (Seitzman and Hanson 1993; Zhao and Ladommatos 1998). This was not a major concern in the 1st stage of ignition as PAHs were formed in the high-temperature fuel-rich regions. The 355 nm PLIF signals mostly detected the HCHO concentration because of the information from the chemical kinetics, which supported the presence of HCHO at these low temperatures (Westbrook 2000; Musculus et al. 2013).

In the conventional diesel HTC, as investigated by Dec and Coy (1996), OH PLIF was used in an optical diesel engine. It was reported that there were no OH radicals detected during the premixed combustion phase. OH was detected during the mixing-controlled combustion phase in the periphery of the jet in the thin layer of diffused flames. The mixture was almost stoichiometric in this layer. There were no signs of OH radicals in the fuel-rich interiors of the jet. This followed our understanding of OH radical formation from its chemical kinetics, as discussed earlier.

In the PPCI combustion, superior premixing was facilitated by the higher ignition delay since the ambient density and temperatures were relatively lower at the time of fuel injection. It generated leaner regions with an almost stoichiometric mixture that otherwise would have been rich. Hence OH radicals were found in larger portions of the jet than the CDC (Musculus 2006). In CDC, there were no OH radicals detected during the premixed combustion phase. However, in the PPC combustion, OH radicals marked the main premixed combustion phase.

In the study conducted by Ganzale et al. (2009), OH PLIF with 355 nm was used to study the 2nd ignition stage in a late injection heavy-duty diesel engine running in the LTC mode. OH was first detected near the 2nd stage ignition heat release peak, consistent with the chemical kinetics. The entire jet was divided into three regions:

1. Regions where HCHO was completely displaced with the OH fluorescence. This was seen in the regions with stoichiometric mixtures. Also, the overlap duration when both the HCHO and OH radicals existed was minimum. This was because of the high temperature in these pockets that allowed the consumption of HCHO before the breakdown of H_2O_2.
2. Regions where HCHO disappeared but showed no evidence of OH fluorescence. After few crank angle degrees, these regions were bright red with fluorescence from the PAHs. These regions were fuel-rich and were the main soot formation zones.
3. Regions where HCHO continued to exist. This was due to the lower temperatures in this region for H_2O_2 decomposition to facilitate OH radical production and subsequent fuel oxidation. These regions were fuel-lean and were the main source of HC and CO emissions.

Figure 5.6 (Musculus et al. 2013) shows the equivalence ratio, 355 nm PLIF, and OH PLIF in a late injection LTC heavy-duty diesel engine with 152° included angle. The injection ended at 3° CA aTDC. The three rows of images show the PLIF results

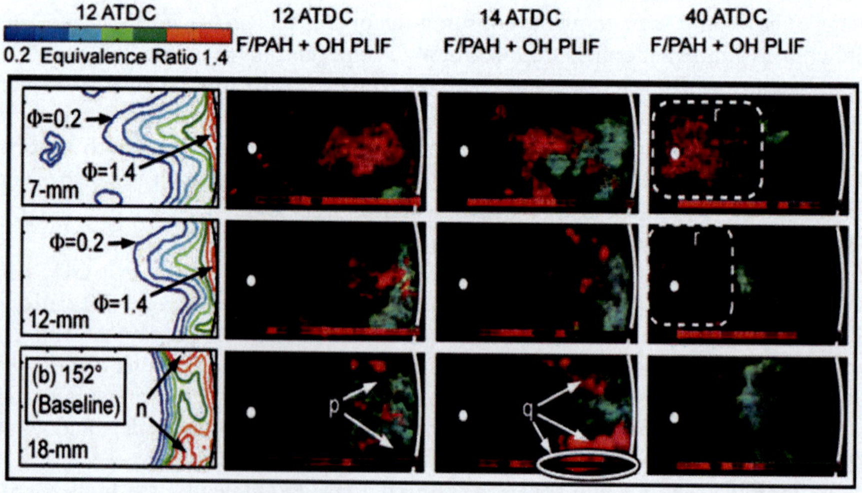

Fig. 5.6 Equivalence ratio, 355 nm PLIF and OH PLIF in heavy-duty late injection diesel PPCI with 152° included angle (Musculus et al. 2013)

of planes located at 7, 12, and 18 mm from the fire deck. The red and green regions indicated the signals from 355 nm and OH, respectively. The regions that are red and are above the red colour in the red–black colour bar in each image correspond to the HCHO signal, and the red regions above the black block correspond to the PAH signal.

The first OH signals were detected in the jet head across the cross-section at 12° CA aTDC, corresponding to region 1. This phenomenon of OH detection in the entire jet cross-section was confirmed by other researchers also using direct chemiluminescence techniques (Singh et al. 2007; Musculus 2006). There was not much overlap in green and red signals. Some black spots marked by 'p' were observed in the 18 mm plane, close to the bowl in the regions where the mixture was rich. Similar dark spots were reported by Genzale et al. (2009). By 14° CA aTDC, these regions showed bright PAH fluorescence marked by 'q.' These regions correspond to region 2. Some regions in the upper part of the cylinder in the 7 mm plane and close to the injector tip (denoted by a white dot) showed HCHO florescence during the entire 2nd stage of ignition. These regions are marked as 'r' in the figure, corresponding to region 3.

From the equivalence ratio studies using tracer PLIF in engines operated in PPCI mode (Genzale et al. 2009), it was seen that the piston bowl size and injection angle play an essential role in mixture formation and subsequent combustion. The relatively smaller piston bowl resulted in quicker spray jet impingement, and richer fuel–air mixture zones were formed in the pockets, where the jets interacted near the jet head. This was due to reduced ambient air entrainment in that region, resulting in richer mixtures near the bowl. However, when the piston bowl was larger, there was no evidence of plume-to-plume interactions, and the fuel-rich regions were at the center of the jet near the bowl. Nevertheless, the takeaway from these studies was that richer regions were formed at the far end of the jet near the bowl, and fuel-lean regions were observed closer to the injector in the engine cylinder center.

Figure 5.7 provides the continuation of the conceptual model (Fig. 5.5) of the late injection heavy-duty LTC combustion (Musculus et al. 2013). It captured different phenomena such as (a) OH appears in the far end of the jet in the entire cross-section, (b) dark spots without HCHO or OH fluorescence, (c) appearance of bright fluorescence from PAHs (soot precursors) or soot, (d) continuous fluorescence from HCHO in regions closer to the injector, indicating that these mixtures have not taken part in the 2nd stage of ignition and hence becoming a source of UHC and CO, and (e) Soot oxidation in the late combustion in the fuel-rich red spots. This was based on the availability of OH radicals and sufficient temperature. The grey zone between the blue and red contains the products of 1st stage combustion such as CO and decomposed fuel. Similar observations were also made in Knox and Genzale (2017). It was reported that the regions near the nozzle were a major source of UHC emissions because the combustion was unable to penetrate the central parts near the nozzle. It was also concluded that the rate of ramp down had an important role in this phenomenon. The quicker was the ramp down, the leaner the fuel was near the nozzle, resulting in higher UHC emissions.

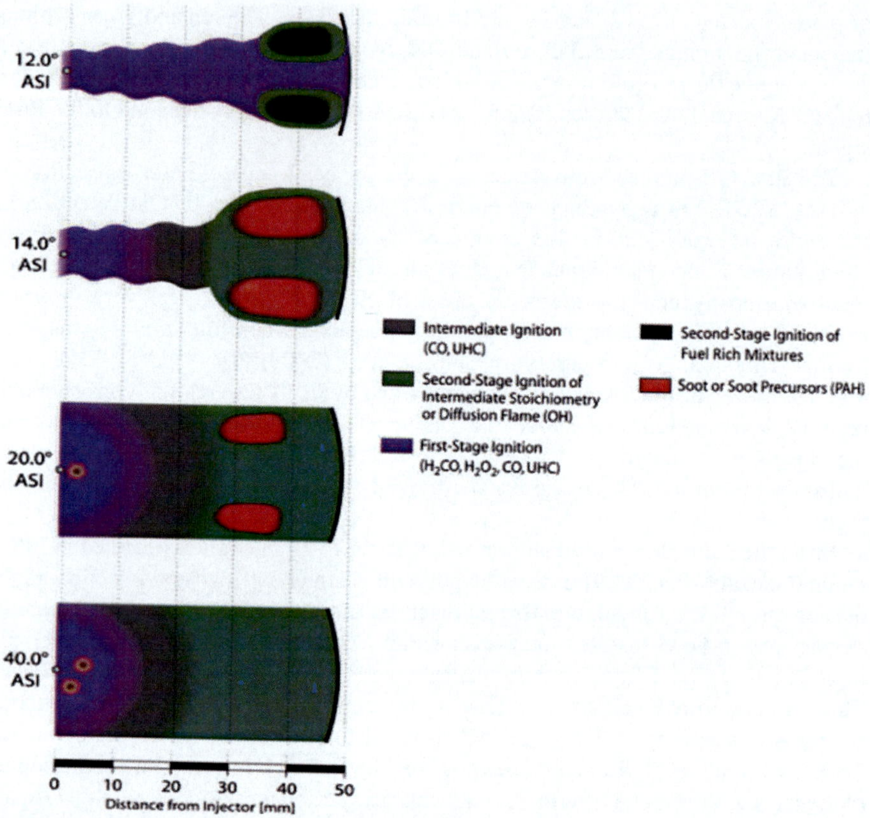

Fig. 5.7 Conceptual model of 2nd stage diesel jet combustion in heavy-duty late injection PPCI (Musculus et al. 2013)

In small piston bowls, the PAH regions were in the piston bowl in the space between the plumes, where the adjacent jets interacted (Genzale et al. 2009). The OH regions were formed around these regions, such as conventional diesel jets. In this case, the fuel-rich regions were pushed towards the cylinder center into the fuel-lean regions, which helped in the late oxidation of the fuel. This was better for the UHC reduction, but soot formation increased (Singh and Agarwal 2021). Studies in early injection heavy-duty diesel engines have also shown similar results (Musculus 2006). The better mixing had resulted in OH radicals across the jet cross-section. The OH radicals were first detected in the jet head and moved upstream. This phenomenon of OH radical distribution appearing simultaneously across a cross-section and broadening with time indicates the volumetric combustion.

Ignition timing was varied to understand the transition from HCCI to PPCI mode in the study performed by Wang et al. (2017) on a heavy-duty optical diesel engine. It was concluded that it is possible to achieve PPC mode when injecting between

46 and 35° CA bTDC. Advancing the injection beyond this range would inject the entire fuel above the piston, resulting in a homogeneous mixture. However, retarding the injection would inject the fuel completely into the piston bowl, decreasing the mixing efficiency. In another study (Lönn 2019), the transition from HCCI to PPCI for gasoline direct injection was studied based on the inlet air temperature required for maintaining the same combustion phasing for a large range of SOI sweeps. It was concluded that an SOI timing between −80° and −45° aTDC resulted in PPCI combustion, and an injection timing before −80° aTDC resulted in HCCI combustion.

The gasoline PPC has positive ignition dwell like diesel PPC (Wang et al. 2019). The fuel vapours interact with adjacent spray plumes after impinging the wall. The eddies from the adjacent plumes interact and produce a recirculating zone where the fuel vapours are pushed back towards the cylinder center. The HCHO also follows a similar trend, i.e., it first appears in the recirculating zone and is later pushed towards the cylinder center. The HCHO formed in the LTC persists until it is consumed in the HTC.

Unlike in the conventional diesel jet, where NO_x is produced in the jet periphery, in PPCI mode, NOx is produced in the entire jet cross-section. But the NO_x emissions in the LTC mode are lower than the CDC mode. Also, in the CDC mode, the entire upstream of the LOL is filled with large soot concentration in the jet head and periphery. In PPC combustion mode, soot is present only in the fuel-rich zones of the jet head or the jet–jet interaction regions (Musculus 2006).

Soot in PCCI mode is produced in the fuel-rich regions formed at the far end of the fuel jet near the bowl, either at the center or at the adjacent zones of the plume-to-plume interactions, as seen earlier. The soot formation starts just after the 1st stage of ignition, where a high-intensity luminescence is observed, which is a few times higher (Johansson 2005a, b). The time for the start of soot formation depends greatly on the EGR (Idicheria and Pickett 2011). Idicheria et al. reported that the soot formation was delayed with an increase in EGR. If the injection ends before the soot formation, there is a huge reduction in soot formation even when the ignition dwell is negative. This information allows bringing the injection closer to the ignition without increasing the soot. A two-colour thermometry study in a heavy-duty optical engine operating in LTC conditions showed that soot formation regions are different for CDC and LTC modes (Bobba et al. 2009). In CDC mode, the soot is formed in the premixed combustion region before the LOL, and it occupies the head of the jet. When the injection ends, this soot region spreads back towards the injector, making the upstream region close to the injector a major source of soot. In the LTC mode, the soot region remains at the far end of the fuel jet near the bowl, especially in plume-to-plume interaction regions.

The effect of EGR on soot formation and subsequent oxidation was also studied (Aronsson et al. 2010; Boiarciuc et al. 2007). These studies using Laser-Induced Incandescence (LII) showed that the increasing EGR reduced the in-cylinder soot formation, but the engine out emissions showed otherwise. With increasing EGR, the mixing of fuels was enhanced due to an increased ignition delay. Still, the lower in-cylinder temperatures decreased the efficiency of soot oxidation, causing higher soot in the engine exhaust.

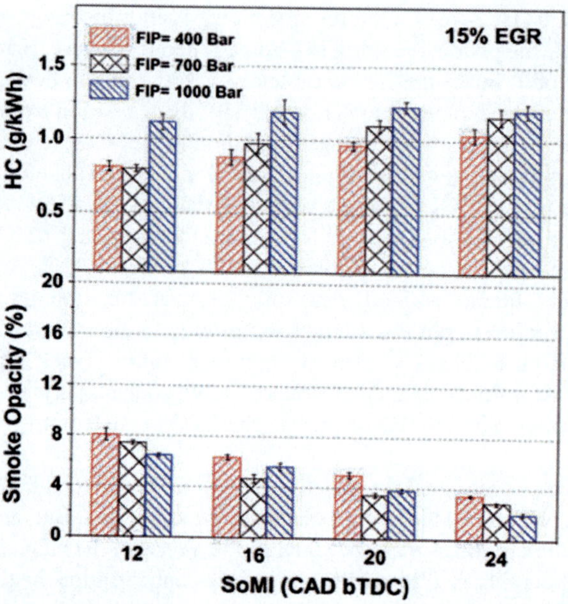

Fig. 5.8 Brake-specific CO, HC, and NO_x emissions and smoke opacity variations w.r.t. start of injection timings at varying fuel injection pressures (Jain et al. 2017)

The effect of injection timing and injection pressure on the exhaust soot and HC emissions were studied (Jain et al. 2017). The results of this study are shown in Fig. 5.8. It can be observed that with injection advance and increased injection pressure, soot decreased. This was because of superior premixing of charge before ignition, but these factors increased the HC emissions. In PCCI mode, the increased mixing led to ultra-lean mixtures near the nozzle, as seen earlier. The increased pressure supplemented the formation of lean mixtures near the nozzle, resulting in higher HC emissions.

The results from PLIF and LII studies in a heavy-duty optical engine operating in LTC mode (Genzale et al. 2009) were seen to be consistent with the natural luminosity images. It can be seen in Fig. 5.9 that the luminosity peaks at 19° CA after the start of injection (ASI) and becomes almost zero by 40° CA ASI. The PAH signals obtained from the PLIF study also indicated large zones of high-intensity PAHs at 19° CA ASI, and it decreased both spatially and in the intensity at 40° CA ASI. Meanwhile, at 40° CA ASI, the OH radical luminosity was still bright and occupied large regions. This indicated that the late oxidation of soot was occurring. These studies concluded that though the PPCI mode reduced the soot formation in the cylinder because of its enhanced mixing, it is important to have adequately high temperatures at the end of combustion to increase the efficiency of soot oxidation. The introduction of EGR in diesel LTC increased accumulation mode particle number concentration and

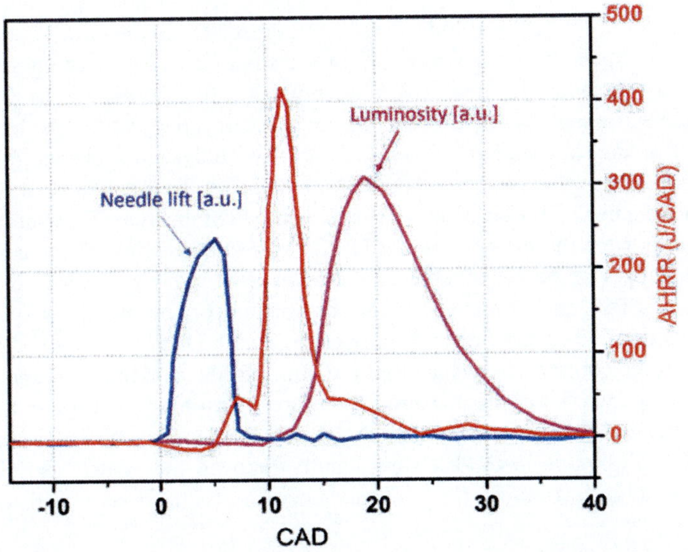

Fig. 5.9 Fuel-injector needle-lift, AHRR, and soot luminosity for low-load EGR diluted PPCI mode operating conditions with late injection. Adapted from Genzale et al. (2009)

decreased nucleation mode particle number concentration, finally resulting in an increased total particulate number concentration (Li et al. 2020).

5.7 Summary and Way Forward

This chapter focuses on the combustion in PPCI engines. It provides comprehensive information on the two-stage ignition strategy in PPCI engines using different optical studies performed in optical engines and constant volume combustion chambers. This chapter introduces the famous Dec model for diesel jet combustion that provided evidence of a very short span of LTC, followed by premixed heat release and diffusion heat release. The LTC phase was not prominent due to the high ambient temperatures in the CDC engines. The Chemical Kinetics studies aimed to understand the effect of premixing in HCCI and PPCI engines exhibited two distinct stages of combustion, wherein the 1st stage was governed by LTC reactions (breakdown of KHP and the formation of HCHO), and the second stage was governed by HTC reactions (breakdown of H_2O_2 and oxidation of CO). It was also concluded that HCHO and OH radicals could successfully mark the 1st and 2nd stages of combustion, respectively. The HCHO and OH PLIF studies also confirmed the two-stage ignition theory, indicated by the chemical kinetics studies. Various studies observed and concluded that the HCHO was detected in the entire fuel–air mixture (rich lean and stoichiometric regions) simultaneously during the 1st stage of ignition. In the 2nd stage of ignition, the jet head indicated OH fluorescence in the entire cross-section with few dark spots that gave rise to bright PAH fluorescence, marking them as the zones of soot formation. These zones were usually found adjacent to the fuel jet in the plume-to-plume interaction regions or jet head center, where the mixing was the least. But late soot imaging/ natural luminosity studies showed enough evidence of soot oxidation, which depended greatly on the EGR. The regions near the nozzle showed almost no OH fluorescence during the entire combustion. They indicated that these regions were the main source of UHC and CO emissions. These conclusions were well supported by the experimental studies done in the engines.

LTC and PPCI modes showed a lot of promise for the reduction of soot and NOx emissions. However, they have to overcome high HRR and high UHC and CO emissions to be commercially successful. Piston profiles and injection models have to be developed to control the mixture formation to eliminate the ultra-lean mixtures near the nozzle. Multiple injection strategies and reactivity-controlled compression ignition (RCCI) mode have also significantly reduced the HRR. Very clean and efficient IC engines running on LTC mode will be realized soon with continued research.

References

Agarwal AK, Singh AP, Maurya RK (2017) Evolution, challenges and path forward for low-temperature combustion engines. Prog Energy Combust Sci 61:1–56. https://doi.org/10.1016/j.pecs.2017.02.001

Alviso D, Costa MW, Backer L, Pepiot P, Darabiha N, dos Santos RG (2020) Chemical kinetic mechanism for diesel/biodiesel/ethanol surrogates using n-decane/methyl-decanoate/ethanol blends. J Braz Soc Mech Sci Eng 42(2):1–14. https://doi.org/10.1007/s40430-020-2186-9

Ando H, Sakai Y, Kuwahara K (2009) Universal rule of hydrocarbon oxidation. SAE Technical Paper 2009-01–0948. https://doi.org/10.4271/2009-01-0948

Aronsson U, Chartier C, Andersson Ö, Johansson B, Sjöholm J, Wellander R, Richter M, Alden M, Miles P (2010) Analysis of EGR effects on the soot distribution in a heavy-duty diesel engine using time-resolved laser-induced incandescence. SAE Int J Engines 3(2):137–155. https://doi.org/10.4271/2010-01-2104

Bobba MK, Genzale CL, Musculus MP (2009) Effect of ignition delay on in-cylinder soot characteristics of a heavy-duty diesel engine operating at low-temperature conditions. SAE Int J Engines 2(1):911–924. https://doi.org/10.4271/2009-01-0946

Boiarciuc A, Foucher F, Mounaim-Rousselle C, Pajot O (2007) Estimate measurement of soot diameter and volume fraction inside the bowl of a direct-injection-compression-ignition engine: effect of the exhaust gas recirculation. Combust Sci Technol 179(8):1631–1648. https://doi.org/10.1080/00102200701244728

Bruneaux G (2008) Combustion structure of free and wall-impinging diesel jets by simultaneous laser-induced fluorescence of formaldehyde, poly-aromatic hydrocarbons, and hydroxides. Int J Engine Res 9(3):249–265. https://doi.org/10.1243/14680874JER00108

Bruneaux G, Auge M, Lemenand C (2004) A study of combustion structure in high-pressure single hole common rail direct diesel injection using laser-induced fluorescence of radicals (spray technologies, mixture formation). In: The proceedings of the international symposium on diagnostics and modelling of combustion in internal combustion engines 2004. The Japan Society of Mechanical Engineers, pp 551–559. https://doi.org/10.1299/jmsesdm.2004.6.551

Cho K, Han M, Sluder CS, Wagner RM, Lilik GK (2009) Experimental investigation of the effects of fuel characteristics on high-efficiency clean combustion in a light-duty diesel engine (No. 2009-01-2669). SAE Technical Paper. https://doi.org/10.4271/2009-01-2669

Cho K, Zhang Y, Cleary D (2018) Investigation of fuel effects on combustion characteristics of Partially Premixed Compression Ignition (PPCI) combustion mode at part-load operations. SAE Int J Engines 11(6):1371–1384. https://doi.org/10.4271/2018-01-0897

Clarke K (2015) Diesel subsidy reform in India: lessons learned. Global Subsidies Initiative and International Institute for Sustainable Development Report. IISD

Colban WF, Miles PC, Oh S (2007) Effect of intake pressure on performance and emissions in an automotive diesel engine operating in low-temperature combustion regimes. SAE Trans 957–977. https://doi.org/10.4271/2007-01-4063

Crimaldi JP (2008) Planar laser-induced fluorescence in aqueous flows. Exp Fluids 44(6):851–863. https://doi.org/10.1007/s00348-008-0496-2

Curran HJ, Gaffuri P, Pitz WJ, Westbrook CK (1998) A comprehensive modelling study of n-heptane oxidation. Combust Flame 114(1–2):149–177. https://doi.org/10.1016/S0010-2180(97)00282-4

Curran HJ, Gaffuri P, Pitz WJ, Westbrook CK (2002) A comprehensive modelling study of iso-octane oxidation. Combust Flame 129(3):253–280. https://doi.org/10.1016/S0010-2180(01)00373-X

Dec JE (1997) A conceptual model of DL diesel combustion based on laser-sheet imaging. SAE Trans 1319–1348. https://doi.org/10.4271/970873

Dec JE, Coy EB (1996) OH radical imaging in a DI diesel engine and the structure of the early diffusion flame. SAE Trans 1127–1148. https://doi.org/10.4271/960831

Fayoux A, Dupré S, Scouflaire P, Houille S, Pajot O, Rolon JC (2004) OH and HCHO LIF measurements in HCCI engine. In: 12th international symposium on applications of laser techniques to fluid mechanics, Lisbon, Portugal. https://doi.org/10.4271/2003-01-3218

Genzale CL, Reitz RD, Musculus MP (2009) Effects of piston bowl geometry on mixture development and late-injection low-temperature combustion in a heavy-duty diesel engine. SAE Int J Engines 1(1):913–937. https://doi.org/10.4271/2008-01-1330

Haraldsson G, Tunestål P, Johansson B, Hyvönen J (2003) HCCI combustion phasing with closed-loop combustion control using variable compression ratio in a multi-cylinder engine. SAE Trans 1233–1245. https://doi.org/10.4271/2003-01-1830

Heywood JB (1988) Internal combustion engine fundamentals. McGraw-Hill Education

Hildingsson L, Persson H, Johansson B, Collin R, Nygren J, Richter M, Aldén M, Hasegawa R, Yanagihara H (2004) Optical diagnostics of HCCI and low-temperature diesel using simultaneous 2-D PLIF of OH and formaldehyde. SAE Trans 1809–1816. https://doi.org/10.4271/2004-01-2949

Idicheria CA, Pickett LM (2011) Ignition, soot formation, and end-of-combustion transients in diesel combustion under high-EGR conditions. Int J Engine Res 12(4):376–392. https://doi.org/10.1177/1468087411399505

Jain A, Singh AP, Agarwal AK (2017) Effect of fuel injection parameters on combustion stability and emissions of a mineral diesel-fueled partially premixed charge compression ignition (PCCI) engine. Appl Energy 190:658–669. https://doi.org/10.1016/j.apenergy.2016.12.164

Johansson B (2005a) HCCI—the optimum combustion process for IC engines? In: Proceedings of the European combustion meeting

Johansson B (2005b) High-load partially premixed combustion in a heavy-duty diesel engine. In: Diesel engine emissions reduction (DEER) conference, Chicago, IL, pp 21–25

Kanda T, Hakozaki T, Uchimoto T, Hatano J, Kitayama N, Sono H (2005) PCCI operation with early injection of conventional diesel fuel. SAE Trans 584–593. https://doi.org/10.4271/2005-01-0378

Kashdan JT, Docquier N, Bruneaux G (2004) Mixture preparation and combustion via LIEF and LIF of combustion radicals in a direct-injection, HCCI diesel engine. SAE Trans 1783–1799. https://doi.org/10.4271/2004-01-2945

Kimura S, Aoki O, Ogawa H, Muranaka S, Enomoto Y (1999) New combustion concept for ultra-clean and high-efficiency small DI diesel engines (No. 1999-01-3681). SAE Technical Paper. https://doi.org/10.4271/1999-01-3681

Klingbeil AE, Juneja H, Ra Y, Reitz RD (2003) Premixed diesel combustion analysis in a heavy-duty diesel engine. SAE Trans 445–459. https://doi.org/10.4271/2003-01-0341

Knox BW, Genzale CL (2017) Scaling combustion recession after end of injection in diesel sprays. Combust Flame 177:24–36. https://doi.org/10.1016/j.combustflame.2016.11.021

Kosaka H, Drewes VH, Catalfamo L, Aradi AA, Iida N, Kamimoto T (2000) Two-dimensional imaging of formaldehyde formed during the ignition process of a diesel fuel spray. In: SAE 2000 world congress. https://doi.org/10.4271/2000-01-0236

Kuwahara K, Hiramura Y, Ohmura S, Furutani M, Sakai Y, Ando H (2012) Chemical kinetics study on effect of pressure and fuel, O_2 and N_2 molar concentrations on hydrocarbon ignition process (No. 2012-01-1113). SAE Technical Paper. https://doi.org/10.4271/2012-01-1113

Kuwahara K, Tada T, Furutani M, Sakai Y, Ando H (2013) Chemical kinetics study on two-stage main heat release in ignition process of highly diluted mixtures. SAE Int J Engines 6(1):520–532. https://doi.org/10.4271/2013-01-1657

Li S, Liu J, Li Y, Wei M, Xiao H, Yang S (2020) Effects of fuel properties on combustion and pollutant emissions of a low-temperature combustion mode diesel engine. Fuel 267:117123. https://doi.org/10.1016/j.fuel.2020.117123

Lönn S (2019) Investigation of PPC in an optical engine: with focus on fuel distribution and combustion characterization. Department of Energy Sciences, Lund University

Lu X, Han D, Huang Z (2011) Fuel design and management for the control of advanced compression-ignition combustion modes. Prog Energy Combust Sci 37(6):741–783. https://doi.org/10.1016/j.pecs.2011.03.003

Mollenhauer K, Tschöke H (eds) (2010) Handbook of diesel engines, vol 1. Springer, Berlin

Musculus MP (2006) Multiple simultaneous optical diagnostic imaging of early-injection low-temperature combustion in a heavy-duty diesel engine. SAE Trans 83–110. https://doi.org/10.4271/2006-01-0079

Musculus MP, Lachaux T (2007) Two-stage ignition and unburned fuel emissions for heavy-duty diesel low-temperature combustion of neat n-heptane (No. SAND2007-0799C). Sandia National Lab. (SNL-CA), Livermore, CA, United States

Musculus MP, Lachaux T, Pickett LM, Idicheria CA (2007) End-of-injection over-mixing and unburned hydrocarbon emissions in low-temperature-combustion diesel engines (No. 2007-01-0907). SAE Technical Paper. https://doi.org/10.4271/2007-01-0907

Musculus MP, Miles PC, Pickett LM (2013) Conceptual models for partially premixed low-temperature diesel combustion. Prog Energy Combust Sci 39(2–3):246–283. https://doi.org/10.1016/j.pecs.2012.09.001

Onda T, Nakamura H, Tezuka T, Hasegawa S, Maruta K (2019) Initial-stage reaction of methane examined by optical measurements of weak flames in a microflow reactor with a controlled temperature profile. Combust Flame 206:292–307. https://doi.org/10.1016/j.combustflame.2019.04.044

Petersen BR, Ekoto IW, Miles PC (2010) An investigation into the effects of fuel properties and engine load on UHC and CO emissions from a light-duty optical diesel engine operating in a partially premixed combustion regime. SAE Int J Engines 3(2):38–55. https://doi.org/10.4271/2010-01-1470

Petersen B, Miles P, Sahoo D (2012) Equivalence ratio distributions in a light-duty diesel engine operating under partially premixed conditions. SAE Int J Engines 5(2):526–537. https://doi.org/10.4271/2012-01-0692

Särner G, Richter M, Aldén M, Hildingsson L, Hultqvist A, Johansson B (2005) Simultaneous PLIF measurements for visualization of formaldehyde and fuel distributions in a DI HCCI engine. In: Powertrain & fluid systems conference & exhibition, San Antonio, TX, USA.https://doi.org/10.4271/2005-01-3869

Seitzman JM, Hanson RK (1993) Planar fluorescence imaging in gases. Instrumentation for flows with combustion, p 405. https://doi.org/10.1007/BF00408770

Shukla PC, Gupta T, Agarwal AK (2018) Techniques to control emissions from a diesel engine. In: Air pollution and control, pp 57–72. Springer, Singapore

Singh AP, Agarwal AK (2021) Performance and emission characteristics of conventional diesel combustion/partially premixed charge compression ignition combustion mode switching of biodiesel-fueled engine. Int J Engine Res 22(2):540–553. https://doi.org/10.1177/1468087419860311

Singh S, Reitz RD, Musculus MPB, Lachaux T (2007) Validation of engine combustion models against detailed in-cylinder optical diagnostics data for a heavy-duty compression-ignition engine. Int J Engine Res 8(1):97–126. https://doi.org/10.1243/14680874JER02406

Singh S, Musculus MP, Reitz RD (2009) Mixing and flame structures inferred from OH-PLIF for conventional and low-temperature diesel engine combustion. Combust Flame 156(10):1898–1908. https://doi.org/10.1016/j.combustflame.2009.07.019

Soid SN, Zainal ZA (2011) Spray and combustion characterization for internal combustion engines using optical measuring techniques—a review. Energy 36(2):724–741. https://doi.org/10.1016/j.energy.2010.11.022

Stanglmaier RH, Roberts CE (1999) Homogeneous charge compression ignition (HCCI): benefits, compromises, and future engine applications. SAE Trans 2138–2145. https://doi.org/10.4271/1999-01-3682

Tagliante F, Malbec LM, Bruneaux G, Pickett LM, Angelberger C (2018) Experimental study of the stabilization mechanism of a lifted diesel-type flame using combined optical diagnostics and laser-induced plasma ignition. Combust Flame 197:215–226. https://doi.org/10.1016/j.combustflame.2018.07.024

Tropea C, Yarin AL, Foss JF (eds) (2007) Springer handbook of experimental fluid mechanics, vol 1. Springer, Berlin, p 237. https://doi.org/10.1007/978-3-540-30299-5

Vanhove G, Petit G, Minetti R (2006) Experimental study of the kinetic interactions in the low-temperature autoignition of hydrocarbon binary mixtures and a surrogate fuel. Combust Flame 145(3):521–532. https://doi.org/10.1016/j.combustflame.2006.01.001

Wang Z, Lonn S, Matamis A, Andersson O, Tuner M, Alden M, Richter M (2017) Transition from HCCI to PPC: investigation of fuel distribution by planar laser-induced fluorescence (PLIF). SAE Int J Engines 10(4):1465–1481. https://doi.org/10.4271/2017-01-0748

Wang Z, Stamatoglou P, Lundgren M, Luise L, Vaglieco BM, Andersson A, Aldén M, Andersson Ö, Richter M (2019) Simultaneous 36 kHz PLIF/chemiluminescence imaging of fuel, CH_2O and combustion in a PPC engine. Proc Combust Inst 37(4):4751–4758. https://doi.org/10.1016/j.proci.2018.06.019

Westbrook CK (2000) Chemical kinetics of hydrocarbon ignition in practical combustion systems. Proc Combust Inst 28(2):1563–1577. https://doi.org/10.1016/S0082-0784(00)80554-8

Yang F, Yao C, Wang J, Ouyang M (2016) Load expansion of a dieseline compression ignition engine with multi-mode combustion. Fuel 171:5–17. https://doi.org/10.1016/j.fuel.2015.12.046

Zhao H, Ladommatos N (1998) Optical diagnostics for in-cylinder mixture formation measurements in IC engines. Prog Energy Combust Sci 24(4):297–336. https://doi.org/10.1016/S0360-1285(98)80026-9

Chapter 6
Gasoline Compression Ignition Combustion Strategies and Recent Engine System Developments for Commercial and Passenger Transport Applications

Yu Zhang, Harold Sun, and Jinsu Kim

Abstract Rising regulatory demand around the world for lower exhaust pollutants and reduced greenhouse gas emissions in the commercial transport and passenger car sectors motivates the development of highly efficient and clean internal combustion engines. In recent years, by harnessing gasoline's low autoignition reactivity and unique spray characteristics, gasoline compression ignition (GCI) has shown the potential as a cost-effective engine technology to achieve high fuel efficiency with low engine-out NOx and soot emissions. This chapter began with an overview of gasoline's autoignition behavior and spray characteristics. Then, building on an-in-depth understanding of the fundamental gasoline combustion knowledge, the basic principles, benefits, and challenges of varying GCI combustion strategies were discussed based on the level of in-cylinder fuel stratification, encompassing homogenous charge compression ignition (HCCI), partially-premixed compression ignition (PPCI), and mixing-controlled compression ignition (MCCI). To reach the full performance potential for GCI, it is essential to develop tailored combustion and air-handling system concepts. Harnessing these learnings, recent efforts towards developing advanced, production-intent heavy-duty and light-duty GCI engines were reviewed. Finally, the benefits and the key technical areas for commercializing GCI engines were reviewed.

Keywords Gasoline compression ignition · Partially-premixed combustion · Mixing-controlled combustion · Combustion system development · Fuel efficiency · Emissions · Commercial transport · Passenger car

Y. Zhang (✉)
Aramco Americas, Aramco Research Center- Detroit, 46535 Peary Ct., Novi, MI 48377, USA
e-mail: yu.zhang@aramcoamericas.com

H. Sun
AVL Powertrain Engineering, 47519 Halyard Dr., Plymouth, MI 48170, USA
e-mail: harold.sun@avl.com

J. Kim
West Virginia University, 1306 Evansdale Dr., Morgantown, WV 26506, USA
e-mail: jk0062@mix.wvu.edu

A. K. Agarwal et al. (eds.), *Advanced Combustion for Sustainable Transport*, Energy, Environment, and Sustainability, https://doi.org/10.1007/978-981-16-8418-0_6

6.1 Introduction

Tightening regulations across transport sectors demand ultra-low vehicle tailpipe exhaust emissions and improved fuel economy. In the commercial transport sector, recent public announcements suggest that the California Air Resources Board (CARB) is considering lowering the on-road heavy-duty vehicle tailpipe NOx emissions by 90% down to 0.027 g/kWh by 2027 (Robertson 2017).[1] Meanwhile, the second phase GHG regulation from the U.S. Environmental Protection Agency (EPA) requires a 3% CO_2 reduction for the heavy-duty vehicles from 2021 to 2027. In the passenger car sector, the U.S. EPA Tier3-Bin 30 and Euro 6d with real driving emissions (RDE) testing demand ultra-low criteria emissions over wide engine operating range while CO_2 and fuel economy standards around the world require 10–15% vehicle fuel economy improvement from 2021 to 2025.

Meeting these stringent standards brings significant challenges to the existing compression ignition (CI) and spark ignition (SI) engines, likely incurring higher engine and aftertreatment system cost along with increased complexity on operation, onboard diagnostics, and maintenance. Therefore, engine combustion strategies that can utilize the existing engine and aftertreatment architectures to produce ultra-low engine-out NOx and soot emissions while achieving diesel-like to better fuel efficiency will become attractive to the engine manufactures. To that end, by harnessing gasoline low autoignition reactivity and favorable physical properties towards enhanced fuel–air mixing, GCI has shown the potential to achieve these goals across a wide range of fuel stratification strategies, including homogeneous or lightly stratified, spark-assisted compression ignition (i.e., HCCI) (Dec et al. 2011a; Yun et al. 2011), moderately stratified PPCI (Kalghatgi et al. 2010; Manente et al. 2010; Ciatti et al. 2013), and heavily stratified MCCI (Zhang et al. 2018a; Paz et al. 2018).

This chapter aims to provide a thorough review of the fundamental aspects of varying GCI combustion strategies and the recent developments in applying the fundamental knowledge towards production-intent technology development. First, gasoline's autoignition reactivity effects on ignition delay and fuel–air mixing under engine-relevant thermal boundary conditions were discussed. Then, gasoline's spray characteristics and the influence on the in-cylinder combustion process were assessed. Subsequently, following the level of in-cylinder fuel stratification, GCI was categorized into HCCI, PPCI, and MCCI. The operating strategy, performance potential, and limitations for each combustion strategy was reviewed in detail. Building on these learnings, a summary of recent industrial efforts on developing tailored GCI combustion and air-handling system concept was provided. Finally, the chapter was concluded with a discussion on the opportunities and challenges for commercializing GCI.

[1] California Air Resources Board (CARB), Heavy-Duty Low NOx program, https:www.arb.ca.gov/msprog/hdlownox /hdlownox.htm.

6.2 Gasoline Autoignition Behavior

It is well documented that the fuel ignition quality measured through cetane number or octane number is influenced mostly by the hydrocarbon oxidation chemistry at low-to-intermediate temperatures (e.g., 700–1200 K), where the onset of ignition is driven by cool flame chain branching and the dissociation of hydrogen peroxide ($H_2O_2 \rightarrow OH + OH$) (Westbrook 2000). Thus, as seen in Table 6.1, gasolines generally have lower autoignition reactivity (i.e., higher RON or lower cetane number) than diesel. However, as the charge temperature migrates into the high temperature regime, the aforementioned cool flame chain branching channels are gradually closed. Instead, the combustion process is initiated kinetically through $H + O_2 \rightarrow OH + H$. Consequently, as reported in ignition delay measurements from shock tube and rapid compression machine experiments, the impact of cetane number or octane number on ignition becomes diminished.

By conducting constant-volume, 0-D combustion simulations (Zhang et al. 2018b), Fig. 6.1 captures the autoignition behavior varying primary reference fuel (PRF) blends under boundary conditions covering both low temperature PPCI and high temperature MCCI. Here, PRF0 (*n*-heptane) was used as a gas-phase surrogate for diesel, while PRF91 was used to kinetically simulate a market-based gasoline. Then, by varying the PRF number from 0 to 91, the impact of fuel ignition quality on ignition delay time (IDT) was investigated.

Figure 6.1a shows that the negative temperature coefficient (NTC) behavior at 900–1050 K became increasingly pronounced as the fuel autoignition reactivity increases from PRF91 to PRF0. As a result, the IDT difference among PRF blends was more notable in the NTC region. In contrast, as the charge temperature increases from the intermediate temperature regime to the high temperature regime (>1050 K),

Table 6.1 Properties of ULSD and gasolines

	ULSD	RON58 Gasoline	RON70 Gasoline	RON80 Gasoline	RON93 Gasoline
FBP [°C]	344	134	169	184	199
Density at 15.56 °C [g/mL]	0.845	0.710	0.717	0.724	0.734
Kinematic viscosity [cSt]	2.49	0.58	0.57	0.56	0.55
Aromatics [−]	27.7	6.7	13.7	19.7	24.9
H/C ratio [−]	1.79	2.11	2.058	1.981	1.854
Cetane number (CN) [−]	46.6	33.9	29.8	25.9	20.5
RON [−]	-	58.0	69.4	80.0	93.2
MON [−]	-	55.7	67	74.9	84.4
AKI [−]	-	56.8	68.2	77.4	88.8
LHV [MJ/kg]	42.87	44.112	43.623	43.58	43.29

Sources Zhang et al. (2018a)

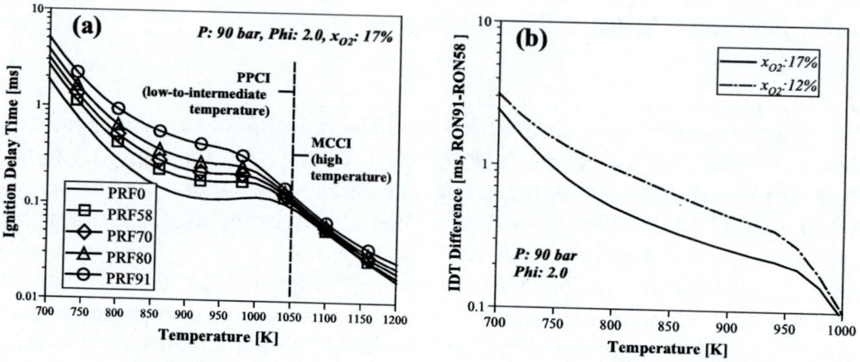

Fig. 6.1 **a** Ignition delay time (IDT) for PRF blends at pressure of 90 bar, equivalence ratio (Phi) of 2, and x_{o2} of 17% and **b** IDT difference between RON91 and RON58 gasoline under x_{o2} of 17 and 12%. *Source* Zhang et al. (2018b)

the IDT difference began to diminish due to the change in chain branching mechanism discussed above. Therefore, as highlighted in Fig. 6.1a, the fuel ignition quality has a notably stronger impact on IDT under PPCI than MCCI. In addition, Fig. 6.1(b) shows that reducing oxygen availability (x_{o2}: 17–12%) strengthened the IDT difference between PRF58 and PRF91 in the intermediate temperature regime, thereby suggesting increasing the EGR dilution level is an effective means to promote PPCI.

The impact of fuel autoignition reactivity on PPCI was then further investigated by performing closed-cycle, 3-D CFD combustion simulations in a 15 L heavy-duty GCI engine (Zhang et al. 2018b). The global combustion behavior was captured in Fig. 6.2 by characterizing cylinder pressure and apparent heat release rate (AHRR)

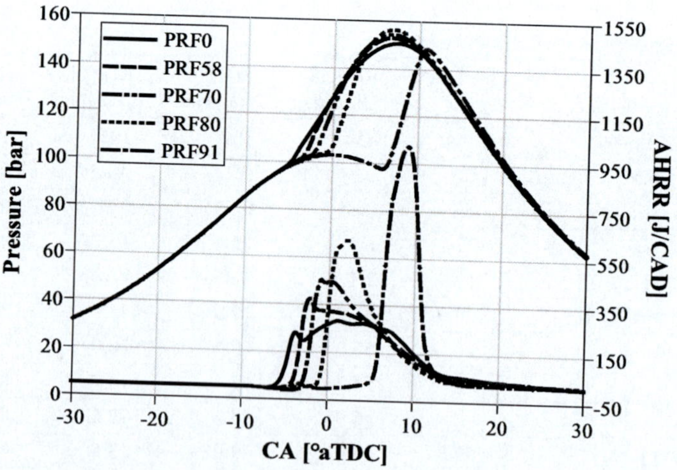

Fig. 6.2 Global combustion behavior for PRF blends under PPCI. *Source* Zhang et al. (2018b)

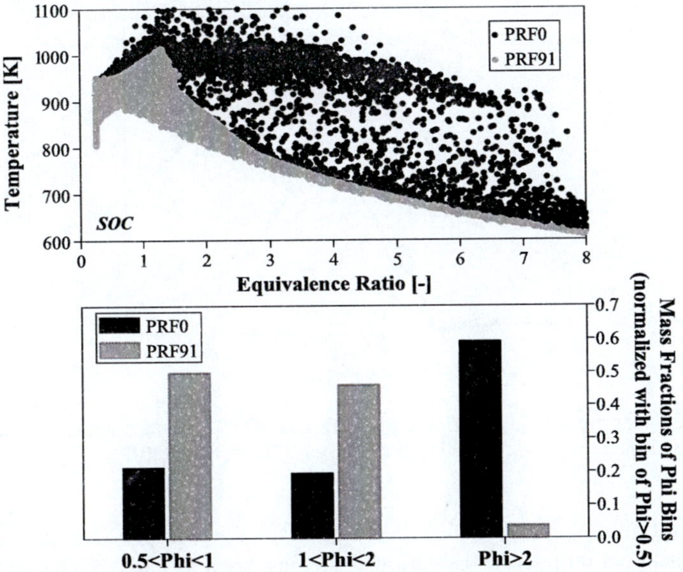

Fig. 6.3 Phi-temperature distribution for PRF blends under PPCI. *Source* Zhang et al. (2018b)

for varying PRF blends. In Fig. 6.2, it is clear that the IDT increased and the combustion phasing was delayed as the fuel autoignition reactivity reduced from PRF0 to PRF91. Consequently, the combustion process migrated from two-stage mixing-controlled combustion to single-stage partially-premixed heat release along with markedly shortened combustion duration. Figure 6.3 provides a comparison of the in-cylinder fuel stratification at the start of combustion (SOC) between PRF0 and PRF91. Due to lengthened ignition delay, PRF91 had a narrower spread of equivalence ratio and notably leaner local fuel–air mixtures. Collectively, Figs. 6.2 and 6.3 suggest that by harnessing gasoline's autoignition reactivity, the in-cylinder fuel–air mixture formation can be tailored by adjusting the fuel injection strategy to achieve low soot emissions and high fuel efficiency while maintaining adequate ignition strength.

In contrast, under high temperature mixing-controlled combustion typically encountered during high load operation in modern heavy-duty diesel engines, cylinder pressure and temperature can be high enough to shift the combustion into the high temperature regime, thereby suppressing the fuel ignition quality influence on ignition delay. As shown in Fig. 6.4, when operated at 15 bar BMEP using an 18.9 geometric compression ratio (GCR), an ultra-low sulfur diesel (ULSD) and a RON91 gasoline exhibited similar global combustion behavior despite a cetane number difference around 20.

Based on the understanding of the sensitivity of gasoline's autoignition reactivity to thermal environment, air-handling and valvetrain strategies can be developed to either promote or suppress gasoline PPCI operation. One approach to expand the

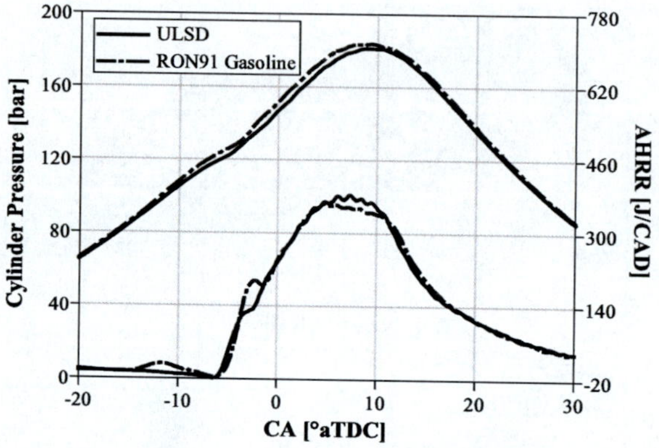

Fig. 6.4 Cylinder pressure and AHRR for ULSD and RON91 gasoline under high temperature MCCI at 1375 rpm/15 bar BMEP and 18.9 CR. *Source* Zhang et al. (2017)

PPCI operation range is to reduce the engine effective compression ratio (ECR) by implementing variable valve actuation strategies, such as early or late intake valve closing (EIVC/LIVC) (Srivastava et al. 2018). Figure 6.5 illustrates that EIVC was an effective means to reduce the engine ECR for GCI operation at a medium load point, thereby lengthening the ignition and enhancing PPCI (Zhang et al. 2020). Further, when combined with a tailored air-handling system, Millerized GCI showed the potential to achieve high fuel efficiency with low engine-out NOx and soot emissions (Zhang et al. 2020; Kumar et al. 2021a).

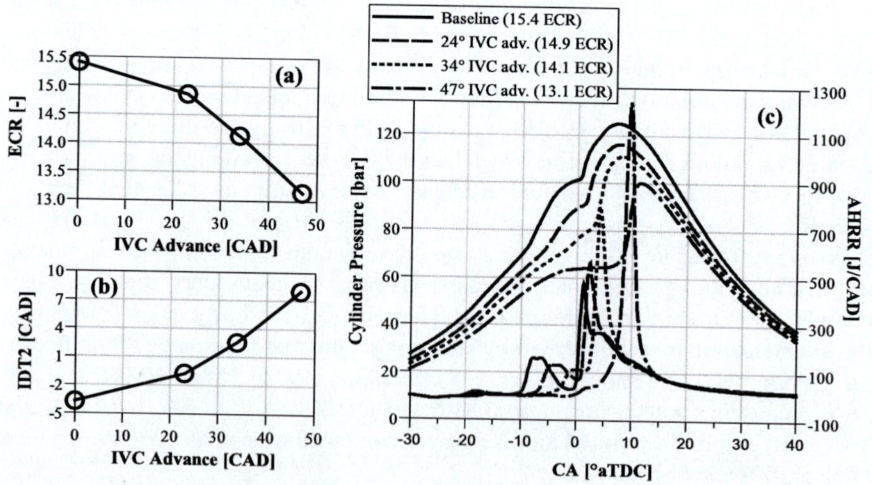

Fig. 6.5 Impact of EIVC on **a** ECR, **b** IDT2, and **c** global combustion behavior for a RON93 gasoline at 1375 rpm/10 bar BMEP. *Source* Zhang et al. (2020)

6.3 Gasoline Spray Characteristics

In addition to the difference in ignition quality, gasoline and diesel have appreciably different physical properties, including lower density, higher volatility, and lower viscosity. It is well established that lower density and high volatility lead to stronger air-entrainment into the spray and faster spray evaporation, while lower viscosity facilitates easier spray breakup. Consequently, through liquid length measurements in a constant-volume combustion chamber and 3-D CFD spray modeling, gasoline spray showed notably shorter liquid penetration than diesel (Fig. 6.6) (Pei et al. 2017). The observation in Fig. 6.6 has significant practical implications for engine combustion system development as the spray-bowl interaction and the subsequent wall-guided fuel–air mixing need to be optimized based on specific fuel spray behavior. Thus, it can be expected that a combustion system design that is developed for diesel may not be ideal for gasoline due to the difference in spray characteristics.

Running both diesel and a RON91 gasoline in a production 15 L diesel engine, Fig. 6.7 provides a comparison of soot evaluation and mass fraction of in-cylinder charge at fuel-rich equivalence ratio bands at a medium load operation point (1375 rpm/10 bar BMEP) (Zhang et al. 2017). GCI clearly demonstrated reduced soot formation and enhanced air utilization (i.e., less pronounced presence in fuel-rich regions). On the other hand, it is worth noting that when combined with gasoline's low reactivity, overly promoted fuel–air mixing from gasoline sprays tend to result in higher HC and CO emissions (Zhang et al. 2017). Mitigation approaches could involve strengthening fuel stratification by tailored spray pattern design along with injection strategy calibration and improving thermal boundary conditions.

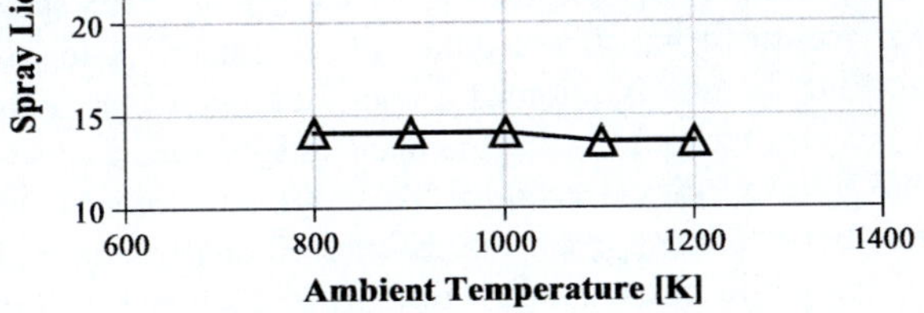

Fig. 6.6 Liquid penetration length for ULSD and RON60 gasoline as a function of ambient temperature at 6 MPa ambient pressure and 150 MPa fuel injection pressure. *Source* Pei et al. (2017)

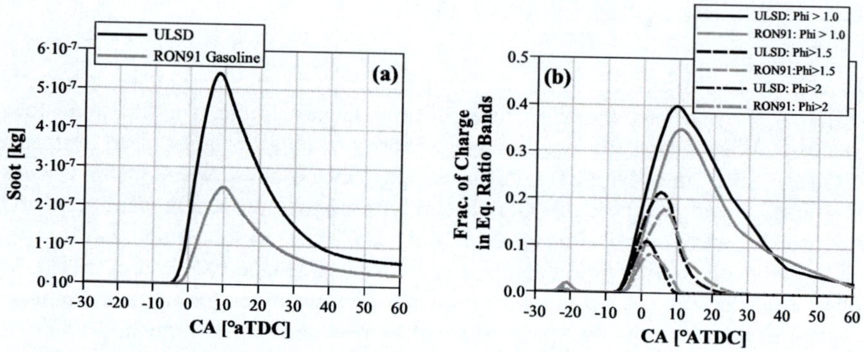

Fig. 6.7 **a** Soot evaluation and **b** in-cylinder air utilization for ULSD and RON91 gasoline at 1375 rpm/10 bar BMEP and 18.9 CR. *Source* Zhang et al. (2017)

6.4 Overview of GCI Combustion Strategies

Harnessing gasoline chemical, physical, compositional properties, a variety of GCI combustion strategies have been proposed and studied extensively. Based on the level of in-cylinder fuel stratification, they can be categorized into homogeneous or lightly-stratified compression ignition (HCCI), PPCI, and MCCI. Note that reactivity-controlled compression ignition (RCCI) of dual fuels encompassing gasoline as ta low reactivity stream has been investigated extensively in the past decade, exhibiting encouraging potential for high fuel efficiency and low engine-out emissions (Kokjohn et al. 2011; Kaddatz et al. 2012; Zhang et al. 2013). However, since this chapter is focused on single-fueled GCI concepts, RCCI is beyond the current scope and not covered.

6.4.1 Homogeneous or Lightly-Stratified GCI (HCCI)

The fundamental principle of implementing HCCI using market-based gasolines (e.g., RON92) is to utilize gasoline's oxidation chemistry at low-to-intermediate temperatures to drive the autoignition of a lean, homogeneous or lightly stratified in-cylinder charge. It is a technology that has been under extensive research and development in the past two decades (Dec et al. 2012, 2017a), aimed at achieving significantly improved fuel efficiency over conventional SI engines through un-throttled, lean, and diluted operation. Since autoignition process under HCCI occurs simultaneously across the engine cylinder, it produces much faster combustion compared to the typical flame propagation process in SI engines. Therefore, combining lean operation with more rapid heat release, HCCI has shown the potential to produce high fuel efficiency with ultra-low NOx and soot emissions.

Despite the performance potential, the kinetics-driven nature of HCCI poses challenges for controlling the onset of combustion over a wide range of engine operating conditions. To address these challenges, research efforts have been dedicated to study the sensitivity of HCCI relative to intake temperature, boost pressure, equivalence ratio, fuel stratification (premixed vs. in-cylinder direct fuel injection), and EGR dilution.

Figure 6.8a shows that partial fuel stratification (PFS) through introducing in-cylinder direct injection (DI) was an effective means to promote faster autoignition and longer burn duration, thereby reducing MPRR and allowing for extended load range (Dec et al. 2011b). In addition, combining PFS and intake temperature control was found to provide enhanced authority over CA50 (Fig. 6.8b). Also, in Fig. 6.8c, the autoignition became more sensitive to equivalence ratio as the boost pressure increased, suggesting the HCCI combustion process can be better tailored and the

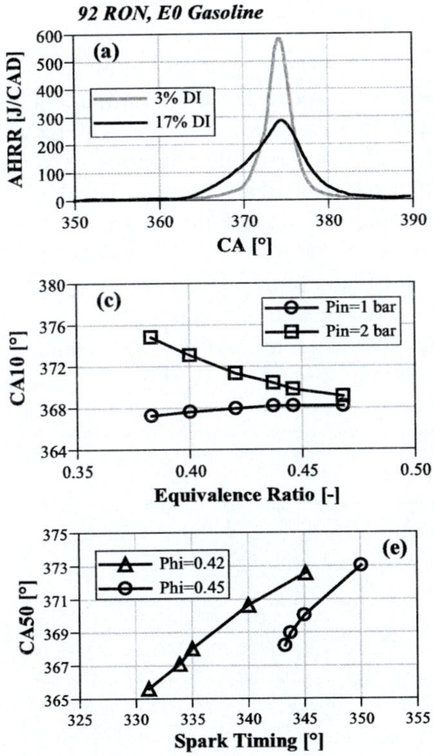

Fig. 6.8 **a** Effect of DI% on heat release rate, **b** comparison between DI and premixed on CA50 control, **c** impact of Pin on CA10-Phi sensitivity, **d** effects of fuel injection strategy and fuel on load range as a function of intake pressure, and **e** impact of spark timing on CA50 control authority. *Source* Dec et al. (2011b, 2016)

load range can be expanded through adequate boost control along with the use of EGR dilution (Fig. 6.8d).

Combatting gasoline's low reactivity to enhance low load HCCI performance has been an important research area. Among various strategies investigated, spark assistance has gained the most attention due to the evolvement of stratified charge lean SI technology (Dec et al. 2016, 2017b; Yun 2018). It is convenient to implement and is synergistic with the aforementioned fuel stratification control. In essence, the strategy utilizes the spark plug to initiate a localized flame kernel and develops a flame propagation process, thereby compressing the remaining charge into autoignition as the flame travels. An example of spark-assisted HCCI is illustrated in Fig. 6.8e (Dec et al. 2011b), where the spark event facilitated an earlier onset of autoignition and exhibited CA50 control authority.

6.4.2 *Partially-Premixed GCI (PPCI)*

The main motivation for developing a partially-premixed GCI concept is to strengthen the in-cylinder fuel stratification while harnessing gasoline's low reactivity to still maintain adequate ignition dwell, thereby attaining enhanced control over the combustion process compared to HCCI. Figure 6.9 shows that tailoring the fuel injection strategy (e.g., SOI) was effective in navigating the combustion mode from HCCI to early and late PPCI (Cho et al. 2019). As the fuel stratification strengthened, CA50 became increasingly sensitive to the start of the second fuel injection (SOI_second). On the one hand, enhancing fuel stratification is beneficial for controlling combustion phasing and combustion noise. On the other hand, excessive stratification can lead to high NOx and soot emissions. Therefore, to reach GCI's full performance

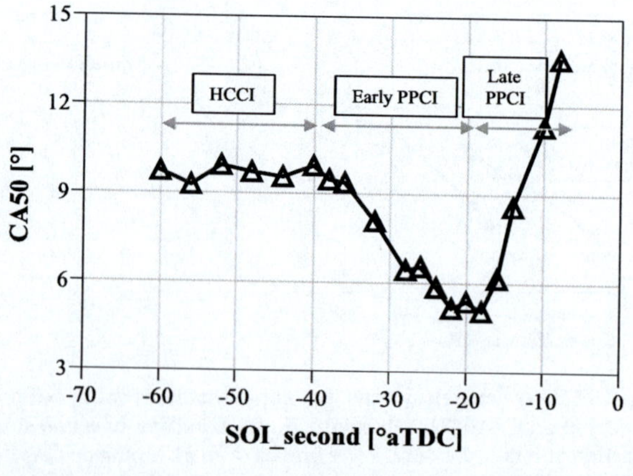

Fig. 6.9 Impact of SOI_second on a CA50 at 1500 rpm/11 bar IMEP in a light-duty GCI engine. *Source* Cho et al. (2019)

potential, it is essential to optimize the fuel stratification distribution across a broad range of engine operating conditions.

Figures 6.10 and 6.11 highlight the experimental GCI results obtained using a heavy-duty single cylinder engine (SCE) with the flexibility to control the thermal boundary conditions (i.e., intake pressure, intake temperature, and EGR dilution level) (Manente et al. 2010). In Figs. 6.10b, c, partially-premixed GCI demonstrated promising potential to produce high fuel efficiency with low engine-out NOx and soot emissions. It is worth noting in Fig. 6.10a that the load range was the widest for a higher reactivity RON70 gasoline. It became narrower as the fuel reactivity reduced. Figure 6.11 shows that high efficiency clean operation at high loads were achieved under idealized conditions with high boost (>3 bar), low intake temperature (303 K), high EGR dilution (>50%), and high MPRR (≥15 bar/CAD). Therefore, although encouraging SCE experimental results were reported, implementing the partially-premixed GCI concept for practical applications using a market gasoline still requires overcoming the technical barriers on developing an adequate air-handling system, a high efficiency and low noise combustion strategy at high loads, and a reactivity-promoting strategy to enhance low load GCI performance.

Recent studies have shown that combining PPCI combustion and heavily-stratified diffusion combustion together (PPCI-diffusion) can be an effective approach to control combustion noise while achieving good fuel efficiency and low exhaust

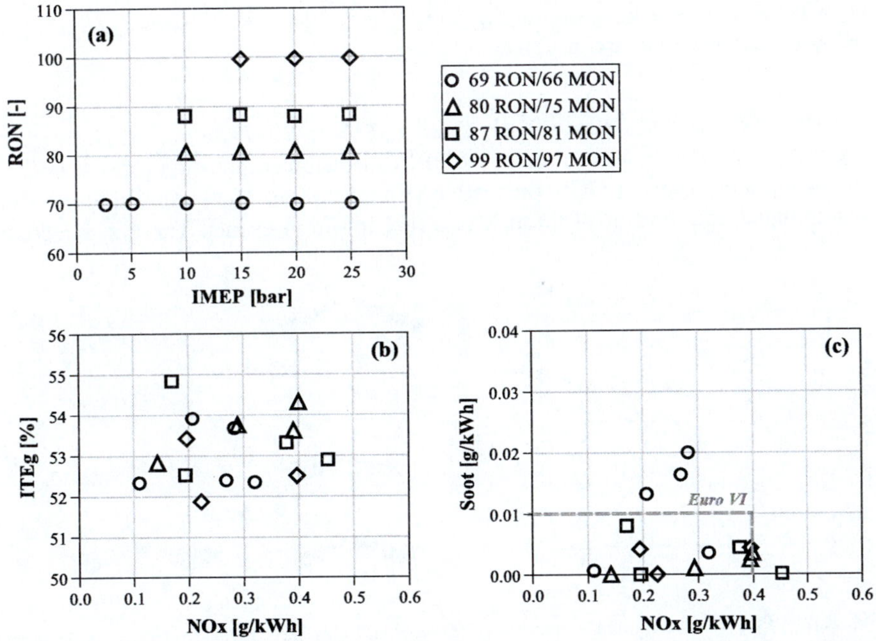

Fig. 6.10 Impact of gasoline fuel reactivity (RON) on **a** load range, **b** NOx and ITEg, and **c** NOx and soot. *Source* Manente et al. (2010)

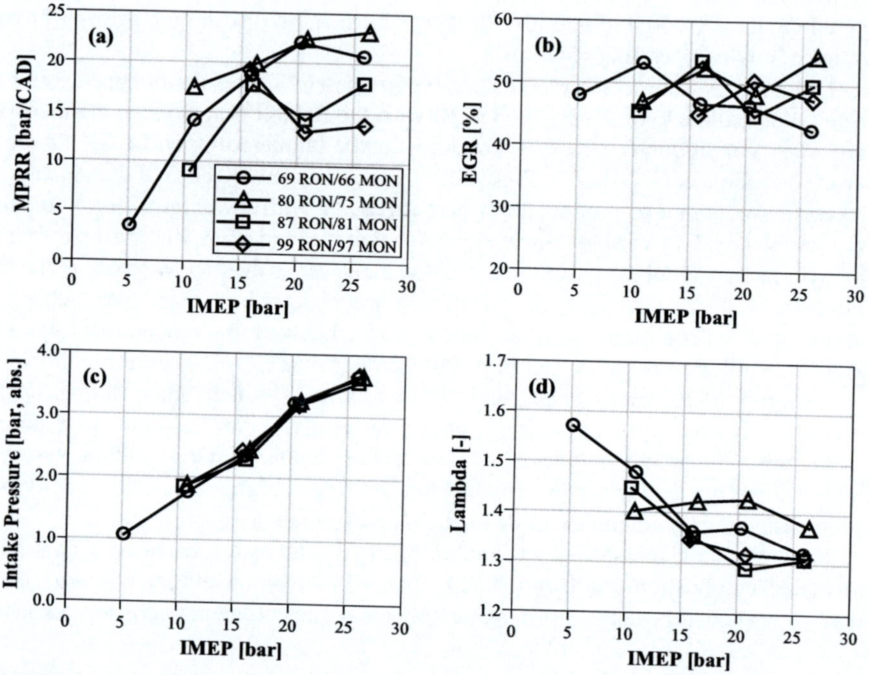

Fig. 6.11 **a** MPRR, **b** EGR, **c** intake pressure, and **d** lambda as a function of IMEPg in a heavy-duty GCI engine. *Source* Manente et al. (2010)

emissions (Zhang and Sellnau 2021; Cung and Ciatti 2017; Sellnau et al. 2019b). Figure 6.12 provides an example of the PPCI-diffusion combustion process generated using a customized GCI combustion system design. The combustion process was initiated and developed through two fuel injection events. The first injection

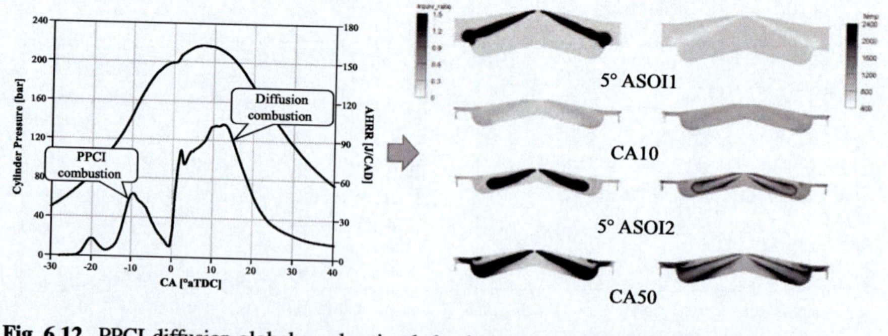

Fig. 6.12 PPCI-diffusion global combustion behavior and localized diffusion-driven combustion process at 2000 rpm/23.5 bar closed-cycle IMEP in a light-duty GCI engine. *Source* Zhang and Sellnau (2021)

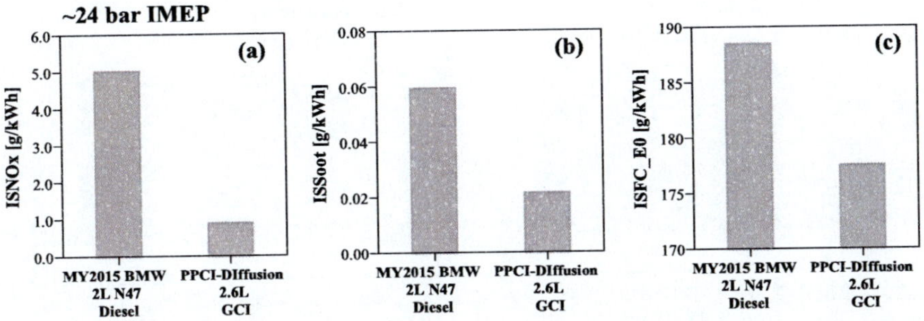

Fig. 6.13 Comparison of **a** NOx, **b** soot, and **c** ISFC_E0 between a MY2015 BM N47 diesel and simulated PPCI-diffusion LD-GCI at 24 bar IMEP. *Source* Zhang and Sellnau (2021)

induced a PPCI combustion process to build an appropriate thermal environment that facilitated effective control over the ignition delay and the MPRR for the main combustion process. Subsequently, the second injection was introduced close to the compression TDC and drove a diffusion combustion process. To maximize the potential for the PPCI-diffusion combustion strategy, it is critical to develop a tailored GCI combustion system that provides favorable spray-wall interaction and geometry-guided fuel–air mixing features. Figure 6.13 shows that optimized spray pattern and fuel injection strategy together had the potential to substantially reduce NOx and soot emission while markedly enhancing fuel efficiency (Fig. 6.14).

On low load performance enhancement, thermal promotion by utilizing variable valve strategy to trap hot residuals is widely documented as an effective approach to enhance charge reactivity (Borgqvist et al. 2013; Zhang et al. 2021). Figure 6.15 shows that by opening exhaust valve a second time during the intake stroke (i.e., exhaust re-breathing), sufficient thermal promotion led to low hydrocarbon (HC) emissions and robust combustion stability while meeting the engine-out NOx and combustion noise targets (Zhang et al. 2021).

6.4.3 *Mixing-Controlled GCI (MCCI)*

In comparison to HCCI and PPCI, MCCI features the heaviest in-cylinder fuel stratification. Under MCCI, combustion typically proceeds at intermediate-to-high temperatures and low-to-moderate EGR dilution levels that are encountered in modern heavy-duty diesel engines equipped with high GCRs (Zhang et al. 2017). Consequently, as discussed previously, gasoline's autoignition reactivity effect is suppressed. Instead, the physical and compositional properties become more impactful, influencing the break-up, evaporation, and penetration of fuel sprays as well as the soot formation and oxidation process. As shown in Fig. 6.15, the global combustion behavior between diesel and gasolines (RON58 and RON91) grew increasingly closer with load increase (Zhang et al. 2017). On the other hand,

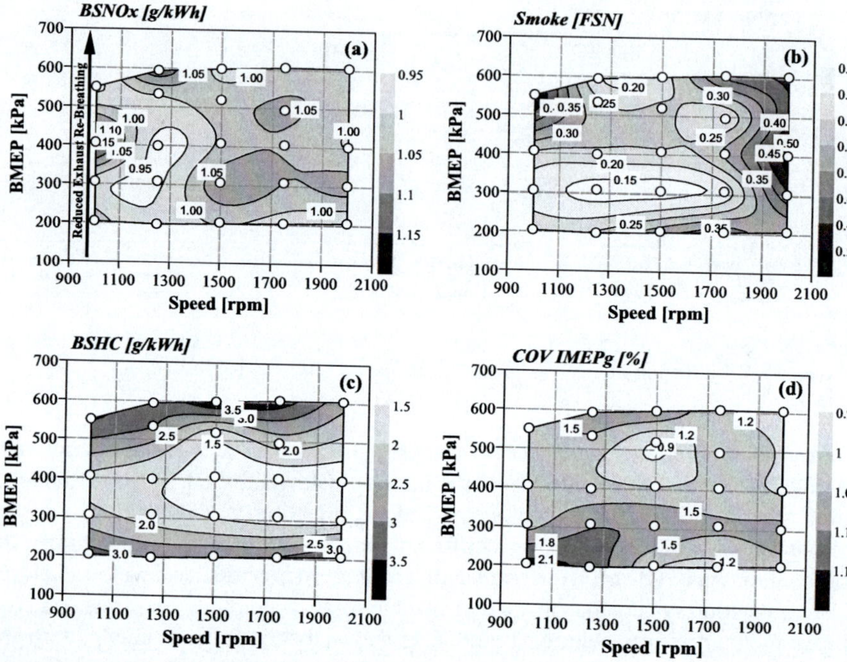

Fig. 6.14 Part-load GCI performance mapping in a light-duty GCI engine. *Source* Zhang et al. (2021)

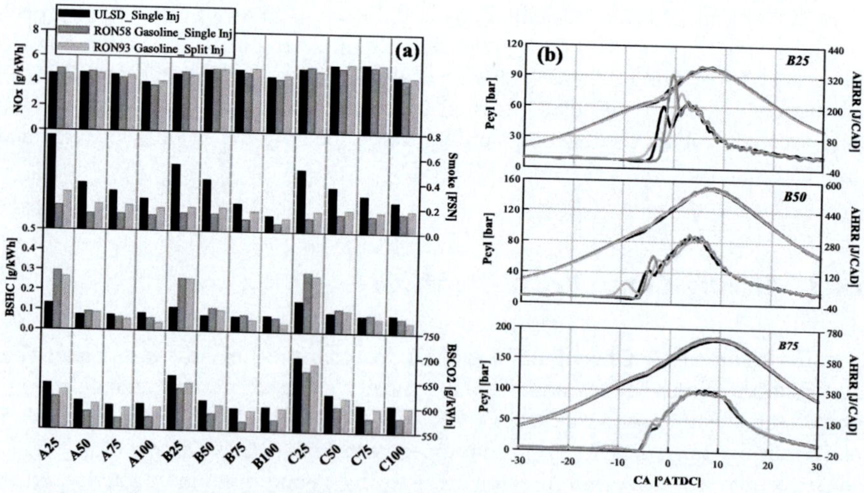

Fig. 6.15 a SET 12-mode performance and global combustion behavior at B50 and B75 for a heavy-duty GCI engine at 18.9 CR. *Source* Zhang et al. (2017)

Table 6.2 Steps and configurations in MCCI GCI combustion system development at 1375 rpm/10 bar BMEP

	CR	Piston Bowl	Design Attributes	Change of ISFC [%]	Change of Soot [%]
Baseline	18.9	-	–	–	–
Step 1	18.9	Bowl A	No. of Noz. Holes: 9 Spray Angle: 146° Total Noz. Area: 1.3 Swirl Ratio: 1.0	– 1.0	– 38.7
Step 2	21	Bowl A	No. of Noz. Holes: 9 Spray Angle: 148° Total Noz. Area: 1.7 Swirl Ratio: 1.0	– 2.3	– 51.6

Source Zhang et al. (2017)

due to the abovementioned differences in fuel physical and compositional properties, both gasolines maintained a consistent soot benefit across the supplemental emission test (SET) 12 modes.

Harnessing gasoline's shorter liquid penetration length, research efforts have been dedicated to designing tailored MCCI GCI combustion system concepts. In Table 6.2, the combustion system development was untaken in two steps, including customized piston bowl design, reducing swirl ratio, increasing injector total nozzle area (TNA), and increasing the engine compression ratio (Zhang et al. 2017). The piston bowls in Fig. 6.16a were narrower and deeper compared to the diesel production design to interact more favorably with gasoline sprays. In addition, leveraging the fuel–air mixing benefit from gasoline sprays due to gasoline's lower density, higher volatility, and lower viscosity, the swirl ratio was modified from the production level of 2.0 down to 1.0 to reduce in-cylinder heat transfer losses. Also, TNA was increased by 70% to shorten the injection duration and strengthen the spray momentum, thereby improving late-stage air utilization and promoting a more rapid diffusion combustion process (Fig. 6.16b). As a result, the analysis predicted and 2.3% improvement of ISFC and 51.6% soot reduction over the baseline production hardware configuration.

6.5 Recent System-Level Developments of GCI Engines

6.5.1 15 L Heavy-Duty GCI Engine for Meeting 0.02 g/hp-Hr Tailpipe NOx

There is an ongoing effort at Aramco to develop a 15 L heavy-duty GCI engine concept that targets an engine-out NOx level of 1–2 g/kWh. It is aimed at achieving California Air Resources Board's (CARB) ultra-low tailpipe NOx standard of 0.027 g/kWh using a production aftertreatment system. Figure 6.17 describes

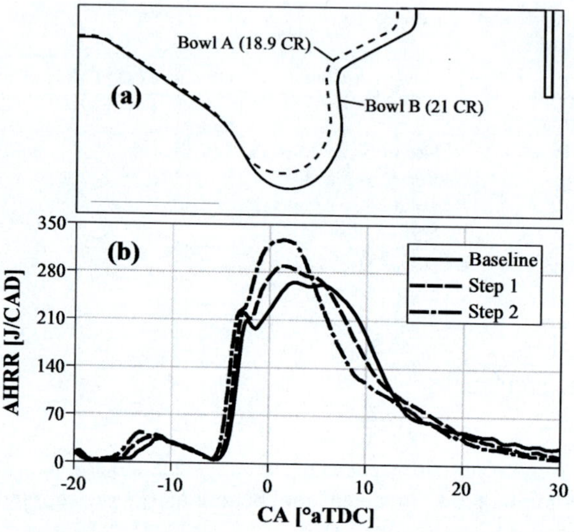

Fig. 6.16 **a** piston bowl geometry and **b** global combustion behavior in MCCI GCI combustion system development at 1375 rpm/10 bar BMEP in a heavy-duty GCI engine. *Source* Zhang et al. (2017)

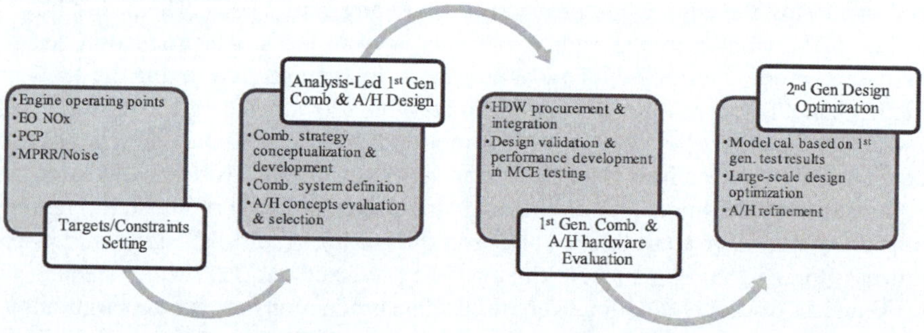

Fig. 6.17 Analysis-led heavy-duty GCI combustion and air-handling development process

the analysis-led design process that is used for the engine concept development, involving close-coupled 3-D CFD-guided engine combustion system design and 1-D system-level air-handling concept evaluation along with proof-of-concept dynamometer engine performance development. Developing high-fidelity analytical tools and streamlined design optimization methodology guided by sound engineering understanding is key in the process (Pei et al. 2019; Kodavasal and Som 2018).

An important initial step in the process is to understand the implications of GCR on fuel efficiency and emissions. Figure 6.18 shows smoke and diesel-equivalent BSFC (BSFC_d) as a function of GCR (15.7–18.9) and engine-out NOx (1–4 g/kWh) for a RON93 E0 gasoline running on a 15 L heavy-duty truck engine (Srivastava et al.

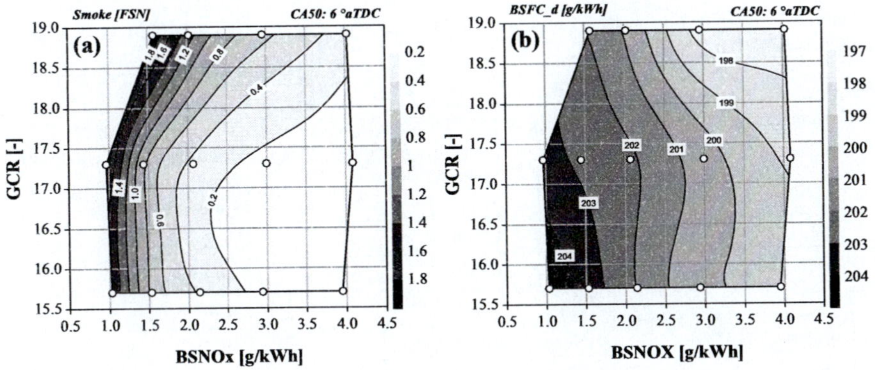

Fig. 6.18 Impact of GCR and NOx on **a** smoke and **b** diesel-equivalent BSFC (BSFC_d) at 1375 rpm/10 bar BMEP in a heavy-duty GCI engine. *Source* Zhang et al. (2020)

2018). Focusing on 1–2 g/kWh NOx, a GCR between 15.7 and 17.3 appears to be provide the best tradeoff between smoke and fuel efficiency. In addition, both smoke and BSFC_d were clearly sensitive to the engine-out NOx, suggesting the production air-handling that was optimized for 4 g/kWh NOx may not be adequate for a low NOx application. Therefore, it is important to optimize combustion and air-handling together to achieve low engine-out NOx and high fuel efficacy simultaneously.

Based on the experimental results in Fig. 6.18, a GCR of 16.5 was selected for the first generation (Gen1) low NOx GCI engine system development (Fig. 6.19). As seen in Fig. 6.19, a new piston bowl was designed with the bowl lip and bowl depth

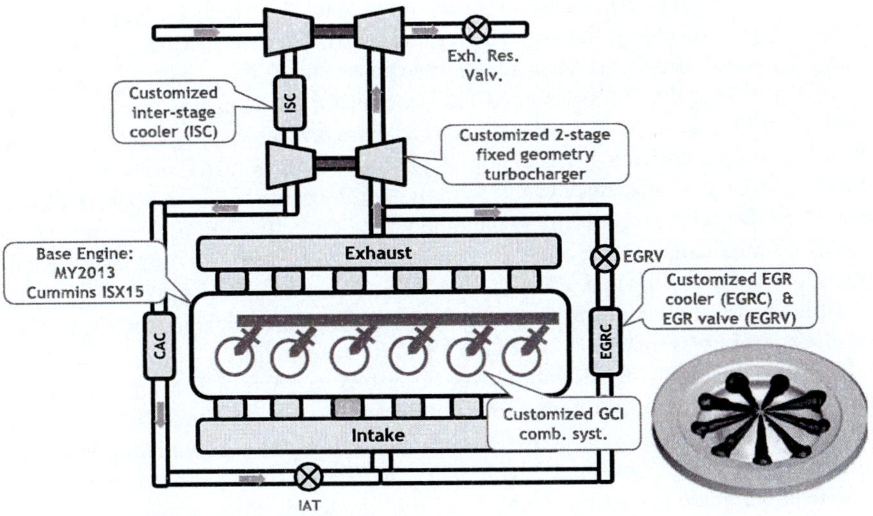

Fig. 6.19 System layout for the Gen1 heavy-duty low NOx GCI engine

Table 6.3 Combustion system configurations for the Gen1 heavy-duty low NOx GCI engine at 1375 rpm/15 bar BMEP

	Config. 1	Config. 2	Config. 3
Combustion Recipe	Baseline	GCR/Piston bowl geometry/ injector spray pattern	Thermal boundary conditions + GCR/Piston bowl geometry/ injector spray pattern
CR [–]	15.7	16.5	16.5
Pivc [bar]	3.05	3.05	3.42
Tivc [K]	399	399	368
Trapped EGR [%]	33.5	33.5	39.2
λ [–]	1.49	1.49	1.65
Number of Nozzle Holes [–]	–	9	9
Spray Inclusion Angle [°]	–	152	152
NOx [g/kWh]	1.40	2.09	1.42
Soot [g/kWh]	0.101	0.066	0.083
ISFC [g/kWh]	174.3	169.1	162.3
ISFC Improvement	–	3.0%	6.9%
MPRR [bar/CAD]	8.7	8.1	10.0

Source Zhang et al. (2018a)

optimized for gasoline sprays. Table 6.3 lists the steps that were undertaken in the design process at 1375 rpm/15 bar BMEP. It is an important high load operating point for the truck engine investigated. From Config. 1 to Config. 2, GCR, piston bowl geometry and injector spray pattern were adjusted. Moving to Config. 3, thermal boundary conditions (Pivc, Tivc, EGR) were customized.

In Fig. 6.20a, the tailor-designed GCI combustion system produced faster and stronger mixing-controlled combustion due to more favorable spray-bowl interaction and enhanced air utilization (Fig. 6.20b). Consequently, it led to a shorter combustion duration and higher effective expansion ratio, thereby contributing to 3% better ISFC and 34.7% lower soot according to the model prediction. In Config. 3, by increasing boost and enhancing EGR cooling (i.e., higher Pivc and lower Tivc), the ISFC benefit was predicted to expand to 6.9% with a lambda increase from 1.49 to 1.65 (Fig. 6.21).

Building on the results from the analysis-led design campaign, the Gen1 GCI engine concept was materialized (Fig. 6.19). The base engine is a model year (MY) 2013 Cummins 15 L diesel engine. As described in Table 6.3, a tailor-designed GCI combustion system was implemented. In addition, several modifications were made to the air-handling system. The production single-stage variable geometry turbocharger (VGT) was replaced with a customized two-stage turbocharging system to deliver adequate boost under high EGR dilution levels. Also, a customized two-stage EGR cooler was used to provide enhanced cooling of the EGR gas.

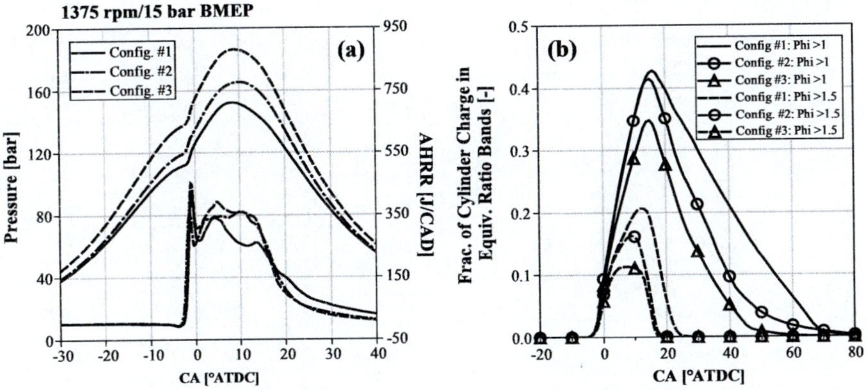

Fig. 6.20 **a** Global combustion behavior and **b** in-cylinder air utilization for the Gen1 heavy-duty low NOx GCI engine at 1375 rpm/15 bar BMEP. *Source* Zhang et al. (2018a)

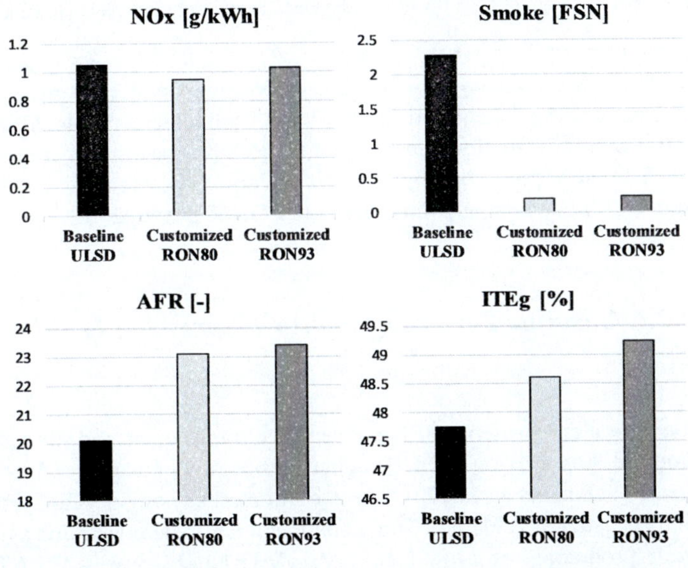

Fig. 6.21 **a** Performance and **b** global combustion behavior for ULSD, RON80 gasoline, and RON93 gasoline at 1375 rpm/10 bar BMEP for the Gen1 heavy-duty low NOx GCI engine

The engine performance results at 1375 rpm/10 BMEP, an operating point representative of the on-highway cruising for the truck engine, were summarized in Fig. 6.22. Three fuels were tested, including an ULSD, a RON80 ethanol-free (E0) gasoline, and a RON93 E0 gasoline. As shown in Fig. 6.22, the customized combustion and air-handling concept collectively promoted leaner and faster mixing-controlled combustion compared to the baseline diesel combustion. As a result, the

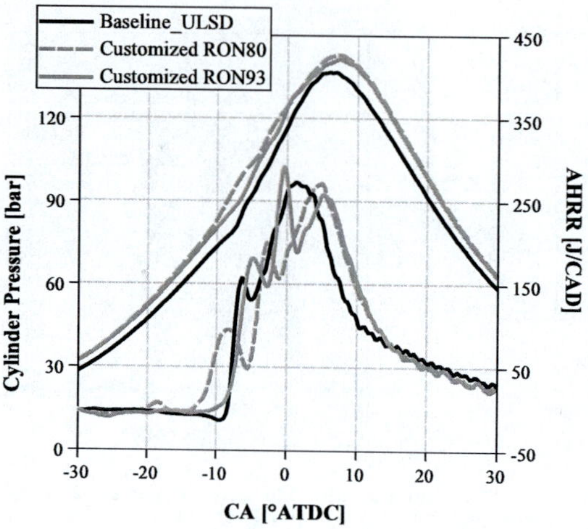

1375 rpm/10 bar BMEP

	RON93 Gasoline
Pinj [bar]	1450
SOI1 [°aTDC]	-30
SOI2 [°aTDC]	-6
Qsplit [-]	30:70
CA50 [°aTDC]	5.5
AFR [-]	23.4
Ext. EGR [%]	36.7
Pintake [bar]	2.25
Tintake [°C]	65

Fig. 6.22 **a** Performance and **b** global combustion behavior for ULSD, RON80 gasoline, and RON93 gasoline at 1375 rpm/10 bar BMEP for the Gen1 heavy-duty low NOx GCI engine

engine-out smoke was reduced by ~ 90% at 1 g/kWh NOx while the gross indicated thermal efficiency (ITEg) was improved by 0.9–1.5 percentage points. Moreover, at 1375 rpm/10 bar BMEP, the low reactivity of the RON93 gasoline enabled the most promoted PPCI combustion, thereby producing a larger ITEg improvement over the diesel baseline (1.5 percentage points) than the RON80 gasoline.

6.5.2 2.2 L Gasoline Direct Injection Compression Ignition Engine

Based upon a modern passenger SI engine, an ultra-efficient and clean gasoline direction compression ignition (GDCI) engine has been under extensive development in recent years (Sellnau et al. 2019b). The evolution on key engine sub-systems and specifications is listed in Table 6.4. The second generation demonstrated a best BTE of 40% while producing ultra-low NOx (<0.2 /kWh) and smoke (<0.1 FSN) over engine operating conditions within typical vehicle drive cycles. Migrating into the third generation, the BSFC was improved by 3–12% over a load range of 2–10 bar BMEP.

The evolution from Gen3 to Gen3X represents a significant change in high load combustion strategy from a single-stage PPCI combustion process to a two-stage PPCI-diffusion combustion process (Fig. 6.12). The engine compression ratio was increased from 16 to 17. As discussed in 4.2, the two-stage combustion process offered enhanced ability to tailor the heat release process, thereby effectively controlling combustion noise at high loads while facilitating a fast and clean diffusion combustion process. Consequently, as seen in Fig. 6.23, the engine operating load

Table 6.4 Evolution of GDCI engines

	Gen1	Gen2	Gen3		Gen3X
Engine type		4-cylinder, 4-valve DOHC			
Bore (mm)		82.0			
Stroke (mm)	85.2	85.2	105		105
Compression ratio	15.0	15.0	16		17
Peak cylinder pressure (bar)		200			
Combustion strategy	PPCI				PPCI-diffusion
Valvetrain	Fully flexible, exhaust rebreathing				2-step exhaust rebreathing
Boost system	2-stage (turbocharger and supercharger)				Single-stage VNT with adv. wide flow range compressor
EGR system		Low pressure loop EGR with EGR cooler bypass			
Swirl		Near Zero			
Fuel injection System		500 bar central-mounted GDI			
Number of nozzle holes	6~8	10	10		10
Nozzle hole diameter (μm)	169	170	170		170
Spray inclusion angle (°)	135	100	130		130
Aftertreatment system (ATS)		GOC + GPF + SCR			

Source Cho et al. (2021)

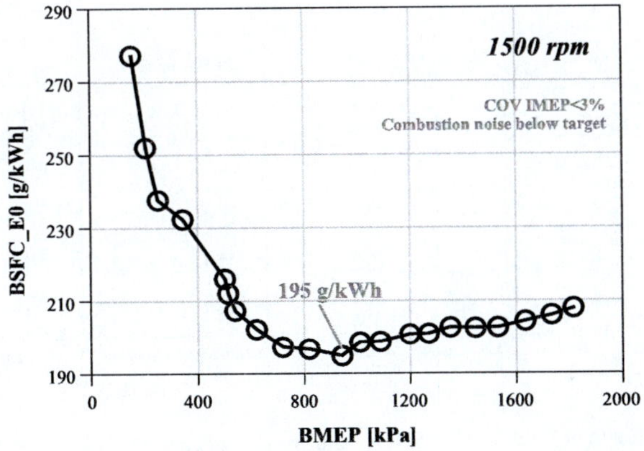

Fig. 6.23 BSFC_E0 over a load sweep at 1500 rpm for Gen3 and Gen3X GDCI engines. *Source* Sellnau et al. (2019b)

was comfortably extended to 18 bar BMEP at 1500 rpm with excellent fuel efficiency (Sellnau et al. 2019b). In addition, BSFC at engine loads over 6 bar IMEP was notably improved due to higher compression ratio, minimized combustion losses, and less parasitic losses by removing the supercharger and replacing boost promotion with thermal promotion. Moreover, the engine performance mapping results demonstrated that a broad operating range exhibited BSFC below 210 g/kWh (40% BTE) and a minimum BSFC of 194 g/kWh (43% BTE) at 1750 rpm/12 bar BMEP (Cung and Ciatti 2017).

Adopting a modified piston bowl design to enhance late-stage air utilization (Fig. 6.12), PPCI-diffusion GCI combustion strategy showed markedly improved performance over a MY2015 BMW N47 2L diesel engine at 1500 rpm/6 bar IMEP (Sellnau et al. 2019a). Figure 6.24 shows that at a targeted NOx level of 1 g/kWh, BSFC was improved by 5.7%, while the smoke was reduced by 92%. Subsequently, a further effort was made to optimize the PPCI-diffusion combustion system at 1750 rpm/20 bar BMEP, involving key design attributes on Pinj, number of nozzle holes, TNA, and spray inclusion angle. As seen in Fig. 6.25, when combining the four steps together, 3-D CFD analysis predicted 14.3% ISFC improvement and 87.7% soot reduction.

Among the four steps, adjusting the number of nozzle holes to 8 (Step 2) was found to be most impactful. Global combustion behavior and air-utilization characteristics (i.e., mass fraction of in-cylinder charge at Phi > 1) are captured in Fig. 6.26. Going through the steps, the heat release progression became increasingly rapid. The trend was well-supported by the enhanced air utilization during the diffusion combustion process as evidenced by reduced presence of Phi > 1 regions in the cylinder. Collectively, faster diffusion combustion with more abundant local oxygen availability during the late combustion stage contributed to the ISFC and soot benefits shown in Fig. 6.25.

Finally, building on the results from Gen3X, a Gen4X was conceptualized to aim for a maximum BTE of 47.6%, featuring reduced heat transfer losses via thermal

	2.2L GDCI
Pinj [bar]	390
SOI1 [°aTDC]	-38
SOI2 [°aTDC]	-3
Qsplit [-]	60:40
CA50 [°aTDC]	9.8
AFR [-]	19.9
Ext. EGR [%]	15.1
haust Re-Breathing	Yes
Pintake [bar]	1.19
Tintake [°C]	65

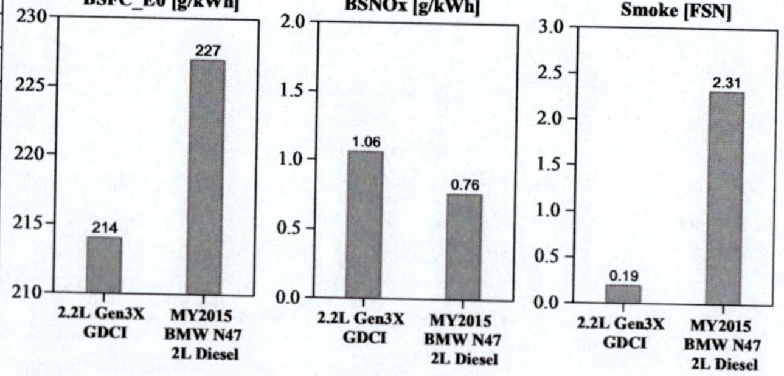

Fig. 6.24 Comparison of BSFC_E0, BSNOx, and smoke between the 2.2 L GDCI engine and two production light-duty diesel engines. *Source* Sellnau et al. (2019a)

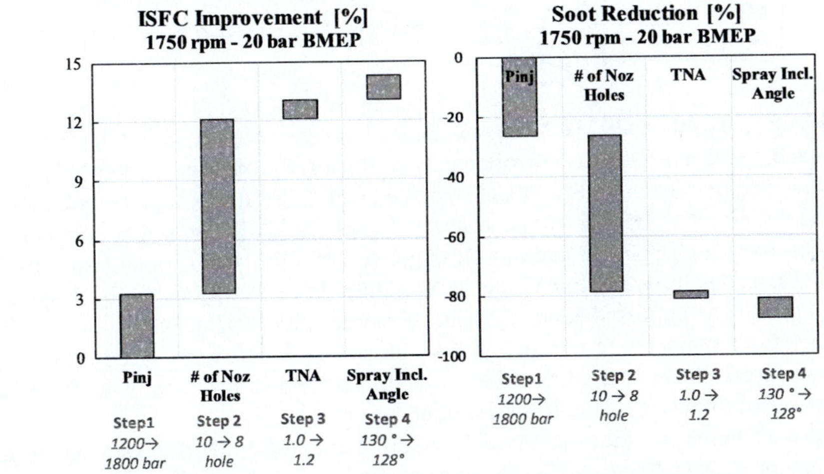

Fig. 6.25 Simulated ISFC, Smoke, and ISNOx for variation of injection pressure, number of nozzle holes, total nozzle area, and spray angle at 1750 rpm/20 bar BMEP. *Source* Sellnau et al. (2019a)

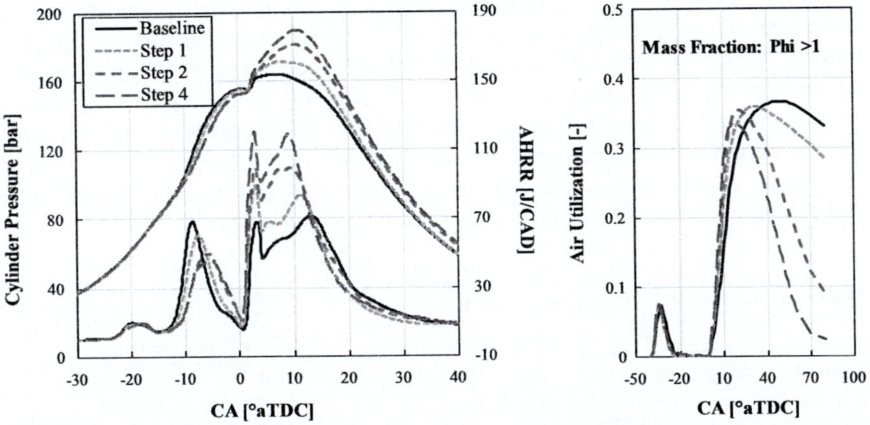

Fig. 6.26 Simulated global combustion behavior and air utilization at 1750 rpm/20 bar BMEP. *Source* Pei et al. (2019)

barrier coating, lower friction, higher gas exchange efficiency, and improved PPCI-diffusion combustion system design.

6.5.3 1.4 L Mixed-Mode Gasoline Low Temperature Combustion Engine

Adopting a multi-mode, HCCI/spark-assisted compression ignition (SACI)/SI, approach, General Motors has reported developing a 1.4 L downsized, boosted gasoline low temperature combustion (LTC) engine (Yun 2019). It was aimed at achieving a specific output of 85 kW/L and meeting CARB SLEV30 emissions regulation without using urea SCR. In addition, it targeted a 15–17% fuel economy improvement over a contemporary stoichiometric SI engine. The enabling technologies under development encompassed downsizing and boosting, lean HCCI combustion system, advanced low temperature plasma ignition, passive ammonia SCR system (PASS), and advanced physics-based controls development and implementation. The key technical attributes are summarized in Table 6.5.

Table 6.6 highlights the mixed-mode combustion strategies developed for gasoline LTC engine (Yun 2019). At low loads, multiple fuel injections were employed during NVO along with a spark event to enable partial fuel oxidation (Combustion Strategy I), thereby enhancing the charge thermal and chemical reactivity prior to the final fuel injection and the spark event. As a result, a more robust HCCI combustion process was achieved. Utilizing this strategy, Fig. 6.27 shows that LTC produced a much faster heat release process, leading to a 16.1% of net indicated fuel consumption (NSFC) improvement at 2000 rpm/2 bar NMEP. The engine-out NOx from the LTC engine was shown at a very low level of 0.2 g/kg_fuel. It was reported that spark ignition under these low load conditions didn't drive the combustion process. Rather, it helped improve the combustion stability by reducing problematic cycles. Moving from Combustion Strategy I to Combustion Strategy II, the multiple fuel injection events were replaced with a single injection event during NVO.

Table 6.5 Gasoline LTC engine specifications

	Base engine	Development engine
Displacement [L]	2.2	1.4
CR [–]	12	
Boosting system	– (naturally aspirated)	Sing-Stage VGT
Fuel injector	Bosch 60-8	Delphi 60/90-10
Ignition system	Conventional spark plug	Groundless barrier discharge igniter (GBDI)
Camshaft	Single-step	Two-step intake (IN) and exhaust (EX)
ATS	TWC	PASS
Targeted specific power output [kW/L]	–	85

Source Yun (2019)

Table 6.6 Gasoline LTC combustion strategy

Strategy I (low load): lean LTC with fuel injection and spark events during NVO to enhance robustness	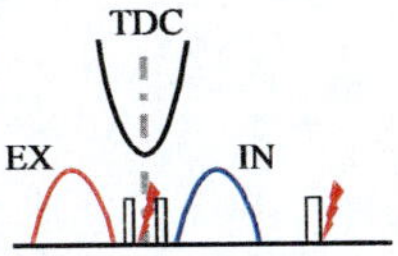
Strategy II (low-to-medium load): Kinetics-controlled, lean LTC with NVO	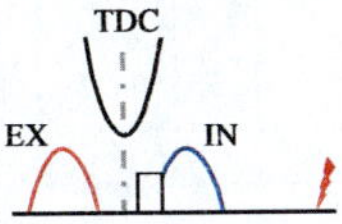
Strategy III (medium-to-high load): spark-assisted, lean LTC with PVO to facilitate smooth transition from LTC to stoichiometric SI	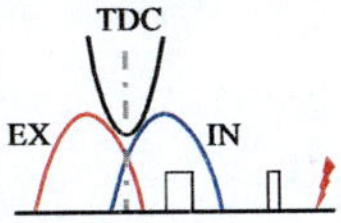
Strategy IV (high load): conventional stoichiometric SI	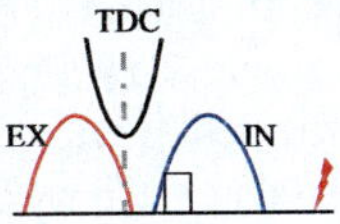

Source Yun (2019)

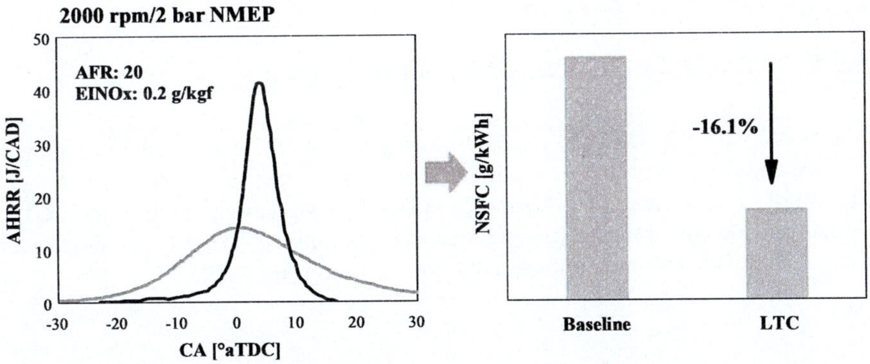

Fig. 6.27 Gasoline LTC engine performance over a load sweep at 1500 rpm. *Source* Yun (2019)

Strategy III served to bridge between lean LTC operation and the conventional homogeneous stoichiometric SI combustion (Strategy IV). This strategy effectively combined spark-initiated flame propagation with the ensuing stratified charge autoignition to control combustion noise, thereby facilitating a smooth mode transition process. Under this strategy, positive valve overlap (PVO) was used to introducing EGR dilution and control NOx emissions. Finally, Combustion Strategy IV was employed after the transition, aimed at achieving the target high specific output through a conventional, boosted, flame propagation process.

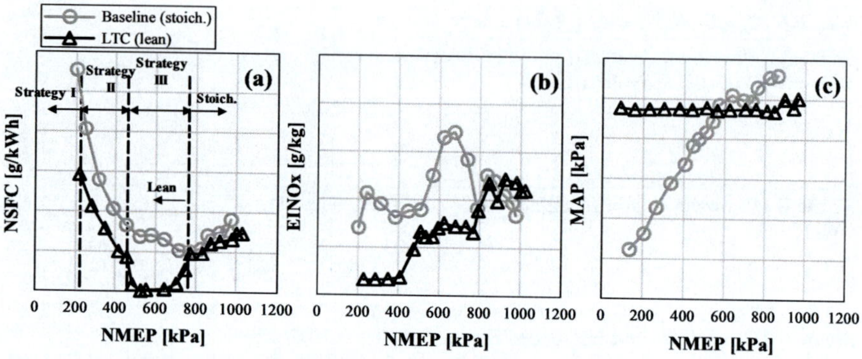

Fig. 6.28 Gasoline LTC engine performance over a load sweep at 1500 rpm. *Source* Yun (2019)

A comparison of engine performance between the LTC engine and the base engine was provided in Fig. 6.28. Across the operating range, the LTC engine was fully un-throttled (Fig. 6.28c). For the LTC engine, the lean-to-stoichiometric transition occurred at ~ 8 bar NMEP. Under lean operation (Modes 1, 2 and 3), the LTC engine consistently produced notably enhanced fuel efficiency and generated considerably lower NOx emissions than the base engine.

6.5.4 2 L Mazda Skyactiv-X Gasoline Engine

Mazda's Skyactiv-X was the first mixed-mode gasoline LTC engine that was commercialized and released to the passenger car market in Europe and Japan. The combustion strategy was named as "Spark Controlled Compression Ignition (SPCCI)", utilizing spark-initiated flame propagation as an "air piston" to compress and trigger the remaining charge into homogeneous autoignition (Fig. 6.29).

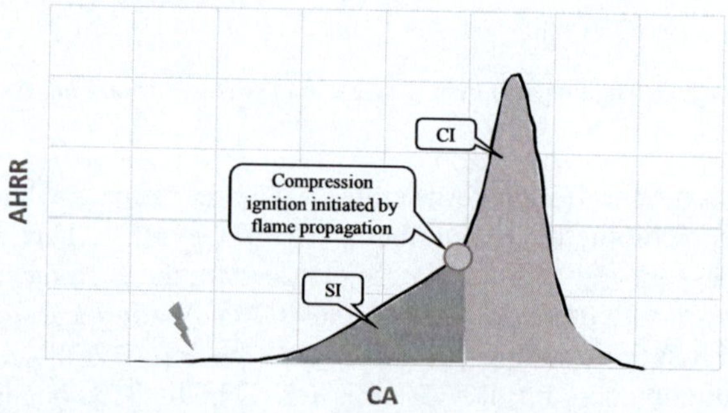

Fig. 6.29 Conceptualized SPCCI combustion strategy for Skyactiv-X. *Source* Isobe et al. (2019)

As seen from Table 6.6 and Fig. 6.27, GM's gasoline LTC engine and Skyactiv-X share similar combustion strategies. However, the performance targets and the implementation approach are different. GM's LTC engine has a higher specific output target (85 kW/L vs. 66 kW/L), so the GCR had to be reasonably low (12) to avoid knock at high loads. In contrast, Skyactiv-X targeted a lower specific output value, thereby allowing the GCR to increase to 16.3. The increased GCR then facilitates achieving robust low load combustion stability and good drive-cycle fuel efficiency. The higher GCR also alleviate the need for complex valvetrain strategies at low loads.

Table 6.7 lists the key engine specifications for Skyactiv-X (Nakai et al. 2019). In addition to the 16.3 GCR discussed earlier, it uses a 700 bar GDI system with centrally-mounted10-hole injectors, thereby leading to enhanced spray atomization, evaporation, and air entrainment. The spark plug is located adjacent to the GDI injector between the two intake ports. The intake ports are designed to introduce both tumble and swirl motion. The level swirl is tailored by adjusting the swirl control valve. The engine is also equipped with cooled low pressure EGR to suppress autoignition. The aftertreatment system (ATS) consists of a three-way catalyst (TWC) and a GPF to meet the emissions regulations in Europe and Japan. Note that a lean aftertreatment system was used for GM's gasoline LTC engine to meet the more stringent SLEV30 HC + NOx regulation.

Figure 6.30 shows that at low loads, a triple injection strategy was combined with introducing swirl motion to organize the in-cylinder fuel–air mixture distribution into a lightly stratified charge. The NOx and BSFC tradeoff was carefully balanced by delivering appropriate fuel–air mixture formation to the spark plug (Sellnau et al. 2019a). At high loads, external cooled EGR, swirl enhancement, and a late injection

Table 6.7 Key engine specifications for Skyactiv-X

	Skyactiv-G	Skyactiv-X
Engine type	In-line 4	
Displacement [L]	2.0	
Bore x stroke [mm]	83.5 × 91.2	
CR	14.0	16.3
Fuel injection	SIDI, 20 MPa	Central DI, 70 MPa
Air-supply system	–	Roots-Type Supercharger
External EGR	–	Cooled Low Pressure EGR
Valvetrain	Dual cam phaser (no VVL)	
Max. power	121 kW (12.1 bar BMEP) /6000 rpm	132 kW (13.2 bar BMEP) /6000 rpm
Max. Torque	210 Nm (13.2 bar BMEP)/4000 rpm	224 Nm (14.1 bar BMEP) /4000 rpm
Specific power output [kW/L]	60.5	66
ATS	TWC + GPF	

Source Nakai et al. (2019)

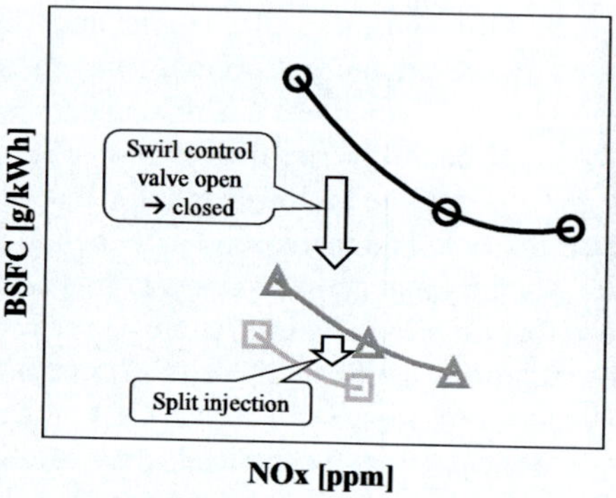

Speed [rpm]	2000
BMEP [kPa]	320
AFR [-]	30
SOI [°bTDC]	336/275/215
Qsplit [-]	0.59/0.31/0.1

Fig. 6.30 NOx/BSFC and in-cylinder fuel–air mixture formation at 2000 rpm/320 kPa for Skyactiv-X. *Source* Nakai et al. (2019)

event in the compression stroke together were found to be effective in mitigating knock and maintaining good combustion stability (Fig. 6.31), thereby enabling to achieve the targeted specific output. Finally, as seen in Fig. 6.32, Skyactiv-X exhibited markedly improved fuel efficiency over the Skyactiv-G gasoline engine. The BSFC improvement was 10% on average and could reach ~ 20% at certain lean burn operating conditions (Sellnau et al. 2019a).

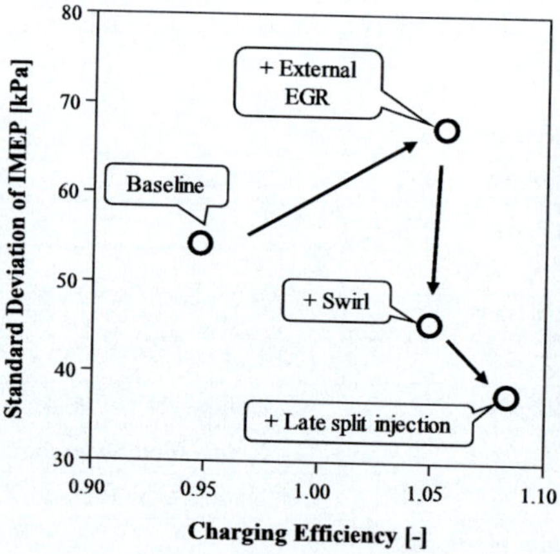

Operating Point	2000 rpm/ full load
Ext. EGR [%]	9
Swirl Control Valve Position [%]	15
Air-Fuel Ratio	12.8
Fuel Inj, Pressure [MPa]	30
SOI [°bTDC]	270/90
Qsplit [-]	0.9/0.1

Fig. 6.31 Charging efficiency/combustion stability and in-cylinder fuel–air mixture formation at 2000 rpm/full load for Skyactiv-X. *Source* Nakai et al. (2019)

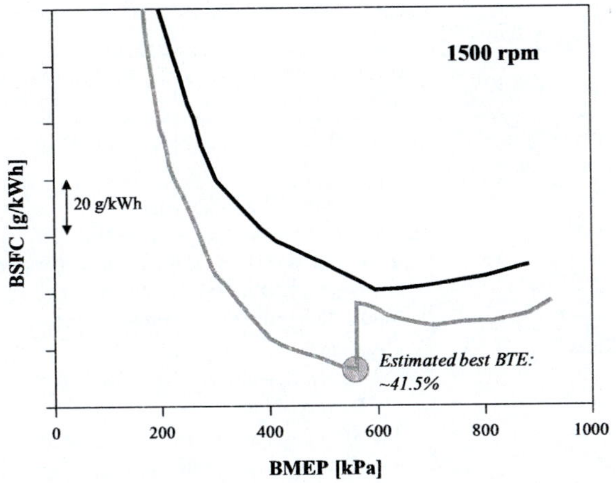

Fig. 6.32 BSFC comparison between Skyactiv-X and Skyactiv-G over a load sweep at 1500 rpm. *Source* Nakai et al. (2019)

6.5.5 Technology Outlook for GCI

Table 6.8 provides a summary on the benefits and key development areas for commercializing the GCI technology. GCI clearly demonstrates enhanced NOx-soot tradeoff over diesel combustion, thereby offering a cost-effective means to achieve future ultra-low emissions regulations in both commercial transport and passenger car sectors. Considering the tremendous challenge faced by today's light-duty diesel engines to meet the stringent Real Driving Emissions (RDE) requirement, GCI's emissions benefit in the passenger car sector may be especially advantageous. Moreover, the PPCI-diffusion combustion strategy combined with tailored combustion and air-handling system design can potentially lead to fuel efficiency that is beyond the state-of-the-art diesel engines. As shown in Pei et al. (2019), it has experimentally reached 43.5% BTE with a technical path laid out towards 48% BTE. In addition, there is a projected shift in transport fuel demand. The demand for diesel likely rises as global commercial traffic continues to expand, while the need for gasoline may reduce as a share of passenger car sector is expected to be electrified. The demand shift could drive a price differential between diesel and gasoline. Consequently, running gasoline in compression ignition engines may offer a potential benefit of reduced total cost of ownership (TCO), which would be particularly attractive to fleet owners.

Key technical areas that need to be developed towards implementing GCI for practical applications include cold start, full-range combustion strategy, and fuel system compatibility. As listed in Table 6.8, each area is presented with good development opportunities. Significant progress has been made to address these areas through tailor-designed engine combustion and air-handling system and by developing customized fuel system components. Overall, GCI is a technology

Table 6.8 Technology outlook for GCI engines

	On-road heavy-duty commercial vehicles	Light-duty passenger vehicles
Typical engine operating load range	10–20 bar BMEP	≤ 10 bar BMEP
Benefits	• Enhanced NOx-soot tradeoff • Diesel-like or better fuel efficiency • Potential to reduce total cost of ownership due to projected price imbalance between diesel and gasoline as future transport energy demand shifts	
Key Technical Areas	• Cold start – Opportunities: spark ignition and/or intake air heating (Zhao et al. 2021) • Stability and performance at low loads – Opportunities: exhaust re-breathing and injection strategy calibration (Borgqvist et al. 2013; Sellnau et al. 2019a) • Maximizing the potential for partially-premixed combustion – Opportunities: air-handling development (boost + EGR) and/or intake millerization (Zhang et al. 2020; Kumar et al. 2021b) • Fuel system compatibility – Opportunities: moderate modifications to high pressure fuel pump and injector components and lubricity additive (Tzanetakis et al. 2019)	

that demonstrated promising performance benefits over today's gasoline and diesel engines. As the implementation challenges continue to be addressed systematically, it may become an important technology in the propulsion system portfolio at engine manufacturers.

6.6 Summary

GCI is a technology the draws significant attention in recent years. Its basic principles and production-intent implementation have been under extensive investigation. Based on the review performed in this study, the main conclusions are summarized as follows.

- Gasoline ignition quality influences ignition delay most strongly at low-to-intermediate temperatures. Under PPCI, gasoline's autoignition reactivity can be harnessed for injection strategy development. In contrast, under MCCI, gasoline's physical and chemical properties are more impactful and can be harnessed for customized combustion system design.

- Based on the level of in-cylinder fuel stratification, GCI can be categorized into HCCI, PPCI, and MCCI. A promising approach towards practical implementation is to develop a PPCI-diffusion combustion strategy. Implementing this strategy, a customized heavy-duty GCI engine has demonstrated drastically enhanced NOx-soot tradeoff and better indicated fuel efficiency over the diesel baseline across the full engine operating range. Similarly, the 2.2L Gen3X GDCI engine achieved a peak BTE of 43% with a wide range of the engine map over 40% BTE.
- Both GM's gasoline LTC engine and Mazda's Skyactiv-X take a mixed-mode approach encompassing HCCI, SI, and a mode transition process through SACI. Depending on the high-level performance targets, the mixed-mode engine can be developed either towards high specific output or favoring high drive cycle fuel efficiency. For mode transition, it is critical to deliver adequate fuel–air mixture formation to spark plug for SACI operation. This can be achieved by properly combining a multiple fuel injection strategy and tailored combustion system design.
- GCI has demonstrated the potential to enhance NOx-soot tradeoff over diesel combustion, thereby offering a cost-effective means to meet future emissions regulations. Fuel efficiency and TCO could also be advantageous for GCI. Significant effort and progress have been made to address the key development areas including cold-start, full-range combustion strategy, and fuel system compatibility.

References

Borgqvist P, Tunestal P, Johansson B (2013) Comparison of negative valve overlap (NVO) and rebreathing valve strategies on a gasoline PPC engine at low load and idle operating conditions. SAE Int J Engines 6(1). https://doi.org/10.4271/2013-01-0902

Ciatti S, Johnson M, Das Adhikary B, Reitz R et al (2013) Efficiency and emissions performance of multizone stratified compression ignition using different octane fuels. SAE Technical Paper 2013-01-0263. https://doi.org/10.4271/2013-01-0263

Cho K, Zhao L, Ameen M, Zhang Y et al (2019) Understanding fuel stratification effects on partially premixed compression ignition (PPCI) combustion and emissions behaviors. SAE Technical Paper 2019-01-1145. https://doi.org/10.4271/2019-01-1145

Cho K, Zhang Y, Sellnau M (2021) Investigation on combining partially premixed compression ignition and diffusion combustion for gasoline compression ignition-part 1: fuel reactivity and injection strategy effects. SAE J. STEEP 2(1):41–58. https://doi.org/10.4271/13-02-01-0003

Cung K, Ciatti S (2017) A study of injection strategy to achieve high load points for gasoline compression ignition (GCI) operation. In: ASME. Internal combustion engine division fall conference, vol 1: large bore engines; fuels; advanced combustion: V001T03A014. https://doi.org/10.1115/ICEF2017-3625

Dec J, Yang Y, Dronniou N (2011a) Boosted HCCI-controlling pressure-rise rates for performance improvements using partial fuel stratification with conventional gasoline. SAE Technical Paper 2011-01-0897. https://doi.org/10.4271/2011-01-0897

Dec J, Dronniou N, Yang Y (2011b) HCCI and stratified-charge ci engine combustion research. US DOE Vehicle Technologies Office, Annual Merit Review Meeting

Dec J, Yang Y, Dronniou N (2012) Improving efficiency and using E10 for higher loads in boosted HCCI engines. SAE Int J Engines 5(3):1009–1032. https://doi.org/10.4271/2012-01-1107

Dec J, Dernotte J, Chunsheng J (2016) Low-temperature gasoline combustion (LTGC) engine research. US DOE Vehicle Technologies Office, Annual Merit Review Meeting

Dec J, Dernotte J, Ji C (2017a) Increasing the load range, load-to-boost ratio, and efficiency of low-temperature gasoline combustion (LTGC) engines. SAE Int J Engines 10(3):1256–1274. https://doi.org/10.4271/2017-01-0731

Dec J, Dernotte J, Chunsheng J, Gentz G (2017b) Low-temperature gasoline combustion (LTGC) engine research. US DOE Vehicle Technologies Office, Annual Merit Review Meeting

Isobe R, Endo K, Sueoka M (2019) New-generation gasoline engine "SKYACTIV-X. Mazda Tech Rev 36:16–23. https://doi.org/10.34338/mazdagihou.36.0_16

Kaddatz J, Andrie M, Reitz R, Kokjohn S (2012) Light-duty reactivity controlled compression ignition combustion using a cetane improver. SAE Technical Paper 2012-01-1110. https://doi.org/10.4271/2012-01-1110

Kalghatgi G, Hildingsson L, Johansson B (2010) Low NOx and low smoke operation of a diesel engine using gasoline-like fuels. J Eng Gas Turbines Power 132(9). https://doi.org/10.1115/1.4000602

Kodavasal J, Som S (2018) Gasoline compression ignition—a simulation-based perspective. In: Srivastava D, Agarwal A, Datta A, Maurya R (eds) Advances in internal combustion engine research. Energy, environment, and sustainability. Springer, Singapore. https://doi.org/10.1007/978-981-10-7575-9_12

Kokjohn S, Hanson R, Splitter D, Reitz R (2011) Fuel reactivity controlled compression ignition (RCCI): a pathway to controlled high-efficiency clean combustion. Int J Engine Res 12(3):209–226. https://doi.org/10.1177/1468087411401548

Kumar P, Zhang Y, Traver M, Watson J (2021) Air-system and variable valve actuation recipe for high load gasoline compression ignition operation in a heavy-duty diesel engine. SAE Technical Paper 2021-01-0516. https://doi.org/10.4271/2021-01-0516

Manente V, Zander C, Johansson B, Tunestal P et al (2010) An advanced internal combustion engine concept for low emissions and high efficiency from idle to max load using gasoline partially premixed combustion. SAE Technical Paper 2010-01-2198. https://doi.org/10.4271/2010-01-2198

Nakai E, Goto T, Ezumi K, Tsumura Y, Endou K, Kanda Y, Urushihara T, Sueoka M, Hitomi M (2019) MAZDA SKYACTIV-X 2.0L Gasoline Engine. In: 28th Aachen Colloquium Automobile and Engine Technology

Pei Y, Torelli R, Tzanetakis T, Zhang Y, Traver M, Cleary D (2017) Modeling the fuel spray of a high reactivity gasoline under heavy-duty diesel engine conditions. In: ASME 2017 ICEF conference, ASME ICEF2017-3664. https://doi.org/10.1115/ICEF2017-3530

Paz J, Staaden D, Kokjohn S (2018) Gasoline compression ignition operation of a heavy-duty engine at high load. SAE Technical Paper 2018-01-0898. https://doi.org/10.4271/2018-01-0898

Pei Y, Pal P, Zhang Y, Traver M et al (2019) CFD-guided combustion system optimization of a gasoline range fuel in a heavy-duty compression ignition engine using automatic piston geometry generation and a supercomputer. SAE Int J Adv Curr Practice in Mob 1(1):166–179. https://doi.org/10.4271/2019-01-0001

Robertson W (2017) California Air Resources Board Heavy-Duty Truck and Engine Plans. In: SAE 2017 Commercial Vehicle Congress, Rosemont, IL, 18–20 Sept 2017

Sellnau M, Cho K, Zhang Y, Cleary D (2019a) Pathway to 50% brake thermal efficiency using gasoline direct injection compression ignition (GDCI). In: 28th Aachen Colloquium automobile and engine technology

Sellnau M, Foster M, Moore W, Sinnamon J et al (2019b) Pathway to 50% brake thermal efficiency using gasoline direct injection compression ignition. SAE Technical Paper 2019-01-1154. https://doi.org/10.4271/2019-01-1154

Srivastava DK, Das A, Singh NK (2018) Variable valve actuation systems. In: Srivastava D, Agarwal A, Datta A, Maurya R (eds) Advances in internal combustion engine research. Energy, environment, and sustainability. Springer, Singapore. https://doi.org/10.1007/978-981-10-7575-9_4

Tzanetakis T, Traver M, Costanzo V, Medina R et al (2019) Durability study of a high pressure common rail fuel injection system using lubricity additive dosed gasoline-like fuel—additional cycle runtime and teardown analysis. SAE Int J Adv Curr Practice in Mob 1(2):654–674. https://doi.org/10.4271/2019-01-0263

Westbrook C (2000) Chemical kinetics of hydrocarbon ignition in practical combustion systems. In: Proceedings of the Combustion Institute, vol 28

Yun H (2018) A high specific output gasoline low-temperature combustion engine. US DOE Vehicle Technologies Office, Annual Merit Review Meeting

Yun H, Wermuth N, Najt P (2011) High load HCCI operation using different valving strategy in a naturally-aspirated gasoline HCCI engine. SAE Technical Paper 2011-01-0899. https://doi.org/10.4271/2011-01-0899

Yun H (2019) A high specific output gasoline low-temperature combustion engine. US DOE Vehicle Technologies Office, Annual Merit Review Meeting

Zhang Y, Sellnau M (2021) A computational investigation of PPCI-diffusion combustion strategy at full load in a light-duty GCI engine. SAE Technical Paper 2021-01-0514. https://doi.org/10.4271/2021-01-0514

Zhang Y, Sagalovich I, De Ojeda W, Ickes A et al (2013) Development of dual-fuel low temperature combustion strategy in a multi-cylinder heavy-duty compression ignition engine using conventional and alternative fuels. SAE Int J Engines 6(3). https://doi.org/10.4271/2013-01-2422

Zhang Y, Sommers S, Pei Y, Kumar P, Traver M, Cleary D (2017) Mixing-controlled combustion of conventional and higher reactivity gasolines in a multi-cylinder heavy-duty compression ignition engine. SAE Technical Paper 2017-01-0696. https://doi.org/10.4271/2017-01-0696

Zhang Y, Kumar P, Pei Y, Traver M, Cleary D (2018a) An experimental and computational investigation of gasoline compression ignition using conventional and higher reactivity gasolines in a multi-cylinder heavy-duty diesel engine. SAE Technical Paper 2018-01-0226. https://doi.org/10.4271/2018-01-0226

Zhang Y, Voice A, Pei Y, Traver M, Cleary D (2018b) A computational investigation of fuel chemical and physical properties effects on gasoline compression ignition in a heavy-duty diesel engine. J Energy Resour Technol 140(10):102202. https://doi.org/10.1115/1.4040010

Zhang Y, Kumar P, Tang M, Pei Y, Merritt E, Traver M, Popuri S (2020) Impact of geometric compression ratio and variable valve actuation on gasoline compression ignition in a heavy-duty diesel engine. In: 2020 ASME ICEF conference, ICEF2020-3035. https://doi.org/10.1115/ICEF2020-3035

Zhang Y, Cho K, Sellnau M (2021) Investigation on combining partially premixed compression ignition and diffusion combustion for gasoline compression ignition-part 2: compression ratio and piston bowl geometry effects. SAE J. STEEP 2(1):59–78. https://doi.org/10.4271/13-02-01-0004

Zhao L, Zhang Y, Pei Y, Zhang A, Traver M, Ameen M (2021) Numerical evaluation of spark assisted cold idle operation in a heavy-duty gasoline compression ignition engine. In: SAE Technical Paper 2021-01-0410. https://doi.org/10.4271/2021-01-0410

Part III
Advanced Combustion Technologies for SI Engines

Chapter 7
Optical Diagnostics for Gasoline Direct Injection Engines

Ankur Kalwar and Avinash Kumar Agarwal

Abstract Advancements in internal combustion (IC) engine technologies have improved both spark ignition (SI) and compression ignition (CI) engines immensely in terms of fuel efficiency and emissions. Gasoline engines are giving tough competition to their diesel counterparts in the light-duty passenger car segment with continuously increasing market share. Apart from lower exhaust emissions, current generation Gasoline Direct Injection (GDI) engines offer higher power output and superior fuel economy. Nevertheless, automotive researchers and engineers have used many tools to evolve state-of-the-art GDI engine technology continuously. The most important and useful tool in engine development is optical diagnostics. Optical access to the engine allows non-intrusive investigations of the in-cylinder processes using optical/laser-based diagnostics, which provides information not available by any other research tool. This chapter discusses applying such techniques to understand various in-cylinder processes such as spray-flow interactions, fuel–air mixture formation, combustion, and pollutant formation in GDI engines. It includes detailed perspectives that help optimize engine design and control parameters adapted to conventional and alternative fuels. It broadly covers studies related to in-cylinder flow characteristics, sprays characterization, fuel spray-air interactions, fuel evaporation, mixture formation, flame propagation characteristics, and pollutant formation, corresponding to different GDI engine operating parameters. This chapter discusses applications of various optical diagnostic techniques in GDI engine investigations such as particle imaging velocimetry (PIV), Mie-scattering, phase Doppler interferometry (PDI), laser-induced fluorescence (LIF), natural flame chemiluminescence, laser-induced incandescence (LII), OH/CH chemiluminescence. The rich information derived from the optical diagnostic techniques helps develop new-generation GDI engines capable of meeting market requirements and emission compliance.

Keywords Gasoline direct injection · Optical diagnostics · In-cylinder flows · Spray · Charge distribution · Pollutant formation

A. Kalwar · A. K. Agarwal (✉)
Engine Research Laboratory, Department of Mechanical Engineering, Indian Institute of Technology Kanpur, Kanpur 208016, India
e-mail: akag@iitk.ac.in

A. K. Agarwal et al. (eds.), *Advanced Combustion for Sustainable Transport*, Energy, Environment, and Sustainability, https://doi.org/10.1007/978-981-16-8418-0_7

7.1 Introduction

This section introduces the state of development in Gasoline direct injection (GDI) engines. Limitations of current GDI engines are also discussed. After this, the need and role of optical diagnostics in GDI engine development are also explained. Since the invention of IC engines, unthrottled engine operation, and efficient control are considered promising. This helps avoid pumping losses and improves combustion efficiency. Thermodynamics conditions suitable for efficient combustion are always attained in such conditions. Heat is released in a mixing-controlled combustion process by auto-ignition of diesel vapours wherever they are in a combustion chamber. In gasoline engines, unthrottled engine operation and predominantly mixing-controlled combustion are difficult due to homogeneous mixture formation and constraints of fixed location of the spark plug. This imposes additional restrictions on the mixture formation process in SI engines. In addition to temporal fuel evaporation requirement, the spatial evolution of repeatable fuel–air mixture distribution is required for a stable stratified mixture formation in every engine cycle. However, homogeneous combustion in the GDI engine does not face any such difficulty related to the charge preparation. With the evolution of the GDI engine technology over time, many measures have been taken to ensure that the combustion system fulfils these requirements (Fraidl et al. 1996). The stratification of the fuel–air mixture is achieved by injecting the fuel into a sub-volume of the combustion chamber, creating well-separated zones with rich, stoichiometric, and lean fuel–air mixtures. This stratified charge combustion engine can be operated with an overall leaner fuel–air mixture, enhancing the fuel economy. The main combustion systems investigated for achieving stratified charge combustion could be air-guided, wall-guided, and spray-guided configurations. Among these, the latter two are the most commonly used, in which spray-guided configuration is considered a new generation GDI engine. Nevertheless, these combustion systems face the problems of wall impingement of sprays, higher HC and soot emissions, limited range of engine operation, and non-repeatability of charge formation. Successful combustion of stratified mixture with lower emissions requires controlled in-cylinder airflow, precise fuel injection, enhanced fuel spray evaporation, optimum fuel–air mixture distribution, and accurate charge motion. Maintaining all these parameters in an optimum range is a challenging task. Several studies have optimized control parameters such as controlled intake flow, combustion chamber design, injector parameters, and fuel injection strategies to achieve desirable engine performance. Our previous chapter describes a detailed analysis of these studies (Kalwar and Agarwal 2020).

Developing sophisticated and new tools for 3D CFD simulations and optical diagnostics has helped obtain newer insights into an engine's complex mixture formation and combustion processes. The availability of these tools helps in further refinement of the overall combustion system of GDI engines. The future development targets for new GDI engines include reducing fuel consumption, improving the engine performance with minimum compromise on fuel economy, and meeting the existing and upcoming emission norms. For emission legislation compliance, exhaust after-treatment technologies such as TWC, particulate trap, DeNOx, etc., are also incorporated in GDI engines. It is expected that cost of GDI vehicles should be comparable with current MPFI vehicles. It is also essential to optimize various parameters for cold-start, transient, and warm-engine conditions and switch between different operating modes of the GDI engine.

In the last couple of decades, the application and functionalities of optical diagnostics of engines have seen dramatic growth in engine research and development. Their crucial role in developing better engines has led to their transition from esoteric research tools to mainstream development tools. Direct observation of flame kernel initiation, flame growth, propagation, and extinction and its correlation with in-cylinder pressure data have rapidly gained importance. The developments in lasers and digital cameras have given a big boost to optical diagnostic tools. It enabled the measurements of various intrinsic processes such as highly spatially and temporally resolved fuel vaporization and multi-dimensional distribution of temperature, flow-fields, and pollutant concentration. These capabilities are of vital importance in modern engine development (Sick 2010). Optimized multi-phase flows, spray-wall impingement, stratification of charge, premixed charge combustion, formation of soot and other pollutants, and other phenomena of importance in carburettor/port injection have become crucial for the success of GDI engines. The cycle-to-cycle variability has become an important aspect of these investigations. The combustion modes involving lean and stratified mixtures result in a high likeliness that the mixtures exceed the flammability limit, eventually leading to misfire (Kalwar and Agarwal 2021). This chapter reviews and summarizes the application of various optical diagnostic techniques to study such aspects for different engine control parameters.

Exploring fundamental physical and chemical processes that control engine operation is essential for improving IC engines. However, it is impossible to evaluate a large number of engine concepts and operation conditions. In addition to being time-consuming and cost-intensive processes, this approach also risks missing the overall optimum. Hence, one of the important areas for optical diagnostics is to aid, develop and calibrate CFD simulation tools for IC engines. Fully predictive numerical models for engine-related processes are gaining significance. Although, considerable differences in the thermal properties of the optical engine with its metal engine counterpart lead to difficulties in matching their combustion characteristics. However, fundamental physics is not much affected by this. These limitations could be avoided by directly applying optical diagnostics on realistic and production-grade engine geometries. This requires greater emphasis on developing micro and small-sized sensors/diagnostics that only need small optical access, e.g., fibre optics could be used for data collection and illumination in some of these techniques.

7.2 Optical Diagnostics in GDI Engines

Over the last couple of decades, several optical diagnostic tools have been developed, applied, and used for GDI engine investigations (Sick 2010). Direct injection of fuel, the multi-phase nature of the in-cylinder charge, and spray-wall interactions pose challenges in implementing optical diagnostic techniques in the engines. This section presents a detailed discussion of all such important phenomena investigated using various optical diagnostic tools. Effect of different engine control parameters, operating conditions, and fuel types are analyzed for the in-cylinder spray characterization, in-cylinder flows, spray-flow interactions, charge formation, flame characterization, and pollutant formation.

7.2.1 In-cylinder Spray Characterization

In-cylinder investigations of spray characteristics provide realistic insights into the fuel injection process and spray quality. Many researchers have successfully attempted high-speed imaging of fuel spray in an optical GDI engine (Catapano et al. 2013; Lee et al. 2020). Mie scattering technique is widely used for visualizing the in-cylinder injection process and characterizing GDI spray.

In a study by Catapano et al. (2013), in-cylinder spray imaging was done by an Intensified Charge-Coupled Device (ICCD) camera with sensitivity in 200–800 nm wavelength covering both UV and visible spectrum in the wall-guided GDI engine. Images were recorded via the piston optical window, and CW halogen lamps were used for illuminating the flow through the optical liner, as shown in Fig. 7.1. The acquired images were post-processed using Labview software and National Instruments hardware. The procedure for obtaining results from the raw spray images is shown in Fig. 7.2. Initially, a raw 12-bit colour image was converted to a 12-bit greyscale image. After this, the image underwent convolution filtering and was binarized by applying a 15% threshold for instantaneous luminosity. Finally, the outline was created by selecting the region of interest. Spray penetration was measured by assessing the maximum distance between the fuel injector and the spray tip.

Fuel injection parameters were the most critical factors affecting the spray characteristics. A substantial number of studies have investigated the impact of fuel injection parameters such as fuel injection timing, fuel injection pressure (FIP), and injection strategies on the in-cylinder spray characteristics. In a study by Lee et al. (2020), the effects of fuel injection timing and FIP on the spray behaviour and wall impingement in a DISI optical engine having a wall-guided configuration were studied. The spray images obtained at 8° CA after the fuel injection were considered for assessing the spray development. The spray collapsed when it reached the optically accessible window, and there were hardly any differences to distinguish the plumes. For injection timing at 330° bTDC or before, spray directly impacted the piston even before

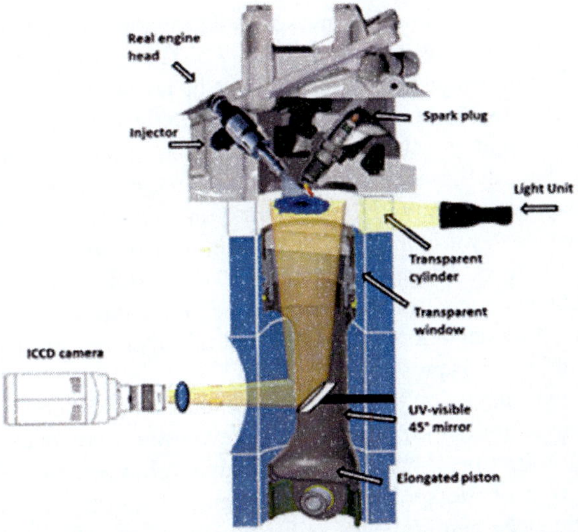

Fig. 7.1 Schematic of the experimental setup of optical GDI engine for digital imaging (Catapano et al. 2013)

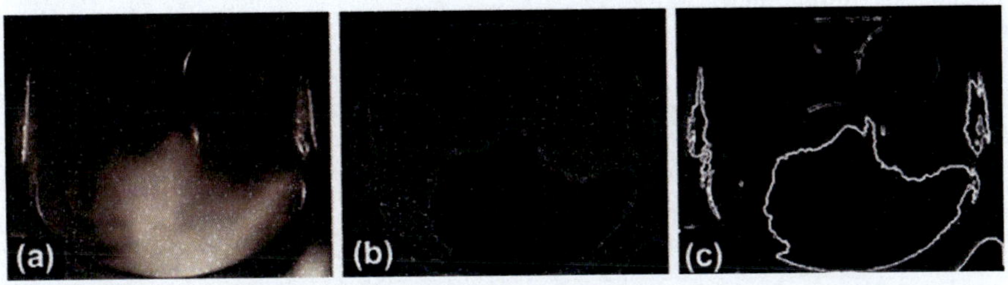

Fig. 7.2 Steps for post-processing of data for calculating the spray penetration (Catapano et al. 2013)

getting completely atomized and evaporated. Due to the shorter distance for injection, in-cylinder flow was quite weak in such a case, and spray behaviour was mainly dependent on the spray momentum. For 300° bTDC injection timing, the spray shape demonstrated bowed trajectory, mainly due to in-cylinder flow in reverse direction w.r.t. the injection during the intake process. This led to the bending of spray towards the injector, causing wetting of the cylinder walls. The spray travelled straight for the injection timing of 270° bTDC or later. These characteristics were in agreement with the variations in the in-cylinder flow during the intake process. At a late injection timing of 210° bTDC, the in-cylinder flow intensity reduced as the maximum valve lift was completed. Hence, injected fuel spray hardly faced any noticeable disruption and travelled straight. Towards the end of injection, the spray tip reached the cylinder wall on the opposite side of the injector, resulting in wall-film formation.

The spray development was further investigated for variations in the FIP at 300° and 210° bTDC injection timing, as shown in Fig. 7.3. The spray penetration was higher for late injection timing than the advanced one, with an increasing FIP. The spray was minimally affected by the ambient flow and was mainly driven by its momentum. Hence, a larger straight movement of the spray led to greater fuel mass reaching the cylinder walls. For fuel injection at 300° bTDC, increasing the FIP enhanced the horizontal movement of spray in the rightward direction. The higher FIP improved the atomization of spray droplets. Hence smaller droplets suffered greater deflection due to in-cylinder flows. The elevated rightward deflection of spray encountered the spray penetration and increased the wall-wetting on the injector side on the cylinder wall. However, spray deviation diminished and inhibited penetration at higher load due to higher ambient gas pressure.

Moreover, spray deviation images were obtained from 50 consecutive injection cycles. The regions of high variations in spray evolution were located for both early and late injection timings. For 300° bTDC, large variations were observed in the lower-right section of the spray, where the spray bending occurred. This region was mainly influenced by turbulence and strong in-cylinder flow movement, which augmented variations in the spray behaviour. These variations were reinforced at higher FIP. For late injection timing of 210° bTDC, the lower part of the spray underwent larger variations. This was mainly because of the turbulence created by the air entrainment in the spray. The low-pressure zone formed on the lower side of

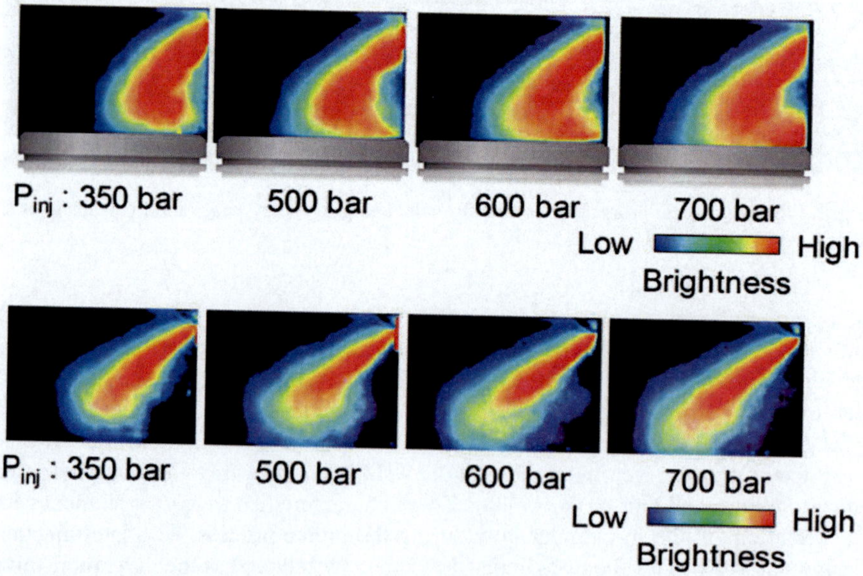

Fig. 7.3 In-cylinder spray behaviour at 300° bTDC and 210° bTDC injection timing with an increased pressure (Lee et al. 2020)

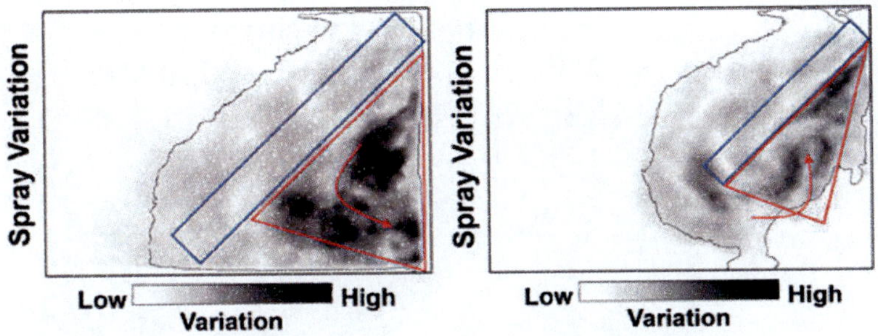

Fig. 7.4 Spray variations for injection timing of 300° bTDC (600 bar pressure) and 210° bTDC (350 bar pressure) (Lee et al. 2020)

the spray by the drag of ambient air due to high-speed droplets resulted in an inflow of air stream. Hence, this region faced larger variations, as shown in Fig. 7.4.

Findings by Song and Park (2015) reported that spray development differed in the engine than the quiescent chamber due to the piston motion. Injecting fuel at 180° aTDC (end of intake stroke) resulted in similar spray morphology to the chamber. Despite turbulence in the engine cylinder, weak intake flow due to the absence of piston motion at BDC didn't affect the spray. In the case of late injection at 300° aTDC, the spray penetration was initially comparatively shorter due to higher ambient pressure (~0.3–0.4 MPa). However, later, the spray plumes were blocked by the upward piston motion causing the wall wetting. For injection at 90° aTDC, at the middle of the intake stroke, the effect of strong intake flow was observed on the spray. It resulted in shorter spray due to dispersion of overall spray along with the surrounding flow. Soon, the spray collapsed as separate plumes were observed. An increase in the FIP resulted in considerable changes in the spray characteristics. It is easily noticeable from Fig. 7.5 that with higher FIP, the spray jet transitioned to an unstable form, resulting in finer spray droplets. Hence, their earlier evaporation was reported as no traces of droplets were observed after 4 ms of fuel injection. Another aspect noticed was related to the shape of the overall spray. Spray turned into rounder and thicker with increasing FIP. This was because smaller droplets at the spray tip experienced higher drag due to ambient, which easily spread out due to the following droplets.

Application of different alternative fuels resulted in variations in the spray breakup, droplet formation, and subsequent spray evolution owing to their different physical and chemical properties. In one such related study (Bao et al. 2014), experiments were conducted to evaluate spray penetration of different fuels, namely, iso-octane, gasoline, and ethanol using Mie-scattering in an optical wall-guided GDI engine. Since the injector was side-mounted, hence image from the bottom presents the side view of the spray. The start of injection could not be viewed due to limited optical access in the piston crown. The spray penetration was obtained by considering the axial penetration of plume C, which was the highest among all the plumes. Spray penetration of ethanol was the lowest at a lower FIP of 40 bar. This was mainly

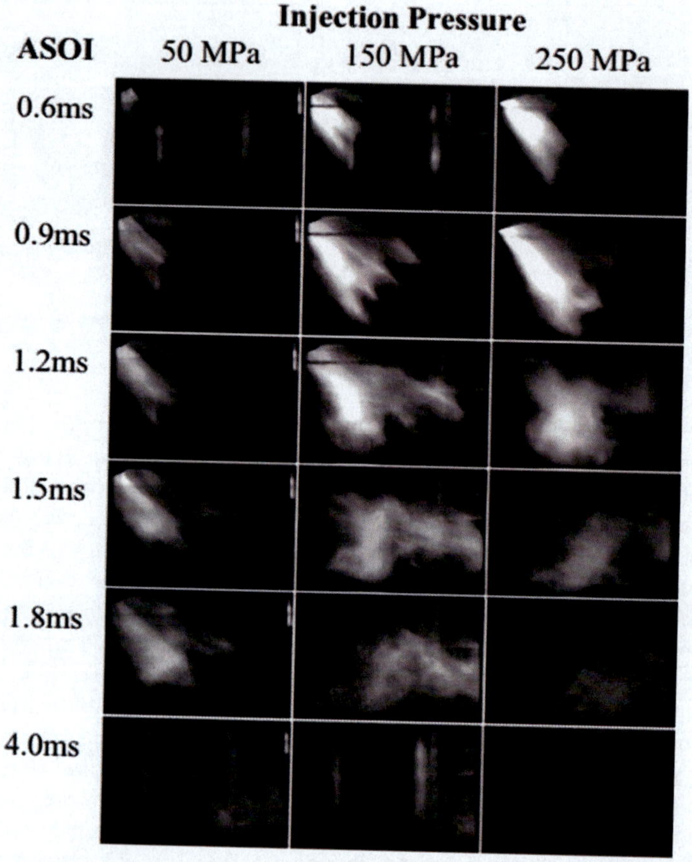

Fig. 7.5 Effect of FIP on spray development at SOI@ 90° aTDC (Song and Park 2015)

due to lower fuel injection velocity, owing to higher nozzle losses. Higher density of ethanol was mainly responsible for this. It can be explained by correlating with the expressions developed by Naber and Siebers (1996) for diesel spray penetration. As per this expression, spray penetration (S) is directly proportional to the velocity of the jet and time after the SOI (t), when time lies before the transition time of the breakup. After the transition time, spray penetration varies proportionally with the square root of the time after the SOI ($t^{0.5}$) and inversely to the ambient gas density. At lower FIP, spray penetration of all three test fuels varied with (t). Hence lower injection velocity for ethanol decreased with its penetration. However, at a higher FIP of 7, 11, and 15 MPa, the crossover between (S) $\propto$ (t) to (S) $\propto$ ($t^{0.5}$) was easily observed for all fuels. Here, spray penetration was mainly governed by the aerodynamic forces over the fuel droplets. Hence, ethanol spray reported the highest penetration for higher FIP due to its larger fuel droplets. In gasoline and iso-octane, smaller droplets suffered higher drag forces, which reduced their penetration.

Another study (Chan et al. 2014) investigated the structural transformation of gasoline and ethanol sprays undergoing flash boiling in an optical GDI engine. The results found that in the case of an early injection (~90° aTDC) when ambient pressure was low (~100 kPa), spray plumes showed higher convergence towards the injector axis, as shown in Fig. 7.6. This was mainly due to the flash boiling of spray as ambient pressure was lower than the vapour pressure of the fuel. On the other hand, for the late injection timing of 300° aTDC, spray plumes showed distinct features due to higher ambient pressure (~4 bar) than the fuel vapour pressure. However, FIP variation affected the structural transformation of the spray in both cases. With increased FIP,

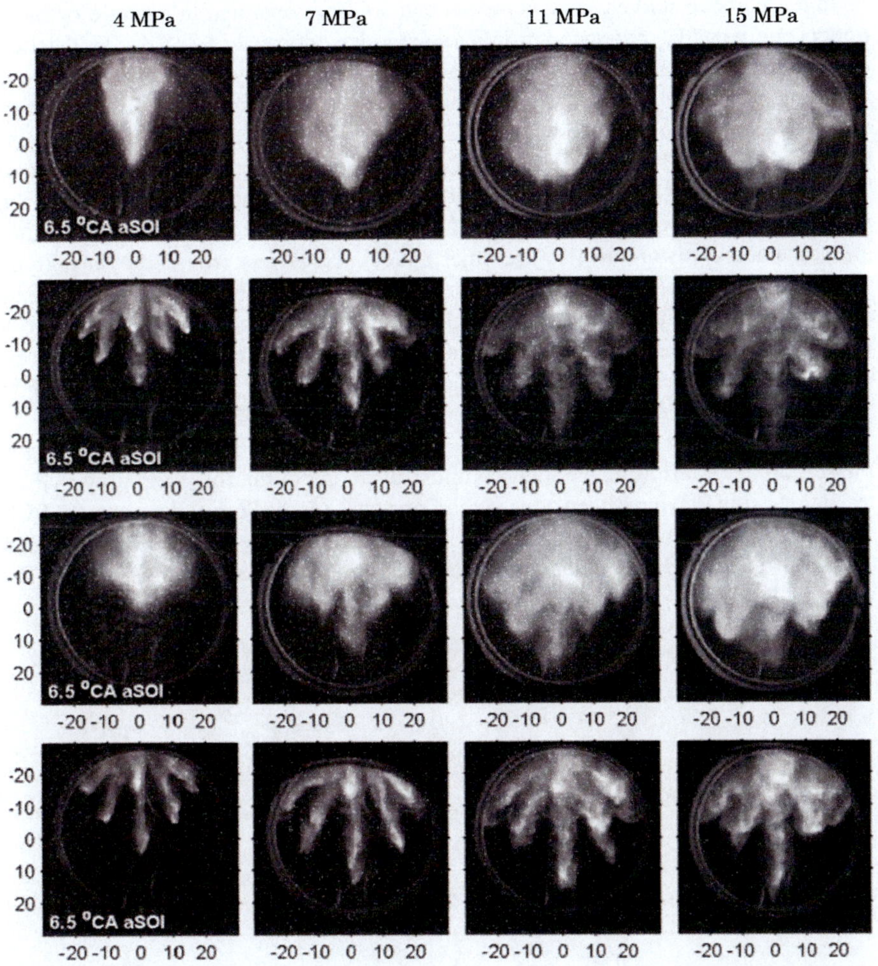

Fig. 7.6 In-cylinder spray images for different injections pressure. **a** Gasoline, SOI@ 90° aTDC, **b** Gasoline, SOI@ 300° aTDC, **c** Ethanol, SOI@ 90° aTDC, **d** Ethanol, SOI@300° aTDC (Chan et al. 2014)

flash boiling was less evident as higher spray momentum outperformed the spray convergence and imparted more directionality to the plumes. In addition, enhanced vaporization due to higher FIP was observed more for the flashing spray. Hence, the study suggested that both FIP and flashing conditions are suitable for providing a homogeneous mixture in the DISI engine. Further, gasoline and ethanol showed flash boiling based on the degree of superheating. For low ambient pressure, flash boiling of gasoline was more predominant at low FIP, while transitional flashing occurred at high FIP. For ethanol, transitional flashing at low FIP and no flashing at high FIP were observed. Overall, FIP on vaporization and flash boiling were more pronounced for gasoline due to higher superheating than ethanol.

In these above studies, optical measurements were obtained in a single-cylinder optical engine with a flat piston profile to maximize optical access to the combustion chamber. Hence, the measurements do not offer a complete picture of a commercial GDI engine. In that respect, a study performed by Catapano et al. used a multi-cylinder commercial GDI engine by converting it to optical configuration (Catapano et al. 2016). The engine had a wall-guided fuel injection system configuration. They incorporated elongated piston arrangement to allow optical access through the sapphire piston window located in the piston bowl. This was done such that it didn't affect the actual piston profile. The other optical access was provided by arranging an endoscopic probe through the cylinder head. This setup ensured minimal modifications to the structure to maintain the thermo-fluid dynamic configuration of a commercial engine. The setup arrangement is shown in Fig. 7.7. The endoscope used had a 70° angle wide field-of-view perpendicular to the tumble plane, including the injector and the spark plug.

For investigating microscopic sprays, the most common technique employed is Phase Doppler Interferometry (PDI). Almost all studies of microscopic spray characterization using PDI have been done in the Constant Volume Spray Chamber (CVSC). This doesn't consider the effects of in-cylinder flows or turbulence on the spray results and fails to simulate the actual engine-like conditions, though. On the contrary, the

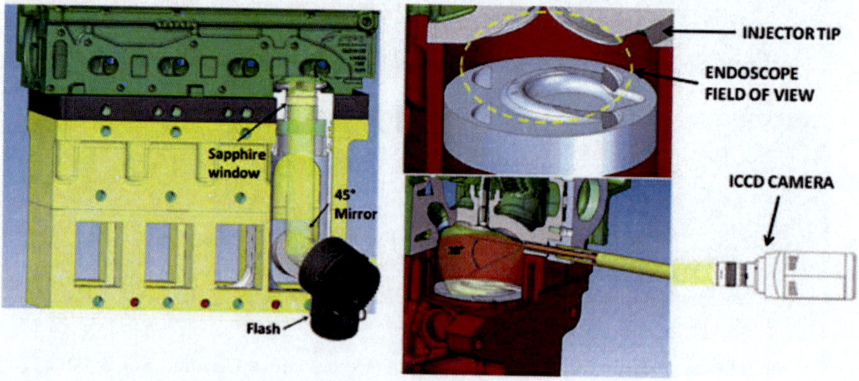

Fig. 7.7 Arrangement of endoscopy for spray imaging (Catapano et al. 2016)

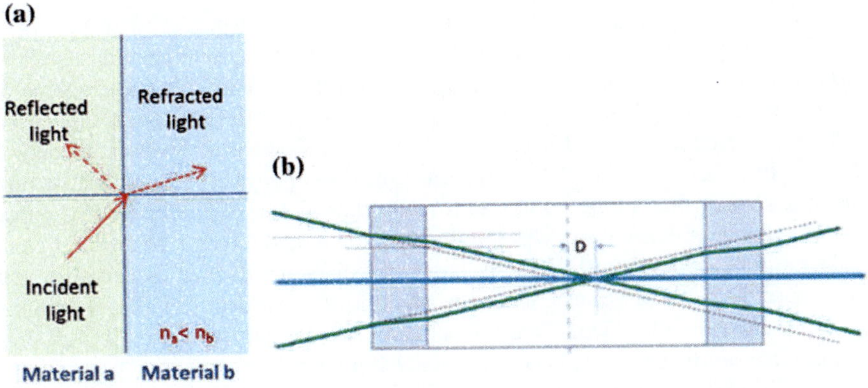

Fig. 7.8 **a** Phenomenon of reflection and refraction of laser beam passing through optical media. **b** Side view showing laser beams propagation (Sharma et al. 2020)

implementation of PDI on firing optical engines is a difficult task. For the first time, PDI was employed in firing optical GDI engine to study spray characteristics in real engine conditions at IIT Kanpur in our group. A comparative analysis was presented vis-a-vis results from the CVSC (Sharma et al. 2020). The most critical aspect in conducting the experiments was the alignment of laser beams through the quartz optical window, which was curved. For measurement accuracy, a pair of laser beams in the orthogonal planes coming from the transmitter must intersect at a common point called 'probe volume.' The procedure of attaining this is called the 'alignment of lasers.' The cylindrical quartz optical window caused difficulty in alignment due to the reflection and refraction of light due to change in medium. Hence, the best possible methodology adopted was to initially align the lasers using a similar surrogate window outside the engine and exact movement to locate the probe volume inside the engine w.r.t. the transmitter and the receiver.

Figure 7.8 shows how the presence of glass media refracts the incoming laser after reflecting a small (~15%) fraction. This is mainly dependent on the curvature, quality of the window, and laser beam quality. Considering a 2-component PDI system, one pair of lasers (blue) traverse horizontally, i.e., orthogonal to the cylinder axis. The other laser beam pair traverses in the vertical plane parallel to the cylinder axis. It was found that the direction of beams travelling in the horizontal plane was affected, while beams in the vertical plane suffered from refraction due to the angle of incidence. This was the case if the measurement point was at the center of the quartz window. However, for any other location, both laser beams showed slight refraction from their path. This study performed in-cylinder spray droplet diameter and velocity investigations in a firing optical GDI engine having a wall-guided spray system. The CVSC and in-cylinder engine experiments showed considerable differences in maximum spray droplet velocity components (V_x and V_y). In the engine experiments, maximum V_x and V_y were reported to be 29.8 m/s and 14.2 m/s, while in CVSC, they were 78.41 m/s and 23.92 m/s, respectively. The presence of dynamic temperature and pressure conditions during engine operation led to a higher drag on the spray droplets, hence enhanced evaporation. Further, in-cylinder turbulence improved the mixing of surrounding air into the spray. In addition, higher ambient temperature reduced the surface tension of the fuel spray droplets. All these factors resulted in a reduction in droplet size, leading to loss of momentum and earlier evaporation before they reached probe volume. Therefore, lower maximum droplet velocity distribution was reported in the engine combustion environment. For a similar reason, spray droplets diameter distribution was also less scattered, and the bulk of droplets were in the range of diameter <10 μm in the engine experiments. This study demonstrated a comparative picture of transient and turbulent flows in the actual engine conditions w.r.t quiescent chamber. This dataset would certainly help improve the CFD models and provide more realistic results after model validation.

In addition to the wall-guided combustion system, several researchers also investigated spray evolution in the spray-guided combustion system. In one such study (Peterson et al. 2014), the laser was used for illumination. Figure 7.9 shows the spray images obtained from the Mie-scattered signals. Geschwindner et al. (2020) also used Mie-scattering for imaging. They used the diffused back illumination (DBI) technique in spray-guided GDI engines to capture the spray images, post-processed to evaluate the spray shape, penetration, and spray cone angle. ECN recommends DBI for liquid spray imaging.

The studies so far discussed used 2D planar imaging for liquid spray visualization in the engine. They have shortcomings when a complex 3D phenomenon is evaluated since out-of-plane motion with gradients are characteristics of the in-cylinder turbulent flows. However, with imaging technologies and optics evolution, 3D imaging techniques have come a long way. Chen et al. (2016) used a plenoptic imaging technique to capture 3D spray structure and its interactions with the surrounding flow under different operating modes in a spray-guided GDI optical engine. Baseline spray images under a quiescent environment were obtained to compare the effect of engine flows on the sprays. Studies were conducted for homogeneous and stratified charge modes, different engine speeds, and swirl ratios of the in-cylinder flows. Under homogeneous mixture conditions (SoI@300° bTDC), a strong intake-air flow was observed on the spray plumes trajectory. The impact of swirl momentum was quite prominent at a higher engine speed of 1300 rpm than 800 rpm. Decreasing the swirl ratio lowered the deviations in the spray plume directionalities. The spray structures varied significantly from cycle to cycle at higher engine speed and with a higher degree of swirl. For stratified charge preparation (SoI@35° bTDC), it was observed that apart from in-cylinder flows, the presence of high pressure and temperature also impacted the spray behaviour. The spray was susceptible to piston interactions. Images showed that the effect of swirl at both engine speeds on spray structures was minimal in stratified conditions. This is mainly due to reduced turbulence intensity and flow momentum as turbulence during the intake stroke dissolves during the compression stroke. Also, cyclic variations didn't affect the spray behaviour considerably. Hence, the spray was minimally affected as it was mostly driven by the spray momentum and not the flow momentum.

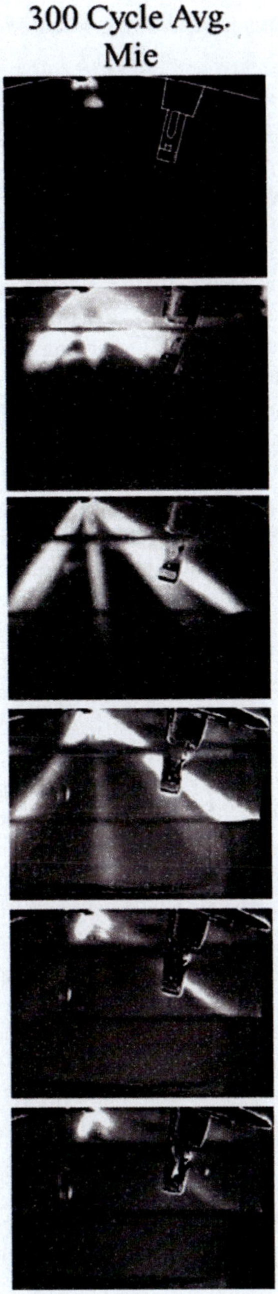

Fig. 7.9 Mie scattering images of fuel spray in a spray-guided DISI engine (Peterson et al. 2014)

7.2.2 *In-cylinder Flows and Spray-Flow Interactions*

Direct injections of high-pressure fuel can substantially change the in-cylinder flow behaviour via momentum exchange. Apart from large-scale bulk flow structures, it can influence small-scale turbulence too. In-cylinder flow mainly influences the mixing of fuel, air, and residual gases to alter the mixture quality. Further, ignition and early flame kernel growth are likely affected by the flow parameters near the spark plug. The rates of combustion and heat transfer also depend on flow turbulence. Hence, measurement of time-resolved flow-related parameters such as velocities, turbulent kinetic energy, dissipation rates, Reynold stresses, etc., are important for the engine design. To understand these phenomena for different fuel injection strategies, Clark et al. (2018) conducted a study to analyze the in-cylinder flows for five different fuel injection timing combinations. Double injection in a 1:1 split ratio was used for all five cases, and combinations were chosen from both timings in intake stroke to both timings in the compression stroke. The mean and fluctuating values of flow-field parameters were quantified using a novel spatial-filtering approach. Two spatial filtering approaches were considered, one based on fixed 8 mm cut-off length and another based on the cut-off filter length according to the mean flow speed at respective crank angle positions. The specifics of the methods used can be referred to in our other work (Clark and Kook 2018).

Experiments were conducted in a single-cylinder optical GDI engine with a multi-hole side-mounted injector. A PIV configuration was set up for swirl plane analysis, as shown in Fig. 7.10. The combinations of fuel injection timings used for double injection are listed in Table 7.1. These timings resulted in a wide range of mixture formation from well-mixed to highly turbulent. The image acquisition for PIV was in intervals of 42° to 28° bTDC, 5 mm below the fire deck. The upper limit was selected to avoid unwanted spray droplet signals, and the lower limit was selected so that the piston did not block the swirling plane.

The results showed that varying the fuel injection timing slightly resulted in significant differences in bulk in-cylinder flow behaviour and magnitude. However, 260–110° bTDC timing showed less bulk flow velocity despite a late injection. Further, it was noticed that the injection timing influenced the magnitude and direction of the tumble. A combination of 300 and 210° bTDC timing showed enhanced tumble flow, while reversal of tumble flow direction was observed at 300 and 110° bTDC and 140 and 110° bTDC timings. This could be a probable reason for low mean flow velocity in the case of 260–210° as flow induced by 2nd injection might be interfering with the flow caused by the first injection. In addition, cases having retarded injection timings resulted in higher TKE except for 140–110° bTDC timings. Cyclic variations in bulk flow structure were estimated by obtaining low frequency fluctuating intensity. A strong correlation was found between these fluctuations and the COV_{imep}. Although, cycle-to-cycle variations in TKE were not as per similar trends. Hence, it was suggested that the bulk flow parameter has a greater influence on the early flame growth than the turbulence. There was a likeliness of experiencing higher heat loss by the flame kernel to the spark plug electrodes in case of higher turbulence or

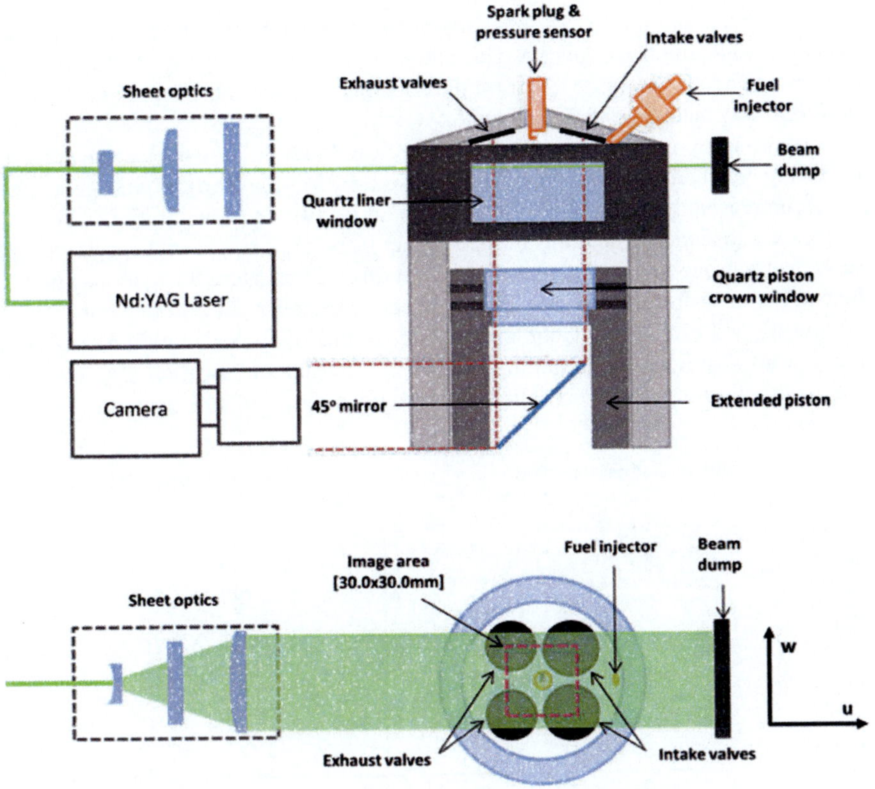

Fig. 7.10 Schematic of PIV setup for swirl plane analysis (Clark et al. 2018)

Table 7.1 Combinations of double injection timings (Clark et al. 2018)

Case 1	300 and 270° CA bTDC
Case 2	300 and 210° CA bTDC
Case 3	300 and 110° CA bTDC
Case 4	260 and 110° CA bTDC
Case 5	140 and 110° CA bTDC

small-scale flows near the spark plug. Also, integral length scales did not experience any significant changes with the injection timings with an 8 mm fixed cut-off filter. However, determining integral scales for engine flows is always a non-trivial task.

Researchers have also explored the potential of the combination of optical tools for a detailed and complete understanding of interdependent processes. In one such study by Stiehl et al. (2013), PIV and high-speed spray imaging were employed to investigate the mutual interaction of sprays and ambient flows in a spray-guided stratified DISI engine. The stratification was achieved by multiple fuel injections. The results showed that the CCVs were higher for subsequent injections while the first injection was quite reproducible. This can be understood from Fig. 7.11, where the distribution of SH was quite larger for the 2nd injection. During the first injection, the flow field was mainly governed by bulk flow structures. The first injection transferred significant spray-induced turbulence to the flow field. As a result of shear forces occurring between spray and gaseous flow, two local toroidal vortices were generated, which increased the turbulence in the flow field locally, as shown in Fig. 7.12.

Hence subsequent injections in such higher turbulence levels with varying length scales were prone to get affected. Apart from this, rail pressure fluctuations after the first injection were also another major reason. Moreover, air-entrainment into the hollow spray during fuel injection caused due to the low-pressure region resulted in the upward airflow. This funnel flow varied in intensity and location, depending on the injection timing. For short delays between injections, the flow structures resulted

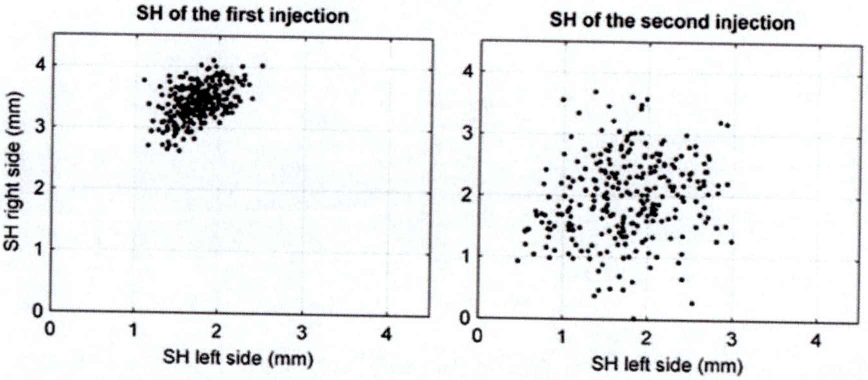

Fig. 7.11 Scatter plot of spray height (SH) of first and second consecutive injections (Stiehl et al. 2013)

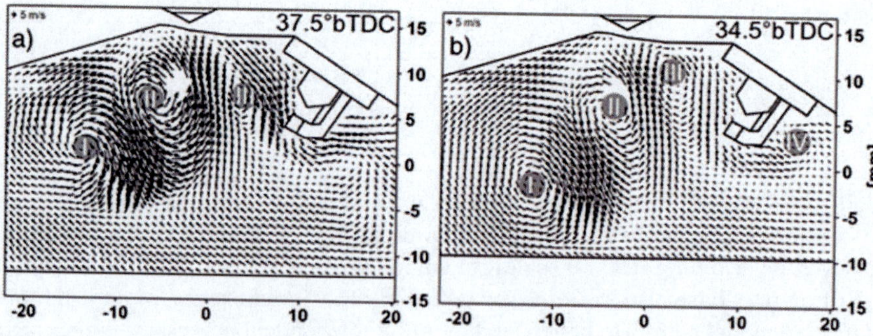

Fig. 7.12 Phase averaged flow fields for two instants between 1st and 2nd injection (Stiehl et al. 2013)

in high velocities, while for longer delays, it led to more stable spray due to early decay. Higher fluctuations were observed for the intermediate delays.

Similarly, another study (Geschwindner et al. 2020) investigated the interactions of bulk in-cylinder flows with late gasoline injection in two optical spray-guided engines. Testing on two optical engines offered different geometric configurations for the parametric study. The operating conditions of the experiment were following spray G standard condition of ECN. In-cylinder flow details were obtained using PIV, and Mie-scattering and DBI techniques were used for spray topology. Experiments were conducted for different inlet valve lifts and engine speeds. With the increase in both parameters, in-cylinder flow velocity increased. For quantifying the effect of flow field interactions with the spray, spatial probability maps for liquid spray were obtained, which showed the likeliness of liquid fuel presence. The results showed that spray outlines depicted broader probability distribution with ongoing injection events, indicating higher cycle-to-cycle variations. Also, cyclic variations increased for higher inlet valve lift and engine speed. It was found that higher in-cylinder flow velocities led to increased fluctuations in the spray G topology. Further, the penetration of spray was marginally affected by increased flow velocity. A slight reduction in spray expansion was observed for higher flow velocity. Also, increasing the engine speed and valve lift enhanced the upward recirculation of flow between the plumes. This restricted the spray collapse as plume-to-plume interactions were minimized. The spray-induced turbulence affected the ambient flow momentum significantly. It was hard to preserve the flow structures during injection in a weak tumble case, which yielded two counter-rotating vortices due to surrounding air entrainment. With high tumble intensity, most flow structures survived the disturbance caused by spray.

The intake manifold design for imparting specific in-cylinder flow structures is also widely studied by several researchers and is considered an important factor for optimizing engine performance. In a study by Stiehl et al. (2016), time-resolved PIV was implemented to investigate the effects of intake geometry on the in-cylinder flows and subsequent interactions with the spray. The experiments were conducted in a single-cylinder optically accessible GDI engine running in stratified mode. The engine had a spray-guided configuration with a hollow cone spray resulting from a piezo-actuated injector. Two modifications at the periphery of the intake port (S1, S2) were studied, as shown in Fig. 7.13.

The history of flow field evolution in the central tumble plane using time-resolved PIV helped determine the origin of flow structures and possible causes for their variations. The investigations were done for a stratified mixture prepared by triple injections in the late compression stroke. It was found that intake geometry variations significantly affected large-scale in-cylinder flow, causing deviations in tumble vortex position and strength. Further, the first spray injection was quite reproducible, while the second injection, which was mainly responsible for forming an ignitable mixture, exhibited significant fluctuations. Fluctuations in the spray shape were quantified by SH, which was defined as the vertical distance from the spray plume boundary to the cylinder head. Variations in SH of 2nd injection were correlated with regions showing large cycle-to-cycle flow field variations. However, the variations in 2nd spray were similar for both port geometries, despite substantial differences in their flow fields.

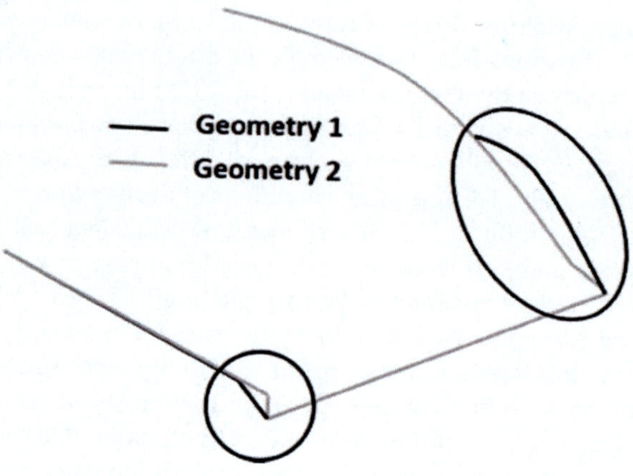

Fig. 7.13 Inner contours of geometric variations of the intake port (Stiehl et al. 2016)

This showed that spray fluctuations were insensitive to geometry changes. For both port geometries, the first injection-induced upward flow field near injector tip due to formation of a low-pressure zone within hollow spray cone. The tracing of the time history of the flow field for geometry G1 using correlation analysis revealed that variation in positions of large-scale tumble vortex from cycle-to-cycle during intake was a major reason behind 2nd spray fluctuations. In addition, turbulence induced by 1st spray injection sought to increase the impact of flow cyclic variations on 2nd spray fluctuations. Further, their correlation decreased in the central tumble plane while tracing backwards towards the intake process. This could be mainly because of the out-of-plane flow, which was highly prevalent in G2 geometry. Tomographic or multi-plane PIV can overcome this limitation of planar or 2D measurements.

Stratified operation in spray-guided engines involves close coupling between the fuel injection and the ignition events. The developing ignition kernel gets subjected to steep gradients of fuel concentration and flow velocity. Hence, in such conditions, the likeliness of variability from cycle to cycle is higher. These unfavourable fluctuations can impact the combustion quality because of misfires and partial burn cycles. Hence, engine control parameters should be optimized for stable and sustained stratified combustion. In this regard, in-cylinder flow evolution and its cyclic variability are critical factors to consider.

Further, the intensity and behaviour of intake flow can impact the turbulence and flow structures resulting from spray-air interactions, impacting the repeatability of fuel–air mixing. In a study by Zeng et al. (2016), intake generated swirl flow resulted in more repeatable and stronger vortex formation in the piston bowl because of the interactions of injected spray plumes from the multi-hole injectors than without swirl generated flows. The injection of liquid fuel spray redistributes the momentum of the gas phase such that it leads to a stronger and repeatable vortex. The formation of this vortex promoted effective combustion with fewer cyclic variations.

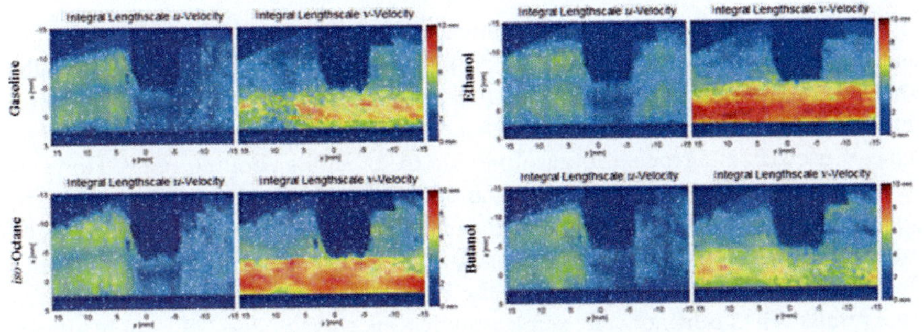

Fig. 7.14 Integral length scales in tumble plane for different fuels (Aleiferis and Behringer 2015)

Studies have also been attempted to categorize the flow characteristics for different kinds of fuel injection. Researchers characterized the in-cylinder flow in one such study and quantified the integral length scales associated with turbulence for different fuels (Aleiferis and Behringer 2015). They considered the effect of gasoline, isooctane, ethanol and butanol injection on the flow field at the time of ignition using PIV. A single-cylinder optical research engine having a side-mounted swirl-type DI injector was used for the study. Experiments were conducted for stoichiometric homogeneous mixture preparation (~60° aTDC) and at a fixed engine speed of 1500 RPM. For without injection runs, mean flow velocities of the order of 3–5 m/s and TKE between 9 and 15 m^2/s^2 were obtained for tumble plane from intake to exhaust. Two counter-rotating vortices were observed beneath the two intake valves in the swirling plane, having flow magnitude similar to that of the tumble.

However, fuel injection during intake didn't cause significant changes in the flow field at the time of ignition. For gasoline, flow velocities were 10% higher, while it was reduced by 30% for butanol and ethanol and isooctane were in-between these. Further, injecting the fuel shifted the highest TKE zone from the exhaust to the intake side. The integral length scales for both fuel-injected and without fuel-injected cycles came in the same order of magnitude for the tumbling plane while showed slightly higher values for injected cases in the swirling plane. Non-dimensionally, these scales were ~20% of the clearance height and 5–12% of the bore. Fuel types didn't result in large differences as average scales were in the range of 4–6 mm in both planes. Specifically, the largest scales were associated with ethanol and the smallest with butanol, as shown in Fig. 7.14.

7.2.3 *Fuel–Air Mixture Formation*

The most intense application of optical measurement in engine research is fuel–air mixing quantification. These are mainly performed by laser-induced fluorescence imaging or absorption techniques via the use of laser diodes. Traditional PLIF works by capturing fluorescence signals emitted by selected tracer upon getting illuminated. Tracers follow the path of fuel droplets, which have to be selected such that they are a close surrogate to the actual fuel and do not emit fluorescence. Otherwise, there would be interference of signals and may affect the gasoline results. Isooctane is used as a surrogate to gasoline due to its similar boiling point. However, isooctane, a single component fuel, cannot represent the actual evaporation characteristics of gasoline, a multicomponent fuel. In the GDI engine, the injected fuel is present in both liquid and vapour phases. Due to differences in emission intensity between liquid and vapour phases of the fuel, it becomes difficult to obtain accurate results. The laser-induced exciplex fluorescence technique (LIEF) uses filtering methods to separate the wavelength spectrum from both liquid and vapour phases to overcome this challenge. Various studies have shown its successful implementation and a detailed experimental setup procedure (Ipp et al. 2000; Kang and Kim 2003; Deschamps et al. 1999). Most studies based on LIF use dopant tracers representing similar evaporation characteristics to the test fuel. However, the real fuel comprises hundreds of distinct hydrocarbon components, with preferential evaporation.

The evaporation of lighter components takes place at different timing compared to heavier ones. For investigating preferential evaporation of multi-component gasoline, 2-tracer LIF is used in many studies. In a study by Bardi et al. (2019), 2-tracer LIF was used, and preferential evaporation was examined for ethanol-blended gasoline. It was observed that with a lower fraction of ethanol blending (E20), evaporation of ethanol took place earlier than iso-octane. This was expected due to the lower boiling point of ethanol. However, for a higher fraction of ethanol blending (E85), ethanol evaporation occurred after isooctane and n-pentane. Moreover, ethanol formed an azeotropic mixture with isooctane and n-pentane due to their higher chemical and physical affinity.

The stratification in wall-guided configuration is very difficult to optimize for a wide range of engine operations. This was mainly due to the shortage of time available for mixture preparation. It involved the transport of fuel in the vicinity of the spark plug via piston cavity. The fuel–air mixture distribution near the spark plug is a critical parameter of the study, as it determined the early flame development and following combustion behaviour. Hence it was crucial to investigate the equivalence ratio of the mixture near the spark plug at the time of ignition. In one such study by Deschamps et al. (1999), two techniques, (i) Combined Catalytic Hot Wire Probe and (ii) LIEF, were used to obtain the mixing characteristics in a stratified mixture. Experiments were conducted for different late injection timings (60–30° bTDC) and different lean mixtures with global Φ varying from 0.3 to 0.6. Both these techniques showed rich mixture cloud at the time of spark near the spark plug. However, it was observed that an increase in the injection duration did not affect the maximum equivalence ratio of the mixture in the vapour cloud. Instead, it caused an increase in the size of the rich vapour cloud. Lean mixture with late injections showed large cycle-to-cycle variations in LIEF measurement compared to homogeneous mixture preparation. Steeper and Stevens (2000) also performed a mixture preparation study using LIF for different injection timings in an optical spray-guided engine. They used gasoline fluorescence for qualitative visualization of vapour phase gasoline.

The measurements were done in a horizontal plane 13 mm below the spark plug at the time of the spark. The results showed an increasing degree of heterogeneity in vapour distribution with respect to the retarded injection timings. A similar trend was evident from the higher CoV of the light intensity distribution. In another study related to the effect of injector fouling, mixture formation was studied using PLIF in a spray-guided DISI optical engine using a fouled injector (Badawy et al. 2018). This study was conducted with isooctane as surrogate fuel and a 3-pentanone (3% w/w) dopant tracer. The PLIF images were captured at 30° bTDC, which was considered ignition timing in the combustion tests. The experiments were performed at a fixed engine speed of 1200 RPM under motored conditions and 150 bar FIP. The PLIF images obtained for choked and clean injectors at varying injection timings are shown in Fig. 7.15. The SOI of 320 and 280° bTDC resulted in a fairly homogenous mixture distribution having Φ in the range of 0.9–1.2. The spatial variations in fuel concentrations were minimum due to greater time availability for almost complete evaporation. Slightly retarding the injection timings to 240 and 200° bTDC resulted in higher inhomogeneity in the mixture ($\Phi \sim 0.9$–1.3), which was further enhanced in the choked injector ($\Phi \sim 1.4$–1.8).

These characteristics were the main reasons for higher particulate emissions. With later injection timings (160 and 120° bTDC), both injectors obtained substantially rich fuel concentrations on the left zone of ROI (near exhaust valves). This was mainly due to the bulk convection of unevaporated fuel in the tumble direction. For further investigating the cyclic variability in charge formation, COV of spatial light intensity distribution (SLID) was measured for each case. Evaluating this showed poor mixing characteristics for choked injectors due to the higher COV of SLID. Also, for both injectors, variations increased with retarded injection timings. Stratification in GDI engines is critical due to the transient nature of the flow field during the engine operation. Design parameters are required to be optimized for the formation of required mixture stratification at different engine operating conditions. Intake manifold geometry and piston shape play a key role in determining the flow pattern in the combustion chamber. The authors investigated different designs for understanding their effects on the mixture formation. In a study by Kang and Kim (2003), different piston shapes and intake flows were investigated for stratified DISI engines having centrally mounted swirl injectors. LIEF was used for obtaining the fuel distribution. Three piston shapes were considered for this study, having flat, bowl and re-entrant shapes. It was found that tumble flow resulted in faster evaporation than swirl flow for early injection timing (270° bTDC). For early injection, a tumble flow caused fuel vapour distribution in the lower region of the combustion chamber. In the case of late injection (90° bTDC), it caused the movement towards the intake valve side. For swirl flow, the mixture was mainly concentrated near the central region in the bowl piston while it was distributed towards the exhaust valve side in a flat and re-entrant piston. A re-entrant piston with injection timing of 60° bTDC was better for stratification due to the smaller bowl diameter. The likeliness of cycle-to-cycle variations was also high in the stratified mode of combustion in DISI engines. These CCVs were mainly due to differences in mixture formation and quality of the combustion cycle. Higher occurrence of partial combustion and misfire generally enhanced the CCV of the

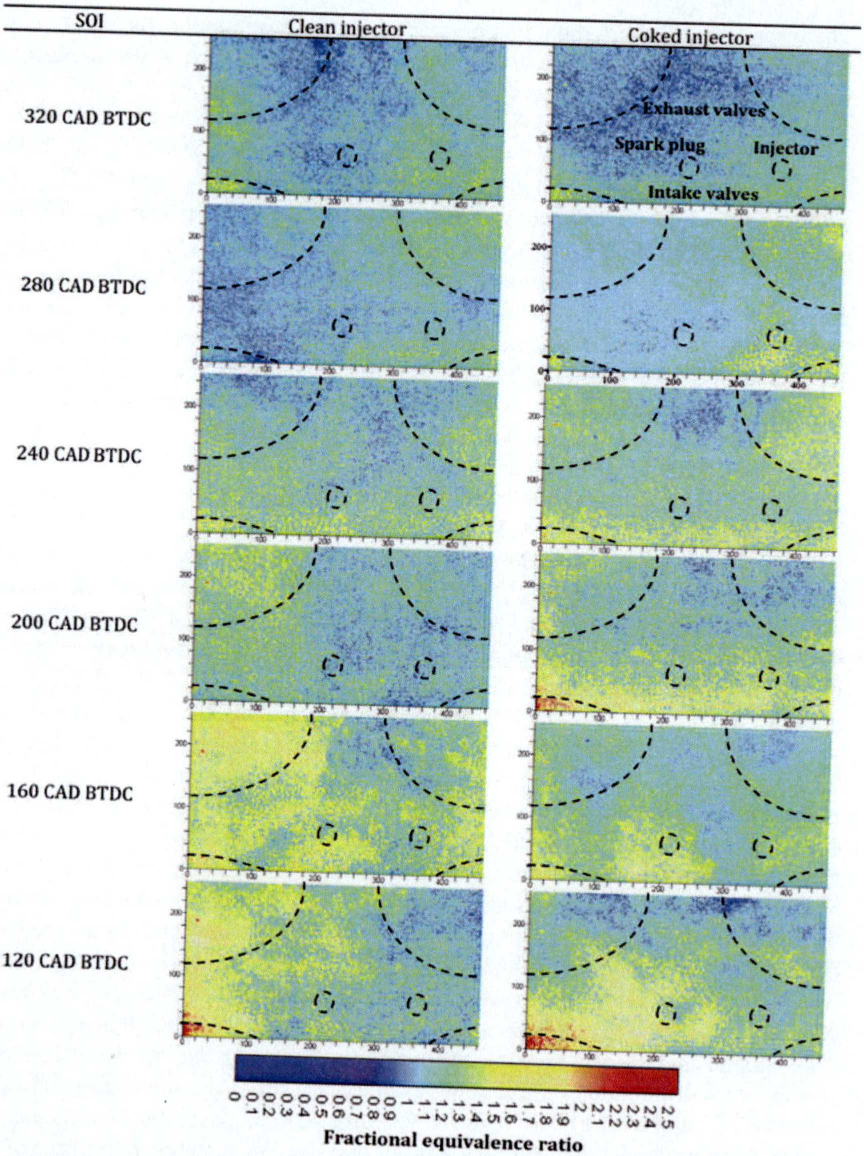

Fig. 7.15 Equivalence ratio distribution using PLIF for injection timing sweep for clean and choked injectors (Badawy et al. 2018)

imep. There is no universal reason for such poor combustion cycles. There could be multiple reasons, such as over-rich or over lean charge, inadequate flow velocities, or fluctuations in ignition energy. Researchers have widely used optical diagnostics for investigating the reasons for misfires and high CCVs. In one such study by Peterson et al. (2011), PIV and PLIF techniques were used for understanding the flow and fuel distribution at the time of ignition for partial burn and misfire cycles in a spray-guided stratified GDI engine operation. The results showed that these cycles were generally associated with lean mixtures and low flow velocities near the spark plug. However, there were well-burning cycles also having low flow velocity and equivalence ratio. Deeper investigations revealed that the ignitable mixture formed near the spark plug, but the flames failed to propagate towards the bowl with fuel-rich zones. Hence, the late arrival of the flame kernel and subsequent surrounding leaner conditions led to failed flame propagation.

Further investigations (Peterson et al. 2014) revealed that major factors responsible for slower flame development were the presence of lean mixtures, dilution of charge by EGR, and unfavourable convection of flame kernel away from the flammable charge due to upward flow. Managing flow to obtain desirable fuel–air mixing is important to enhance combustion stability in stratified DISI engines. Studies have found that intake-generated swirl is beneficial in spray-guided configurations for repeatable mixture formation during early frame growth (Zeng et al. 2016, 2015). However, on the other hand, a strong swirl has shown a detrimental effect (Drake et al. 2005). The authors also investigated the interactions of swirl and tumble dominated flow fields with sprays. These findings are described in our previous chapter (Kalwar and Agarwal 2021). In another study (Zeng et al. 2017), the authors evaluated its impact on gasoline vapour distribution using high-speed infrared (IR) imaging of C-H band emitting wavelengths. It was found that both swirl and tumble dominated flows led to asymmetric distribution of gasoline vapours in the later part of the compression stroke. These regions were also correlated with high soot luminosity. Symmetric and uniform distribution of fuel vapour was observed for the no-swirl case.

To limit the exhaust CO_2 emissions, downsizing of GDI engines is essential. At the same time, boosting is done to compensate for the power loss by engine downsizing. However, the performance of these engines is limited by increased knock or pre-ignition tendency. Researchers investigated Octane on demand (OOD) concept to resolve this, where two fuels with different RON are injected in the dual-injection mode. Base fuel with low octane numbers such as Naptha and the octane boosters such as ethanol is injected using optimized injection strategies to maintain the desired RON, depending on the engine operating conditions. Pilla et al. (2016) conducted experiments to evaluate the fuel concentration and mixing processes for three different dual-injection strategies (port + lateral), (port + Central) and (Central + lateral) using 2-colour PLIF. The measurements were done in motored engine conditions, having SOI at 260° bTDC.

The PLIF images were captured at 60° bTDC in the vertical plane. They reported that the injection strategies did not affect the fuel repartition in the combustion chamber. Hence it was mainly dependent on a single injection having negligible interaction between the spray plumes. The single injection from the central injector accumulated the fuel near the piston top and on the right side of the liner near the exhaust valves. For injection from the lateral injector, the fuel presence was mainly on the intake side of the liner and lower right zone near the walls. Direct fuel injection in the SI engine also benefits the charge cooling, an important parameter considered during mixture formation. It enhances volumetric efficiency and avoids preignition in the downsized engine. In a study by Attar et al. (2014), the thermometry technique was successfully demonstrated by tracer-based 2-line PLIF to evaluate the charge cooling due to fuel injection. The results obtained were in agreement with the calculations of temperature from the pressure data. The study suggested that both injection quantities and timings influenced the charge cooling. Charge cooling happens when the injected fuel evaporated by absorbing heat from the surrounding air. Hence, increasing the fuel injection quantity led to an increased cooling effect. Late injection timing (250° aTDC) reduced the in-cylinder temperature by a higher amount than earlier (90° bTDC). This was mainly due to the opening of the intake valve during the early injection. Hence, the enthalpy of vaporization allowed greater air mass to enter the cylinder. On the other hand, for later injection timings, the evaporation occurs for a fixed air mass as the inlet valve closes. The authors reported a 50% increase in the charge cooling when injection timing shifted from early to late injection for a low fuel injection quantity. Although for higher fuel injection quantity, the rise was only 10% due to the fuel impingement. Though in-cylinder temperature measurement (thermometry) is performed by various techniques such as LIF, Coherent anti-stokes Raman scattering (CARS), or Rayleigh scattering, these methods result in low accuracy. The authors demonstrated a laser-induced thermal grating system (LITGS) for more precise measurement of in-cylinder temperature in a range of 0.1–1% (Williams et al. 2014). This can also resolve the issue of high CCVs in the temperature measurements. LITGS uses signal's frequency, whereas other techniques use signal intensity. Hence, it remains unaffected by noise.

7.2.4 *Flame Evolution and Pollutant Formation*

Optical combustion and flame development investigations are generally performed using various techniques, such as natural flame luminosity, chemiluminescence imaging, emission spectroscopy, and LIF to detect plasma and unburnt charge characteristics, etc. In homogeneous charge preparation, combustion processes are similar to carburetted or port-injected SI engines. While for late injection, heterogeneous and stratified mixture presence near the spark plug offers difficulties in combustion due to highly rich or highly lean mixture. A highly turbulent flow can affect ignition stability. Hence, it is important to study the combustion in GDI engines as a possible combination of premixed, partially premixed, or even non-premixed charge. The following paragraphs discuss research studies that investigated the effect of fuel injection parameters on combustion and flame characteristics via high-speed imaging. In a study by Lee et al. (2020), combustion images for various fuel injection

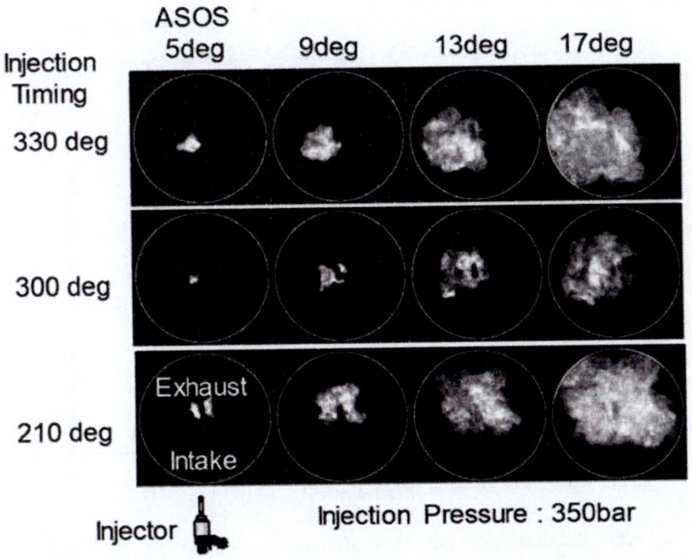

Fig. 7.16 Images of flame propagation for various injection timings (Lee et al. 2020)

timings were obtained from an optical GDI engine with the side-mounted injector. Initial flame growth was slower for early injection timing (~300° bTDC) than in the other cases, as seen in Fig. 7.16. This slower flame development was mainly due to weaker in-cylinder flows due to injected fuel and the in-cylinder airflow in the opposite directions. Further, the flame center movement was obtained for different injection timings, indicating bulk in-cylinder airflow. The movement of the flame center in the direction of fuel injection took place as the injection timing was retarded, as shown in Fig. 7.17.

In another study, it was observed that compared to early injection, late injection showed faster initial flame propagation due to presence of a locally rich mixture around the spark plug (Clark et al. 2014). Hence, in addition to global equivalence ratio or mixture properties, local conditions also play a significant role in the subsequent combustion processes. They also reported that advancing the ignition timing increased the flame speed. At early ignition timing, higher piston speed maintained a higher TKE of the ambient airflow. The overall combustion process of the stratified mixture was observed to have both lean premixed and mixing controlled combustion (Oh and Bae 2013). Non-luminous flames for early injections mainly resulted from over-mixing of charge, while sooty luminous flames for late injections were due to under mixing of charge. In a study by Song et al. (2018), it was found that increasing the FIP from 50 to 500 bar reduced the fuel film over the piston for injection timing of 210° bTDC, as shown in Fig. 7.18. They reported that at a lower FIP of 5 MPa, a change of injection timing from 330° bTDC to 270° bTDC resulted in a considerable reduction in particulate emissions. In contrast, at 50 MPa FIP, this change was negligible.

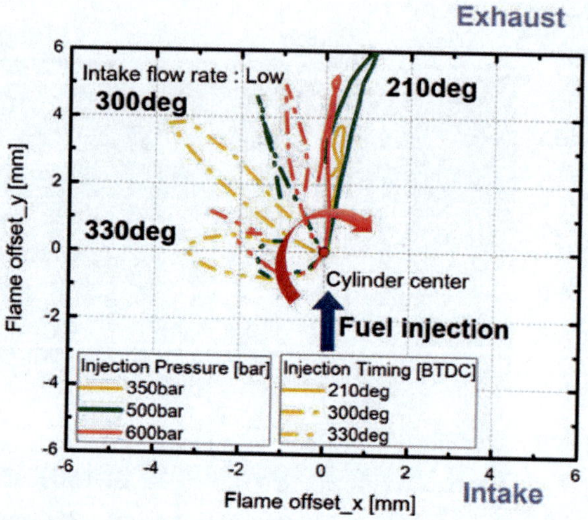

Fig. 7.17 Variations of flame center travel for different injection timings (Lee et al. 2020)

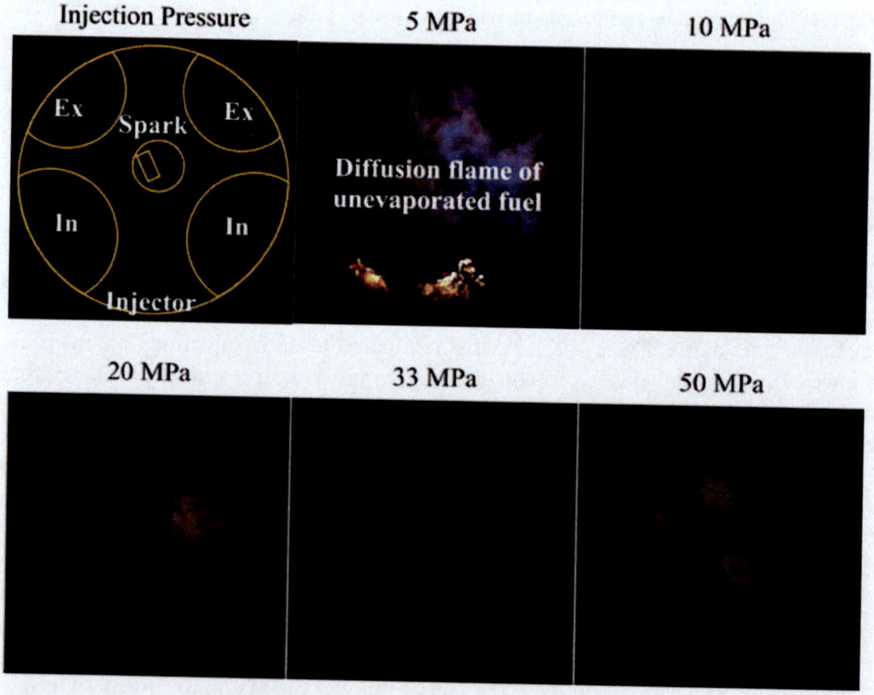

Fig. 7.18 Flame images for different FIP after 52° SOS (Song et al. 2018)

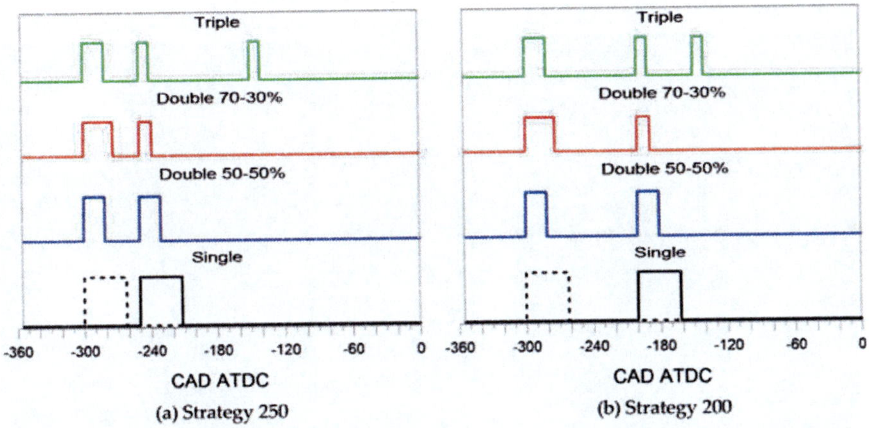

Fig. 7.19 Details of injection strategies adopted w.r.t. 2nd injection position. **a** 250° bTDC, **b** 200° bTDC (Martinez et al. 2019)

Injection strategies are important tools that define mixture formation and optimize the engine for higher performance with lower emissions. Several authors investigated multiple injection strategies. In a study by Martinez et al. (2019), the effect of single, double and triple injections was investigated for different split ratios and injection facing the wall guided optical engine, as shown in Fig. 7.19. and soot emissions obtained with the splitting of injections.

Figure 7.20 shows flame images from the double injection strategy. It can be concluded that splitting the injection with retarded second injection resulted in a lower luminosity than baseline early injection. However, with the late second injection, flame growth was delayed for both single and double injections. This shows that the combustion is improved for the split cases. Similarly, triple injection also exhibited similar behaviour. Overall, double injection with a 70:30 split ratio gave the best results. The authors concluded that flame propagation became less sensitive to injection facing than a single injection with split injection. The Heywood circularity factor (HCF) was obtained, which indicated distortion in the flame shape. Figure 7.21 shows that the HCF increased with retarded injection facing and decreased with multiple injections. Also, single injection led to the preferential direction of flame propagation towards the exhaust valve, while, for multiple injections, flame propagation was more uniform. Hence, these observations presented the basis for lower cyclic variations.

In another study, the authors investigated the effect of injection strategies on flame characteristics and propagation for a side-mounted injector equipped GDI engine (Song et al. 2015). The authors analyzed a single injection strategy with different injection timings during the intake stroke and a double injection with a short second injection at the end of the compression stroke. Figure 7.22 shows the flame images for the single injection strategy. It was observed that flame speed slowed down with the advancing injection timings.

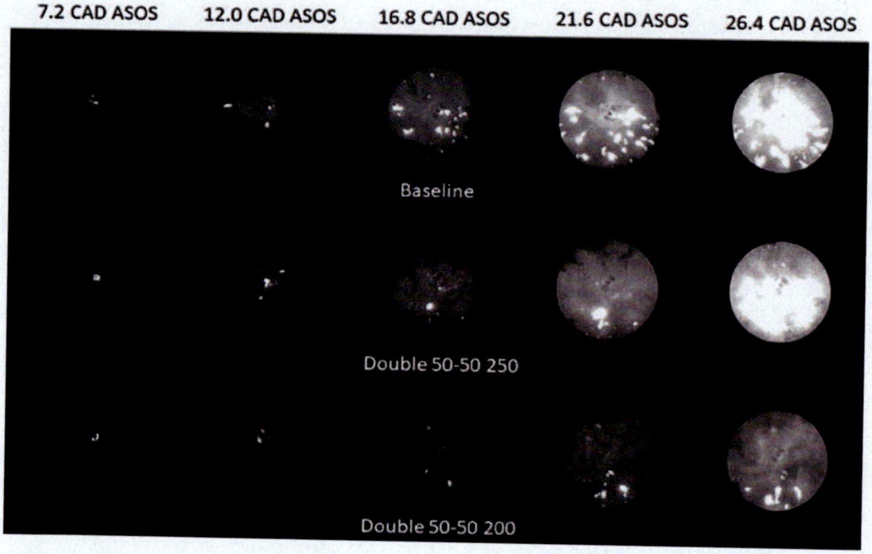

Fig. 7.20 Flame images for single (baseline) and double injection strategy (Martinez et al. 2019)

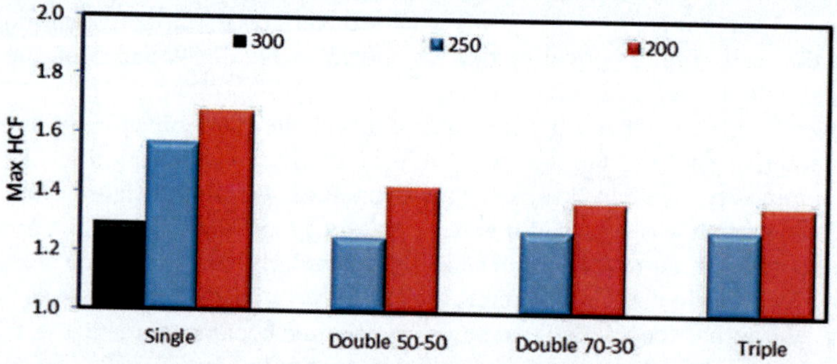

Fig. 7.21 Heywood Circularity Factor (HCF) for different injection strategies (Martinez et al. 2019)

Late injection timings led to higher peak pressure and faster flame propagation due to enhanced in-cylinder turbulence. Injection timing of 180° aTDC resulted in a flame speed of 20 m/s (55 times that of gasoline laminar burning speed), while 0° aTDC injection timing resulted in the slowest combustion with the flame speed of 10 m/s. 80% of fuel was injected during the intake stroke (~45° bTDC) in the double-injection strategy, while 20% was injected later at different timing. From Fig. 7.23, it can be observed that injection timing of 35° and 45° bTDC resulted in similar flame characteristics.

For both injection timings, bright yellow diffusion flames were observed during early frame growth due to the combustion of liquid fuels. Further flames interacted with fuel impinged on the piston, which was observed as a pool fire. Piston wetting

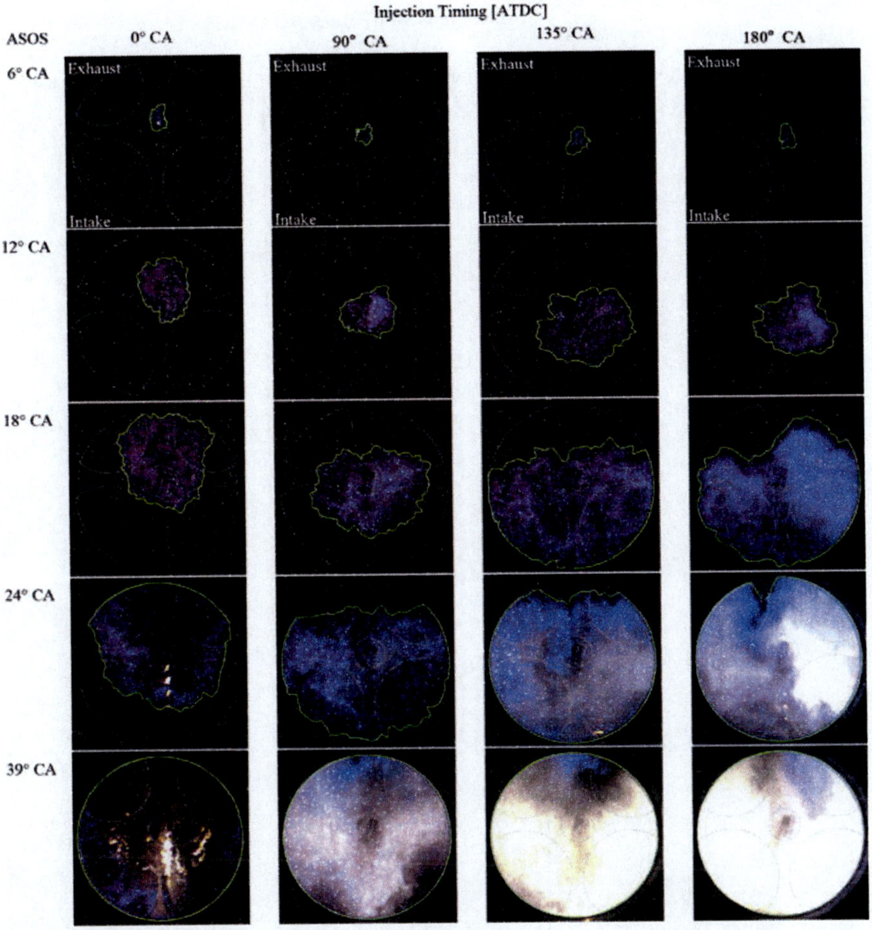

Fig. 7.22 Flame propagation images for single injection strategy having different injection timings (Song et al. 2015)

is unavoidable in the wall-guided configuration, and it is a major source of soot formation. However, 50° bTDC injection timing showed very slow flame propagation due to lack of combustible mixture. These observations correlated with the peak combustion pressure as 35° and 45° bTDC resulted in higher peak pressure than 50° bTDC. Flames propagated towards the exhaust valve for early injection, while the flames moved towards the intake valve for late injections. These were mainly dependent on the presence of combustible mixture.

Similar observations were reported in the study conducted by Clark et al. (2016). They found that double injections resulted in more spherical flame while single injection led to elliptical flames. The eccentricity of the flame front was not affected by the equivalence ratio variations. Close alignment between the semi-major axes of

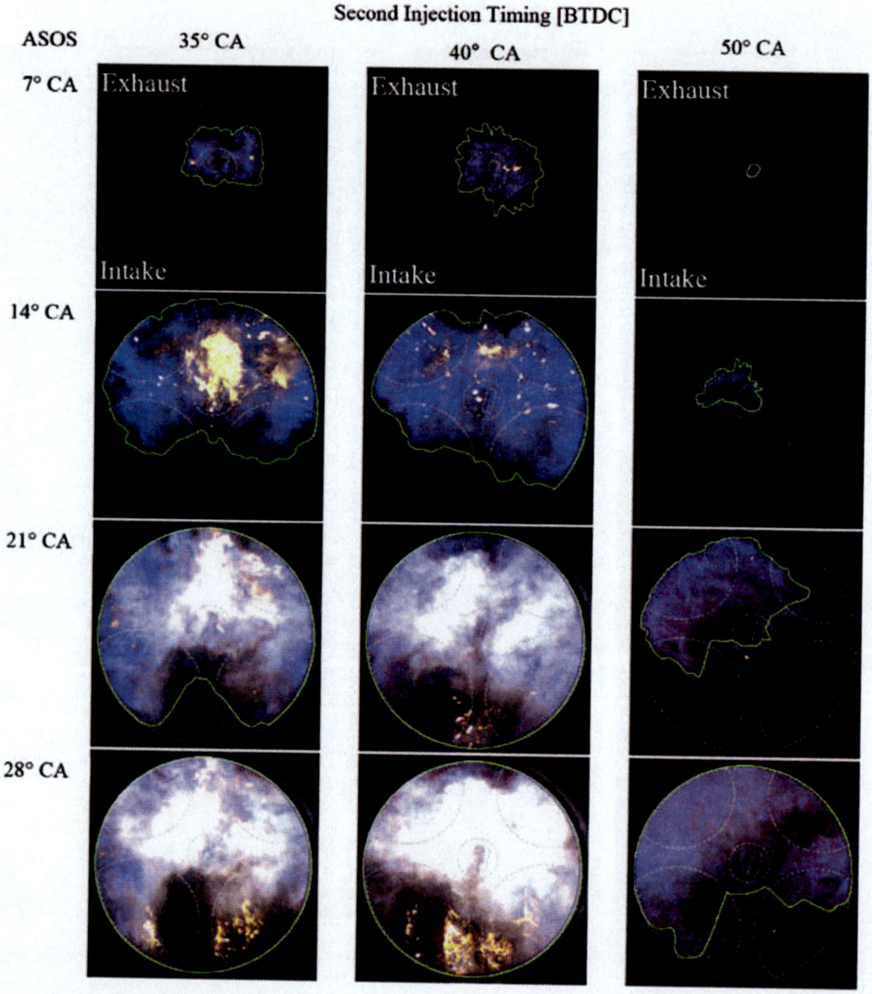

Fig. 7.23 Flame propagation images for double-injection strategy having different 2nd injection timings (Song et al. 2015)

the elliptical-shaped flames and the fuel jet path in the direction of the tumble was observed for both strategies. Wrinkling of the flame front was observed for both injection strategies, which implied that wrinkling was mainly an outcome of turbulence effects and was not affected by the richness or leanness of the mixture. Regardless of the second injection was positively correlated with the eccentricity of flames due to decreased charge homogeneity (Clark et al. 2017). Marseglia et al. (2017) observed that the main flame front exhibited asymmetric and heart-shaped geometry. For single injection, the concavity of flame front was towards the intake valve side, while that for split injection, it was towards the exhaust valve side. The highest knock occurrences were concurrent with the zones having residual charge susceptible to auto-ignition before consuming the flame front. CFD simulation results supported the observations from optical combustion engine experiments. TKE distribution was higher for single injection, asymmetrically, along with the flame front propagation. In addition, equivalence ratio distribution also guided the flame speed and flame movement. Overall, the split injection was advantageous in reducing cyclic variations and knock propensity, improve combustion efficiency, and reduce undesirable pollutants.

Lean mixture combustion in DISI engines is widely researched due to its thermal efficiency and fuel economy benefits. Its advantages are mainly obtained during part-load conditions due to reduced pumping losses. However, leaning the mixture is limited by acceptable cycle-to-cycle variations due to combustion instability. Flow control for proper fluid motion can be beneficial in compensating the reduced mixture reactivity by enhanced turbulence level. Stratification of the mixture by late injection offers combustion sustainability at the overall leaner mixtures. Though, it leads to diffusion flames due to wall impingement, resulting in higher PM emissions. For greater insights into lean mixture combustion and their flame characteristics, Irimescu et al. (2018) conducted flame imaging for lean mixtures at different equivalence ratios in a wall-guided optical engine. This provided more valuable information about the spatial distribution of burned species and interactions with the in-cylinder airflow. They also analyzed the effect of the start of injection and engine speed on the combustion characteristics. The crankshaft rotational speed of 1000 rpm allowed maximum dilution ($\lambda \sim 1.6$), and 2000 rpm allowed the least dilution. Results showed that flames were generally directed towards the exhaust valves with similar overall flame intensity for stoichiometric mixture at different engine speeds. Higher wall temperature region towards that part along with tumble could have led to this. Slower flame evolution ($\lambda \sim 1.3$) with inherent offset for the engine speed showed a prolonged initial phase as the underlined factor. Although, combustion velocity was quite similar in fully established turbulent flame propagation. Higher luminous intensities were observed during late combustion, attributed to chemiluminescence of premixed combustion and soot particles formation during diffusion-controlled combustion. It was reported that diffusion flames mainly originated from two sources. First, from the oxidation of liquid fuel film over the piston, the second was associated with the unburned gas returning from the piston top (jet-like flames). Diffusion flames were for all cases, where luminosity at the intake region was quite repeatable. This could be due to the presence of fuel impingement from a vertical plume from the injector. Speed of 2000 rpm led to more distributed intensity due to the tumbling effect. However, the time dependence for both cases was different, which affected the chemical kinetics. The liquid film over the piston was consumed by propagating a premixed combustion flame for the first case.

In contrast, flame propagation was almost over for the second case when the unburned gas returned from the top land region. Hence, the latter type of flames were found near the exhaust valve side with lesser repeatability. Also, the diffusive flame area for $\lambda \sim 1.3$ was smaller than for $\lambda \sim 1$, which correlated well with the smoke opacity. For further details about the chemical nature of reactive flame species, spectroscopic analysis was done at the intake valve and exhaust valve side.

At 60° CA after ignition, detection of OH radicals (~310 nm) indicated fuel oxidation even during late combustion. However, their presence was found where diffusive flames are absent, which could be associated with a high-temperature zone in the premixed combustion. At 70° CA after ignition, high wavelength emissions were observed for both locations (2 and 7), which indicated the presence of a carbonaceous structure. The engine speed of 1000 rpm recorded higher soot compared to 2000 rpm. For different injection phasing (340°, 300°, 260°), it was observed that both early and late injection timing resulted in the higher displacement of flames. These were mainly a result of charge stratification and flow motion generated due to interactions of spray jets. 300° CA was found to be optimum with a reduced flame shift. The authors concluded that the fuel injection timing for lean limit would be optimized for stoichiometric combustion. Martinez et al. (2017) investigated flame characteristics of fuel–air mixture with different air–fuel ratios (from $\lambda = 1$ to $\lambda = 1.6$) in a wall-guided GDI engine with early injection. The engine operated at $\lambda \sim 1.6$ with acceptable COV_{imep} slightly greater than 5%. Leaning the mixture prolonged the flame evolution process with overall lower flame intensity. The reacting species obtained from flame spectroscopy such as OH, CO_2, O, and soot spectroscopy signals were in lower concentrations for the mixture having a higher air–fuel ratio. Thus, lean mixtures resulted in reduced polluting species and ensured much less particle formation. Also, flame circularity reduced as the equivalence ratio was decreased. Strongly induced fluid flow such as swirl and tumble effectively extends the lean limit or reduces combustion variability. Studies have shown their role in improving engine performance (Zhang et al. 2014; Matsuda et al. 2017). However, these advantages are obtained at the cost of reduced volumetric efficiency in a naturally aspirated engine. Boosting along with organized flow can result in desired engine performance. In a study performed by Yang et al. (2019), different tumble deflectors were incorporated in the intake manifold, which augmented the tumble ratio of fluid motion from 0.5 to 1.5 and 2.2 in spray-guided DISI engines. The study reported an increase in flame development rate by 13.9% and 34.5%, increasing the tumble ratio of 1.5 and 2.2. The TKE of in-cylinder flow increased with a higher tumble, enhancing the fuel–air mixing and intensifying mass transfer between the burnt and unburnt zones. A more blueish flame was observed for a higher tumble. The propensity of yellow flames reduced and delayed, as shown in Fig. 7.24. Calculation of presence probability images of flames showed that for a tumble ratio of 0.5, the flame development was slower with higher cyclic variations. Repeatability and faster flame propagation were observed for a tumble ratio of 2.2. Hence, optimizing the tumble ratio for improved engine performance and reduced cyclic variations can be effective for GDI engines, especially for leaner mixture combustion.

The highly organized flows were also investigated to reduce the variability in the in-cylinder conditions and provide good combustion stability. For stratified operation in a spray-guided engine, fuel injection and ignition are closely coupled. Therefore, the high flow velocities and turbulence could affect the flames. Further, the presence of steep gradients in the fuel concentration and flow velocities may induce unfavourable conditions for early flame growth. In a study by Zeng et al. (2016), the deterioration of the combustion performance of stratified mixture caused by increased

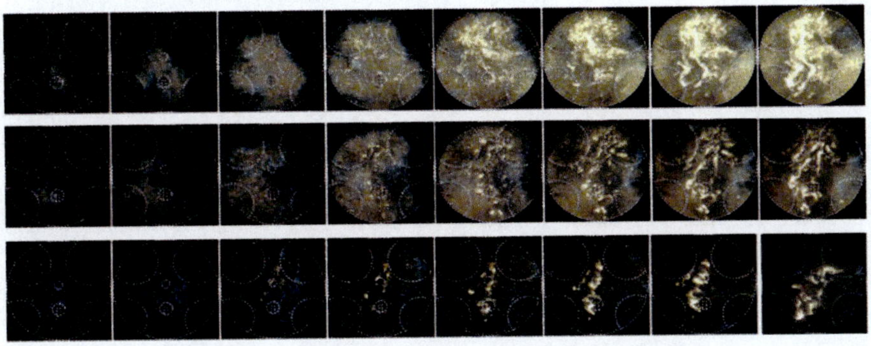

Fig. 7.24 Combustion images for different tumble ratios (Yang et al. 2019)

speed was prevented by swirl enhanced intake airflow. The authors mentioned that the poor burning cycles were dependent on variability in the early flame growth. The regression of burning rate in the later combustion period was also a factor affecting the combustion cycles. The spray-flow interaction analysis revealed that spray from a multi-hole injector redistributed the swirl-enhanced flow momentum and better-centred vortex formation with reduced variability. The flame imaging results showed that the kernel randomly stretched without swirl based on the ignition of two spray plumes stranding on either side of the spark plug. While, in case of enhanced swirl, flame kernel repeatedly convected towards the specific spray plume, resulting in repeatable ignition. Hung et al. (2014) conducted a study to examine the effect of swirl (swirl ratio of 0.55 and 5.68) on the early flame development in the spray-guided DISI engine. Highly directed swirl flow (5.68) advanced the initial frame kernel origin by 4° CA compared to low swirl. Flame kernel propagation speed during early flame growth was enhanced by two times w.r.t. the low swirl case. Hence, peak pressure location also advanced, which improved the engine power output. The location of flame centroid was also highly repeatable for the higher swirl case, which resulted in lower cyclic variations in the combustion characteristics.

Though direct injection of fuel offers the advantage of in-cylinder charge cooling, which helps reduce knock compared to port-injected engines, the complexity involved in proper fuel–air mixture distribution, charge inhomogeneity, fuel impingement and boosting leads to favourable conditions for engine knock. With the help of optical diagnostics, the in-situ processes responsible for knocking and its relationship with engine variables can be understood. In a study by Catapano et al. (2017), UV–visible flame imaging was done for a high spatial and temporal description of knocking characteristics in a multi-cylinder turbocharged wall-guided GDI engine. It was observed that split-injection strategies were effective in suppressing the knocking. Flame visualization allowed tracking of the normal flame propagation, end-gas auto ignition, and residual flames under knock, which correlated well with the knocking parameters obtained via in-cylinder pressure measurements. Close correlations were found between the pressure oscillations due to knock onset

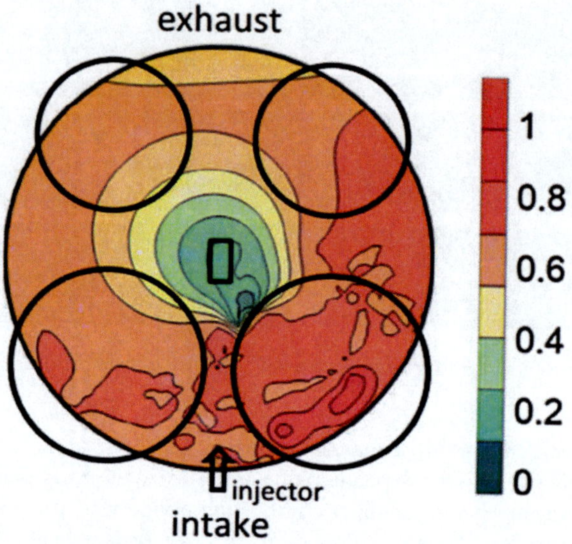

Fig. 7.25 Probability maps for secondary flames (Merola et al. 2016a)

and negative flame front curvature. The authors worked on a novel way of determining knock onset's temporal and spatial location by analyzing centroid luminosity dynamics. Merola et al. (2016a) investigated the hot-spot induced pre-ignition and secondary flame analysis leading to abnormal combustion in wall-guided GDI engines. Knocking cycles were observed with the secondary flames near the cylinder walls. The knock intensity was found to be correlated with the area of secondary flames. In Fig. 7.25, it can be seen that the higher probability regions of secondary flames near the cylinder wall were towards the injector location. It inferred that the fuel deposits were mainly responsible for abnormal combustion, as the engine was devoid of oil droplets. Further, the initiation of abnormal combustion from hot-spot induced pre-ignition was less evident.

The engine researchers further need to consider the diversification of fuels since bio-components can reduce global CO_2 emissions. However, the biofuels differ in chemical and physical properties, affecting the mixture formation and combustion characteristics. Hence, it is essential to understand the in-cylinder combustion phenomena for different biofuels to manage fuel flexibility and fuel utilization better. Catapano et al. (2013) optically investigated the combustion process of ethanol and gasoline fuelling in a small DISI engine having a side-mounted injector. Chemiluminescence imaging was done via ICCD camera for flame visualization, and soot formation analysis was done using a CCD colour camera by 2-colour pyrometry. The intensity of diffusive flames lowered for higher FIP due to improved mixture homogeneity. Ethanol combustion resulted in considerably lower diffusive flames. The KL factor obtained for soot analysis was lower for ethanol combustion owing to its chemical properties. In another study (Catapano et al. 2016), authors attempted endoscopic imaging of a multi-cylinder turbocharged engine. E50 and E85 showed higher integral flame luminosity during the early combustion phase due to enhanced flame speed. However, gasoline and E10 exhibited higher diffusive flames in the later phase, when the flame front devoured the piston fuel film. These observations correlated well with lower UHC and smoke emissions for E50 and E85. Quicker fuel vaporisation and higher oxygen content in ethanol-blended gasoline improved homogenization, increased flame speed, and reduced soot emissions. In a study by Sementa et al. (2012), it was found that in ethanol combustion, diffusive flames were less visible, which was also evident from low soot obtained from the 2-colour pyrometry. Apart from methanol and ethanol, butanol was also investigated as a fuel of interest due to recent developments in its bio-based production pathways. Its fuel properties are comparable to gasoline, such as calorific value and corrosiveness. However, butanol has four isomers, and their molecular structure could affect combustion. In a study by Han et al. (2020), different butanol isomers, namely n-butanol, sec-butanol, iso-butanol and tert-butanol, were blended with gasoline surrogates, PRF/TPRF in a 3:7 ratio and investigated for flame characteristics. Butanol addition resulted in slower flame development as compared to gasoline surrogates due to longer ignition delay. Higher viscosity and surface tension of butanol resulted in larger spray droplets which slowed down the fuel–air mixing. Higher LHV also led to reduced mixture reactivity. Among all isomers, n-butanol showed the fastest flame propagation due to a straight-chain structure, while tert-butanol exhibited the slowest flame. However, their PN emission reduction capability ranked in order of sec- > iso- > n- > tert-. Merola et al. (2016b) performed spatially resolved UV–visible flame visualization for butanol as a complete gasoline replacement in a wall-guided optically accessible GDI engine. Both fuels showed macro-distortion in flames during throttled operation. However, at WOT, butanol flames showed higher Heywood Factor values at both stoichiometric and lean operation. Gasoline flames were more luminous, with continuous strong emission between 300 and 550 nm wavelength, including OH and CH bands. Higher fluctuation in intensity luminous integral signals for gasoline flames indicated higher cyclic variations. Koupaie et al. (2019) reported that faster combustion was observed with alcohol addition. Flames tend to propagate towards the exhaust valve side. Placing the spark plug slightly biased towards the intake valve side could resolve the issue of exhaust migration characteristics and symmetric flame development. However, the maximization of inlet valve size could be a possible constraint in this.

Natural gas-fuelled engines also get attention due to lower carbonaceous emissions despite reduced engine power output and thermal efficiency. Though, late direct injection of natural gas offers favourable conditions for improved engine performance. In one such study by Chen et al. (2019), methane combustion in leaner conditions was investigated for varying injection timings. The latest injection timing (~100° CA) reported the highest flame propagation speed (increment of 50%). Late injection promoted in-cylinder turbulence and volumetric efficiency (due to injection at IVC), resulting in faster combustion. Di Iorio et al. (2016) investigated the effect of hydrogen enrichment of methane on combustion characteristics. Hydrogen addition enhanced the flame speed due to the higher diffusivity of hydrogen. The improved mixing enhanced the homogeneity, leading to uniform flame propagation. Hence, lower carbonaceous emissions were reported for hydrogen-methane blends. Di Iorio et al. (2016) performed flame analysis for different fuels, namely methane, ethanol, butanol, isooctane and gasoline, in a spray-guided optical DISI engine. Engine encompasses a multi-hole injector with a FIP of 150 bar. Experiments were done at a fixed engine speed of 1500 RPM and 0.5 bar intake pressure. For methane, the injection was done in the intake plenum to achieve a homogeneous mixture, which was considered the baseline for other fuels injected earlier into the cylinder. Effect of equivalence ratio ($\varphi = 1$ and 0.83) and coolant temperature (20 and 90 °C) were also investigated for flame characteristics. The study reported that post ignition at $\varphi = 1$, flame growth was the fastest for ethanol, followed by butanol, gasoline and isooctane. Methane reported slower flame growth during the initial flame development stage. However, after 20–25° aSOI, its flame grew at a speed of 10–13 m/s, similar to alcohols. Trends of flame speed were similar at both coolant temperatures, where 90 °C showed 5–10% higher values. Calculations of turbulent flame speeds revealed that the burning velocity of butanol was ~8% lower than ethanol but ~5% faster than gasoline. The flame centroid path for all test fuels was similar towards the exhaust valves. Global flame centroid displacement speed was common and fairly stable for all test fuels of the order of 2 m/s. However, there were differences in flame texture among different test fuels. Alcohols and methane flames showed comparatively lower luminosities than isooctane and gasoline. In another study (Aleiferis and Behringer 2015), authors attempted chemiluminescence imaging and double-pulsed laser-illuminated flame tomography to characterize flame characteristics and quantify the flame front topology, respectively. Variations in flame speed for different test fuels were similar to the previous study. However, flame speed obtained from the flame tomography was independent of equivalent flame radii, while from the chemiluminescence imaging, flame speeds were higher for the higher flame radii. Flame roundness was of the order of ~10–15%, with ethanol having the highest, followed by butanol, gasoline and iso-octane. This showed that flames from hydrocarbon fuels exhibited a higher level of distortion and multiple cluster levels than alcohols.

Despite GDI engine technology offering many advantages in terms of engine performance, it results in high particulate emissions. The formation of ultrafine or fine particles in the exhaust becomes unavoidable due to the availability of locally rich zones due to incomplete fuel–air mixing, stratified mixture, and fuel film formation on the combustion chamber walls. Researchers focus on reducing particulate emissions from the engine as it poses a serious risk to the environment and human health. Hence, it is important to understand the soot formation mechanism in greater detail. Several researchers have studied the in-cylinder soot formation process via optical diagnostics techniques such as the 2-colour pyrometry (2C), LII and Laser Extinction Method (LEM) (Pastor et al. 2016; Wang et al. 2018). In one such study by Xie et al. (2021), 2-colour LII was applied to evaluate soot volume fraction, soot evolution, and distribution in a single-cylinder optical GDI engine having a side-mounted injector. It was found that with an increase in the fuel injection quantity, soot volume fraction in a measured plane increased and then decreased. It also led to increased non-uniformity of soot distribution. The soot formed peaked at around 42° CA aTDC, and afterwards, it decreased due to the oxidation. A dominant factor behind soot formation was the pool fire which was evident till 52° CA aTDC. Later, the produced soot gathered around a 10 mm plane. The produced soot reached a balanced state after 82° CA aTDC and was distributed throughout the combustion chamber following the in-cylinder flows. In a study by de Francqueville et al. (2010), soot volume fraction was obtained using a coupled application of LII and LEM in a stratified spray-guided combustion system. The soot levels were measured in a vertical plane for different injection strategies and EGR. Double injections decreased the soot while EGR promoted the soot formation. The results supported a commonly understood hypothesis. Rapid formation of soot took place with combustion. The fuel-rich zones were mainly responsible for soot production. Soot oxidation and soot formation were competing phenomenon occurring during the expansion stroke. There was a sharp reduction in soot due to oxidants such as OH radicals. The oxidation process was detected significantly till 90° aTDC, after which the in-cylinder temperature became too low to sustain these reactions. Hence, air–fuel mixing, local fuel-rich zones and oxidant concentrations are the key factors determining the soot formation. Potenza et al. (2020) demonstrated a neural network two colours (NNTC) method for quantifying the soot concentration in a GDI engine. The values obtained from NNTC were in good agreement with exhaust soot measurement. Apart from this, several studies used the KL factor and 2-colour method to determine the soot concentration. The study found that SOI variations had a profound effect on the soot concentration compared to rail pressure variations.

Hence, from the above discussion, it can be summarised that optical diagnostics techniques have immensely contributed to understanding different complex in-cylinder phenomenon and their interactions in very short time scales in the DISI engines. Although implementing these tools helps obtain optimized engine design features and control parameters in a wide range of engine operating conditions, several scarcely understood or completely unknown aspects of the in-cylinder processes remained. Further development of these diagnostic tools will enable the scientific community to utilize their potential in developing advanced GDI engines beyond the current limitations.

7.3 Summary and Way-Forward

Optical diagnostics is a well-established research tool for advanced combustion development. Future advancements require developing more sophisticated high-speed imaging capabilities to study coupled, non-linear processes and generate accurate data for critical model validation. Demand for multiple signal acquisitions and their synchronization with optical instruments is bound to increase in future. Simultaneous deployment of multiple optical tools is of great interest for researchers to understand the coupled and complex relationships in engine processes. Greater focus is required on generating time-resolved three-dimensional engine data. Data acquisition and processing systems would have to be more powerful to handle an enormous amount of data acquired during experimental measurements. Multiphase flow measurement capabilities need to be further improved to characterize spray-flow interactions and processes involved in the formation and destruction of soot. In-situ measuring equipment working in high pressure and temperature conditions need to be improved. More work is required for higher resolution measurements for studying the liquid–gas interactions and related surface properties measurements. More work should focus more on developing endoscopic tools than research studies, which would boost their availability and capability. Optical diagnostics has been a crucial component in developing superior engines and would continue to play an increasingly more important role. A growing interest in direct injection technologies has resulted in an increased number of in-cylinder investigations. Phenomenons such as multiphase flow, spray-wall impingement, spray-flow interactions, charge stratification, partially premixed combustion and soot formation that were not of much importance in port-injected SI engines have become very important in GDI engines. The application of optical diagnostic tools to investigate the effect of engine control parameters on the in-cylinder flows, mixture formation, and subsequent combustion processes would help optimise the engine for obtaining superior performance and reducing emissions.

References

Aleiferis PG, Behringer MK (2015) Flame front analysis of ethanol, butanol, iso-octane and gasoline in a spark-ignition engine using laser tomography and integral length scale measurements. Combust Flame 162(12):4533–4552

Attar MA, Herfatmanesh MR, Zhao H, Cairns A (2014) Experimental investigation of direct injection charge cooling in optical GDI engine using tracer-based PLIF technique. Exp Thermal Fluid Sci 59:96–108

Badawy T, Attar MA, Xu H, Ghafourian A (2018) Assessment of gasoline direct injector fouling effects on fuel injection, engine performance and emissions. Appl Energy 220:351–374

Bao Y, Chan QN, Kook S, Hawkes E (2014) A comparative analysis on the spray penetration of ethanol, gasoline and iso-octane fuel in a spark-ignition direct-injection engine (No. 2014-01-1413). SAE technical paper

Bardi M, Di Lella A, Bruneaux G (2019) A novel approach for quantitative measurements of preferential evaporation of fuel by means of two-tracer laser-induced fluorescence. Fuel 239:521–533

Catapano F, Sementa P, Vaglieco BM (2013) Optical characterization of bio-ethanol injection and combustion in a small DISI engine for two wheels vehicles. Fuel 106:651–666

Catapano F, Sementa P, Vaglieco BM (2016) Air-fuel mixing and combustion behaviour of gasoline-ethanol blends in a GDI wall-guided turbocharged multi-cylinder optical engine. Renew Energy 96:319–332

Catapano F, Sementa P, Vaglieco BM (2017) Characterization of knock tendency and onset in a GDI engine by means of conventional measurements and a non-conventional flame dynamics optical analysis. SAE Int J Engines 10(5):2439–2450

Chan QN, Bao Y, Kook S (2014) Effects of injection pressure on the structural transformation of flash-boiling sprays of gasoline and ethanol in a spark-ignition direct-injection (SIDI) engine. Fuel 130:228–240

Chen H, Lillo PM, Sick V (2016) Three-dimensional spray–flow interaction in a spark-ignition direct-injection engine. Int J Engine Res 17(1):129–138

Chen L, Wei H, Zhang R, Pan J, Zhou L, Liu C (2019) Effects of late injection on lean combustion characteristics of methane in a high compression ratio optical engine. Fuel 255:115718

Clark LG, Kook S (2018) Correlation of spatial and temporal filtering methods for turbulence quantification in spark-ignition direct-injection (SIDI) engine flows. Flow Turbul Combust 101(1):161–189

Clark LG, Kook S, Chan QN, Hawkes E (2018) The effect of fuel injection timing on in-cylinder flow and combustion performance in a spark-ignition direct-injection (SIDI) engine using particle image velocimetry (PIV). Flow Turbul Combust 101(1):191–218

Clark LG, Kook S, Chan QN, Hawkes ER (2014) Effects of injection timing and spark timing on flame propagation in an optically accessible spark-ignition direct-injection (SIDI) engine. In: 19th Australasian fluid mechanics conference, pp 350–356

Clark LG, Kook S, Chan QN, Hawkes ER (2016) Multiple injection strategy investigation for well-mixed operation in an optical wall-guided spark-ignition direct-injection (WG-SIDI) engine through flame shape analysis (No. 2016-01-2162). SAE technical paper

Clark LG, Kook S, Chan QN, Hawkes ER (2017) Influence of injection timing for split-injection strategies on well-mixed high-load combustion performance in an optically accessible Spark-Ignition Direct-Injection (SIDI) engine (No. 2017-01-0657). SAE technical paper

de Francqueville L, Bruneaux G, Thirouard B (2010) Soot volume fraction measurements in a gasoline direct injection engine by combined laser-induced incandescence and laser extinction method. SAE Int J Engines 3(1):163–182

Deschamps B, Ricordeau V, Depussay E, Mounaïm-Rousselle C (1999) Combined catalytic hot wires probe and fuel-air-ratio-laser induced-exciplex fluorescence air/fuel ratio measurements at the spark location prior to ignition in a stratified GDI engine. SAE Trans 1631–1641

Di Iorio S, Sementa P, Vaglieco BM (2016) Analysis of combustion of methane and hydrogen–methane blends in small DI SI (direct injection spark ignition) engine using advanced diagnostics. Energy 108:99–107

Drake MC, Fansler TD, Lippert AM (2005) Stratified-charge combustion: modelling and imaging of a spray-guided direct-injection spark-ignition engine. Proc Combust Inst 30(2):2683–2691

Fraidl GK, Piock WF, Wirth M (1996) Gasoline direct injection: actual trends and future strategies for injection and combustion systems. SAE Trans 543–559

Geschwindner C, Kranz P, Welch C, Schmidt M, Böhm B, Kaiser SA, De la Morena J (2020) Analysis of the interaction of spray G and in-cylinder flow in two optical engines for late gasoline direct injection. Int J Engine Res 21(1):169–184

Han D, Fan Y, Sun Z, Nour M, Li X (2020) Combustion and emissions of isomeric butanol/gasoline surrogates blends on an optical GDI engine. Fuel 272:117690

Hung DL, Chen H, Xu M, Yang J, Zhuang H (2014) Experimental investigation of the variations of early flame development in a spark-ignition direct-injection optical engine. J Eng Gas Turbines Power 136(10)

Ipp W, Egermann J, Engermannn J, Wagner V, Leipertz A (2000) Visualization of the qualitative fuel distribution and mixture formation inside a transparent GDI engine with 2D MIE and LIEF techniques and comparison to quantitative measurements of the air/fuel ratio with 1D Raman spectroscopy. SAE Trans 931–941

Irimescu A, Merola SS, Martinez S (2018) Influence of engine speed and injection phasing on lean combustion for different dilution rates in an optically accessible wall-guided spark ignition engine. SAE Int J Engines 11(6):1343–1370

Kalwar A, Agarwal AK (2020) Overview, advancements and challenges in gasoline direct injection engine technology. In: Advanced combustion techniques and engine technologies for the automotive sector. Springer, Singapore, pp 111–147

Kalwar A, Agarwal AK (2021) Lean-burn combustion in direct-injection spark-ignition engines. In: Alternative fuels and advanced combustion techniques as sustainable solutions for internal combustion engines. Springer, Singapore, pp 281–317

Kang JJ, Kim DJ (2003) Effects of piston shapes and intake flow on the behaviour of fuel mixtures in a GDI engine. KSME Int J 17(12):2027–2033

Koupaie MM, Cairns A, Xia J, Vafamehr H, Lanzanova T (2019) Cyclically resolved flame and flow imaging in an alcohol-fuelled SI engine. Fuel 237:874–887

Lee Z, Kim D, Park S (2020) Effects of spray behaviour and wall impingement on particulate matter emissions in a direct injection spark ignition engine equipped with a high-pressure injection system. Energy Convers Manag 213:112865

Marseglia G, Costa M, Catapano F, Sementa P, Vaglieco BM (2017) Study about the link between injection strategy and knock onset in an optically accessible multi-cylinder GDI engine. Energy Convers Manage 134:1–19

Martinez S, Irimescu A, Merola SS, Lacava P, Curto-Riso P (2017) Flame front propagation in an optical GDI engine under stoichiometric and lean-burn conditions. Energies 10(9):1337

Martinez S, Merola S, Irimescu A (2019) Flame front and burned gas characteristics for different split injection ratios and phasing in an optical GDI engine. Appl Sci 9(3):449

Matsuda M, Yokomori T, Iida N (2017) Investigation of cycle-to-cycle variation of turbulent flow in a high-tumble SI engine (No. 2017-01-2210). SAE technical paper

Merola SS, Tornatore C, Irimescu A (2016a) Cycle-resolved visualization of pre-ignition and abnormal combustion phenomena in a GDI engine. Energy Convers Manag 127:380–391

Merola SS, Tornatore C, Irimescu A, Marchitto L, Valentino G (2016b) Optical diagnostics of early flame development in a DISI (direct injection spark ignition) engine fueled with n-butanol and gasoline. Energy 108:50–62

Naber JD, Siebers DL (1996) Effects of gas density and vaporization on penetration and dispersion of diesel sprays. SAE Trans 82–111

Oh H, Bae C (2013) Effects of the injection timing on spray and combustion characteristics in a spray-guided DISI engine under lean-stratified operation. Fuel 107:225–235

Pastor JV, Garcia-Oliver JM, Garcia A, Mico C, Möller S (2016) Application of optical diagnostics to the quantification of soot in n-alkane flames under diesel conditions. Combust Flame 164:212–223

Peterson B, Reuss DL, Sick V (2011) High-speed imaging analysis of misfires in a spray-guided direct injection engine. Proc Combust Inst 33(2):3089–3096

Peterson B, Reuss DL, Sick V (2014) On the ignition and flame development in a spray-guided direct-injection spark-ignition engine. Combust Flame 161(1):240–255

Pilla G, Kumar R, Laget O, De Francqueville L, Dauphin R, Solari JP (2016) Simulation and optical diagnostics to characterize low octane number dual fuel strategies: a step towards the octane on-demand engine. SAE Int J Fuels Lubr 9(3):443–459

Potenza M, Milanese M, Naccarato F, de Risi A (2020) In-cylinder soot concentration measurement by Neural Network Two Colour technique (NNTC) on a GDI engine. Combust Flame 217:331–345

Sementa P, Vaglieco BM, Catapano F (2012) Thermodynamic and optical characterizations of a high-performance GDI engine operating in homogeneous and stratified charge mixture conditions fueled with gasoline and bio-ethanol. Fuel 96:204–219

Sharma N, Bachalo WD, Agarwal AK (2020) Spray droplet size distribution and droplet velocity measurements in a firing optical engine. Phys Fluids 32(2):023304

Sick V (2010) Optical diagnostics for direct injection gasoline engine research and development. In: Advanced direct injection combustion engine technologies and development. Woodhead Publishing, UK, pp 260–286

Song J, Kim T, Jang J, Park S (2015) Effects of the injection strategy on the mixture formation and combustion characteristics in a DISI (direct injection spark ignition) optical engine. Energy 93:1758–1768

Song J, Lee Z, Song J, Park S (2018) Effects of injection strategy and coolant temperature on hydrocarbon and particulate emissions from a gasoline direct injection engine with high-pressure injection up to 50 MPa. Energy 164:512–522

Song J, Park S (2015) Effect of injection strategy on the spray development process in a single-cylinder optical GDI engine. Atomization Sprays 25(9)

Steeper RR, Stevens EJ (2000) Characterization of combustion, piston temperatures, fuel sprays, and fuel-air mixing in a DISI optical engine (No. 2000-01-2900). SAE technical paper
Stiehl R, Bode J, Schorr J, Krüger C, Dreizler A, Böhm B (2016) Influence of intake geometry variations on in-cylinder flow and flow–spray interactions in a stratified direct-injection spark-ignition engine captured by time-resolved particle image velocimetry. Int J Engine Res 17(9):983–997
Stiehl R, Schorr J, Krüger C, Dreizler A, Böhm B (2013) In-cylinder flow and fuel spray interactions in a stratified spray-guided gasoline engine investigated by high-speed laser imaging techniques. Flow Turbul Combust 91(3):431–450
Wang Y, Makwana A, Iyer S, Linevsky M, Santoro RJ, Litzinger TA, O'Connor J (2018) Effect of fuel composition on soot and aromatic species distributions in laminar, co-flow flames. Part 1. Non-premixed fuel. Combust Flame 189:443–455
Williams B, Edwards M, Stone R, Williams J, Ewart P (2014) High precision in-cylinder gas thermometry using laser-induced gratings: quantitative measurement of evaporative cooling with gasoline/alcohol blends in a GDI optical engine. Combust Flame 161(1):270–279
Xie F, Zhang M, Wang Y, Su Y, Hong W, Cheng P (2021) Soot development in an optical direct injection spark ignition engine fueled with isooctane. Int J Automot Technol 22(2):455–463
Yang J, Dong X, Wu Q, Xu M (2019) Effects of enhanced tumble ratios on the in-cylinder performance of a gasoline direct injection optical engine. Appl Energy 236:137–146
Zeng W, Sjöberg M, Reuss DL (2015) PIV examination of spray-enhanced swirl flow for combustion stabilization in a spray-guided stratified-charge direct-injection spark-ignition engine. Int J Engine Res 16(3):306–322
Zeng W, Sjöberg M, Reuss DL, Hu Z (2016) The role of spray-enhanced swirl flow for combustion stabilization in a stratified-charge DISI engine. Combust Flame 168:166–185
Zeng W, Sjöberg M, Reuss DL, Hu Z (2017) High-speed PIV, spray, combustion luminosity, and infrared fuel-vapour imaging for probing tumble-flow-induced asymmetry of gasoline distribution in a spray-guided stratified-charge DISI engine. Proc Combust Inst 36(3):3459–3466
Zhang Z, Zhang H, Wang T, Jia M (2014) Effects of tumble combined with EGR (exhaust gas recirculation) on the combustion and emissions in a spark-ignition engine at part loads. Energy 65:18–24

Part IV
Dual-Fuel Combustion Technology

Chapter 8
Dual-Fuel Internal Combustion Engines for Sustainable Transport Fuels

Babalola Aisosa Oni and Samuel Eshorame Sanni

Abstract The quest for a sustainable future for transport-fuels has led to the consideration of advanced methods of admixing fossil fuels without compromising their qualities; this is aimed at improving/complementing the properties of these fuels for high engine performance. Owing to the high abundance of biomass from which alternative fuels can be sourced for blending or improving the properties of conventional gasoline and diesel fuels, it has become pertinent to consider their use as complementary fuels towards ensuring high sustainability of the fuels as transport-fuels. The synergistic effects offered by these fuels helps to improve the properties of the fuels better than the individual components that make up the fuel mix. Hybrid gasoline-biofuel fuels offer these improved properties as a result of the complementary effects of either or both components offer in the hybrid fuels such that there is a boost in the fuel's combustion potential owing to the degree of homogenization attained during blending. Furthermore, despite ensuring high compatibility of the individual fuels that make up the biofuel-diesel fuel mix, it is also pertinent to emphasize the need to obtain an optimum blend of the dual fuel system for the engine performance because, for specific dual fuel systems, beyond the optimum mixture composition, the performance of the engine begins to wane owing to the alteration in the properties of the fuel mix beyond favorable conditions for complete/near complete combustion of the fuels. Therefore, in this chapter, the properties of different dual fuel systems will be discussed alongside the degree of homogenization that can improve the atomization/combustion potential of the fuels towards attaining high engine BTEs, high engine power, moderate heat release rates as well as good air–fuel ratios.

Keywords Alternative fuels · Combustion · Conventional fuels · Dual fuel systems · Homogenization · Sustainable transport fuels

B. A. Oni (✉) · S. E. Sanni
Department of Chemical Engineering, Covenant University, Ota, Ogun State, Nigeria
e-mail: Babalola.oni@covenantuniversity.edu.ng

B. A. Oni
Department of Chemical Engineering, China University of Petroleum, Changping, Beijing, People's Republic of China

A. K. Agarwal et al. (eds.), *Advanced Combustion for Sustainable Transport*, Energy, Environment, and Sustainability, https://doi.org/10.1007/978-981-16-8418-0_8

8.1 Introduction

Internal combustion engines (ICEs) convert produced heat into work from the combustion chamber. Two main ICE types such as diesel and petrol engines, contribute to environmental pollution from the release of emission from the exhausts of these engines (Lyng et al. 2018). There are about 25% of global greenhouse gas (GHG) emissions from the transportation sector as a result of higher demand in automobile (Xue et al. 2020).

Gasoline and diesel are the widely adopted fuels in ICEs/compression ignition engines (Boisen 2008). In SI engines, flame propagation occurs after an initial spark event ignites the homogeneous air–fuel mixture (Rimkus et al. 2018). Despite several decades of consistent supply and market position for diesel and gasoline as regular automotive fuels, the search for alternative fuels began to emerge in the 1980s (Das et al. 2020).

The definition of alternative fuels varies; however, it depends on the context. Alternative fuels are those fuels other than diesel and gasoline fuels, which cover a wide range of final forms and manufacturing sources (Ozcanli et al. 2018). Ethanol is an example of alternative fuel for use in spark ignition (SI) engines, irrespective of its source either from biomass or crude oil/fossil fuel. According to Arnoersson (2011), alternative fuels also include a wide class of other fuels such as non-conventional fuels (methanol, ethanol, butanol, LPG (liquified petroleum gas), Hydrogen, coal-derived liquid fuels (CTL) and pure biodiesel (B100). Some of the importance of alternative fuels can be ascribed to the following:

(i) trailing sustainable energy via increasing the consumption of alternative fuels from biomass while justifying the concerns associated with the exhaustion of energy recovered from fossil fuels (Santoso et al. 2013).
(ii) improving the efficiencies/reducing the emissions of engines by employing alternative fuels of high quality that are comparable with conventional fuels, in terms of their physicochemical properties (Manoharan et al. 2019)
(iii) .reducing the unbalanced use of traditional petroleum-based fossil fuels (Chong and Hochgreb 2011).

Despite projections that conventional fuels will continue to dominate transportation fuels for the next 5 decades, an urge for energy sustainability and a desire to cutdown greenhouse gas (GHG) emissions have caused an increase in biofuel consumption (Antunes et al. 2008). Fossil fuels significantly contribute to the emission of CO_2 with approximately 27 billion tons of CO_2 generated from anthropogenic activities per year. The use of biofuels informs a lower CO_2 life cycle which constitutes a significant drop in CO_2 emissions from a social standpoint (Kukharonak et al. 2017). Based on a life cycle assessment that was conducted to compare the greenhouse gas impacts of biofuels and gasoline/diesel fuels, the results showed that biofuels from vegetable oil gave reduced greenhouse gas emissions compared to those obtained for conventional fuels (Chintala and Subramanian 2017). The analyses considered all GHG emissions associated with their transportation, production, and storage, as well as emissions from vehicles.

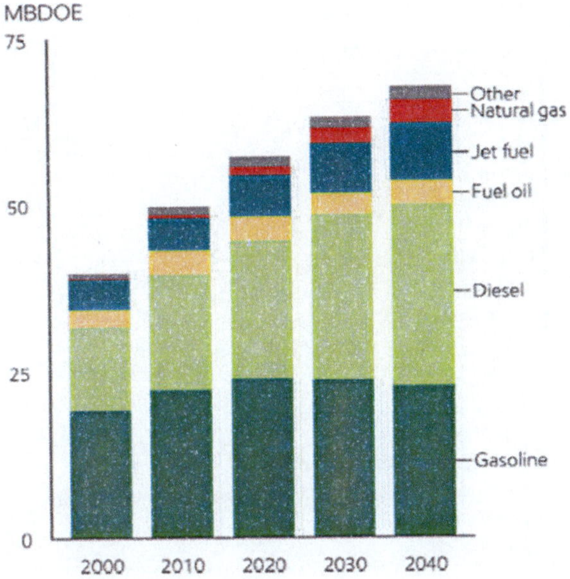

Fig. 8.1 Global transportation demand of different fuels. Reprinted from Bae and Kim (2017) with copyright permission obtained from Elsevier

In 2012, biofuels accounted for about 4% of the total global transportation fuel. Figure 8.1 illustrates the increase in energy demand in the transportation sector where it is also noticed that the trend in the demand for biofuel also increased (Huyskens et al. 2011) which in the long run may lead to mitigation of net greenhouse emissions, thus giving rise to low-carbon transportation fuels; this also informs an expected reduction in the emission of GHG in the nearest future (Heffel 2003).

In addition, it is critical to understand whether a fuel is renewable energy-based such as those sourced from crops, because when they burn, they only marginally increase the levels of CO_2 in the atmosphere, hence the need to understand the clear distinction in the origin of these fuels as to whether they are generated from biofuels or fossil. Figure 8.2 is an illustration of several fuels that can be sourced from different origins.

8.2 Different Biofuels and Their Blends for Transportation

8.2.1 Dual Fuel System

A dual fuel engine operates on diesel (B100) or the mixture of natural gas and diesel, distributing the same engine torque curve, transient response and power density as the

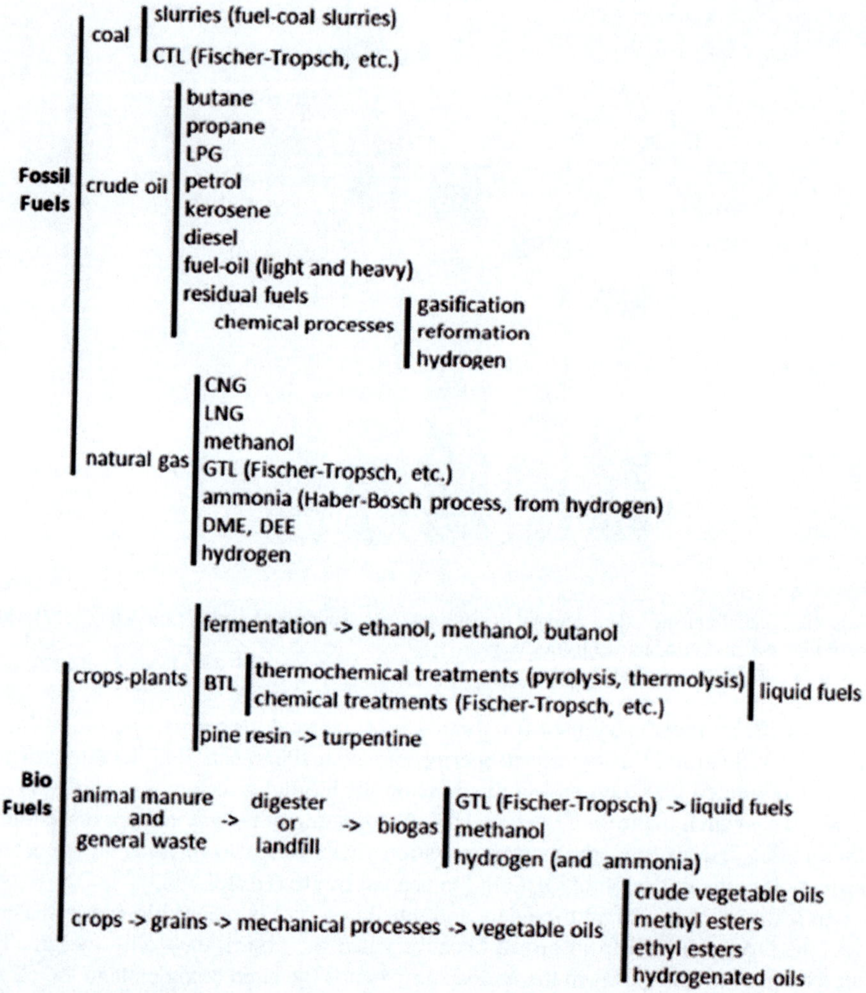

Fig. 8.2 Types of fossil and biofuels that can be used in ICEs. Reprinted from Martins (2020) and Martins and Brito (2020)

base diesel engine. The transportation sector is the largest oil consumer accounting for about 53% of total oil/fuel-consumption (Åhman 2010). Hence, there is a crucial demand for sustainable fuel in the transportation sector. The current trend of fossil fuel as means of transportation increases the tendencies for high GHG crisis. Several studies are being currently conducted on the aforementioned and much emphasis is being placed on alternative fuels that would provide similar effects as gasoline on a long-term/large-scale basis as transport-fuel (Banerjee et al. 2017).

8.2.1.1 Recent Works on Dual-Fuel System Based-Engines

Biofuel is a vehicle alternative that has the potential to significantly minimize engine emissions and improve air quality. Biofuel production and their uses have increased dramatically in recent years around the world. Some countries have made biofuel consumption a goal and mandate. Elfasakhany (2018) carried out an experiment on bio-methanol blended with gasoline and iso-butanol and gasoline (A), and bio-ethanol blended with the admix of gasoline and n-butanol (B) at 3–10 vol% biofuels. The 2-ternary fuels (A and B) were compared with gasoline in terms of emissions and efficiencies in a gasoline-fueled engine. The results show that the A and B fuel blends showed appreciable efficiencies and better emissions than the gasoline fuel. Oni et al. (2021) blended neem and camelina biodiesel methyl esters at different ratios in a 1.9 Multijet diesel engine. The experimental result shows an increase in BTE with lesser carbon monoxide, hydrocarbons and CO_2 emissions with higher NOx emission as compared to the conventional diesel fuel.

Elfasakhany (2021) blended methanol, acetone, i-butanol-n-butanol, and ethanol biofuels, in single/dual modes in SI engine. All the analyzed biofuels were compared with each other as well as with neat gasoline in a SI engine. The improvement of methanol in the SI engine was similar to that of ethanol, although methanol is poisonous, which makes it not compatible with SI engine. I-butanol and n-butanol biofuels, showed less engine performance with higher emissions of CO and HC. Methanol and ethanol showed the highest engine performance, with higher carbon dioxide emissions. The engine performance of acetone was moderate, with very low emissions of HC and CO when compared with other biofuels. Different dual biofuel blends of ethanol–methanol was found to be more superior in terms of engine efficiencies and low emissions than the blends of i-butanol–n-butanol and ethanol-i-butanol.

8.2.1.2 Projects Underway for Dual Fuel-Based Systems Engine Applications

Fuel pilot-injection, which includes Dual-fuel technology and a High Pressure Direct Injection (HPDI) method, is a well-known type of ignition assist. Dual-fuel engines, have high NOx and CH_4 emissions as well as low efficiency, especially at 50% load condition, while the HPDI technology has particulate matter emissions from natural gas diffusion flames. This HPDI injector can be used to inject (CNG) and H_2 compressed natural gas (Juknelevičius et al. 2019). This (HPDI) injector consists of two-concentric spring-loaded needles, thus allowing gas and liquid fuels to flow simultaneously.

Port injection can be used to inject methanol into the intake port, then directly injecting diesel near the TDC to ignite the mixture of methanol/air. Wu and Han (2020) studied the effects of intake-air temperature and the mass fraction of methanol on a six-cylinder turbocharged diesel engine. There was a decrease in the intake-air temperature which in turn reduces the EGT and indicated thermal efficiency, thus increasing the methanol energy fraction in a dual-fuel operation mode.

8.2.2 *Biomethane CNG Hybrid*

One of the several alternatives for the transportation sector is biomethane-CNG hybrid. Methane is the main composition of biomethane. It is produced by anaerobic decomposition of organic matter via biomass gasification or by bacterial decomposition. To supply vehicles with biomethane, the gas must be compressed to about 200 bars (Boesel 2009) prior being mixed with CNG. This combined hybrid fuel is one of the most promising fuels for achieving a carbon–neutral environment, with a significant reduction in the 'Well to Wheel' emissions of any wheel Baldwin system; this mixture is also referred to as the green energy equivalent of methane (Arnoersson 2011).

Raw biogas is primarily composed of CH_4 and CO_2, with average concentrations ranging from 60–75 and 35–60%. Upgrading raw biogas to a highly purified biomethane can give about 98% CH_4. Pressure swing adsorption, water scrubbing, and amine absorption are the common approaches adopted in attaining high purity of the gas (Zhang et al. 2019a). Upgraded biomethane can be applied in two ways i.e., by compressing it to 200 bars or cooling it to a liquid at 162 °C (liquefied biogas-LBG). To liquefy biomethane, an additional polishing step is frequently required, in which the biomethane is purified further. Kaiser and White (2009) compared various upgrading methods for compressed biogas and liquified biogas, and their method of distribution, the result showed that liquified biogas needs more energy when compared to compressed biogas, however, liquid biogas (LBG) has lower global warming impacts relative to compressed biogas (CBG) due to their reduced CH_4 slips (Obermair et al. 2010). Biogas produced from waste matter is a good potential alternative gaseous energy fuel for ICEs, and it has the potential to replace fossil fuels.

When compared to gasoline, the use of biomethane-compressed natural gas hybrid in the transportation sector can contribute to lowering GHGs by about 80% (Bordelanne et al. 2011); biomethane is essentially biogas that has been purified to make it of high-quality natural gas. Biomethane is a non-CO_2-emitting biogas that is 98% CH_4 with mixed proportions of C_2H_6 which helps to increase its calorific value (Zheng et al. 2012).

It is impossible to use a biogas-CNG blend in a CNG vehicle unless the biogas quality is improved. Biomethane is an improved biogas that is used as a fuel in CNG vehicles. CNG vehicles emit approximately 17% less GHG than gasoline vehicles (Zhang et al. 2019b). When compared with gasoline, vehicles running on biomethane-CNG blends can give reduced emissions of about 51%. Toyota Prius CNG Hybrid prototype powered by biomethane reduces emissions by 87% when compared to gasoline vehicles (Wellinger and Lindberg 2000).

Energy crops, residential, industrial waste, solid waste, forest residue, agricultural residues etc., are raw materials used to produce short-term synthesis of biomethane (You et al. 2020). Energy crops are good options for medium term production of these fuels, while biomass gasification from lignocellulosic resources is usually employed for long term purposes (Bordelanne et al. 2011; Stenlåås et al. 2004). Biomethane needs additional processes to work in a vehicle so as to prevent engine corrosion.

Biomethane can be produced from—wet materials (as in anaerobic fermentation) or dry materials (as seen in gasification processes) (Bordelanne et al. 2011; Salazar et al. 2009). The production of biomethane via biogas is susceptible to pH, type of feedstock, temperature, moisture content, digester technologies, type of bacteria used etc. (Liu and Dumitrescu 2019). Biomethane can also be produced from gasification of woody biomass. New gasification technologies are currently ongoing to improve the production of biomethane. The first is indirect gasification which is suitable for small scale (< 100 MW) (Lee et al. 2001, 2013). The second is the use of fluidized bed gasification (FB) which is suitable for large scale (> 100 MW) plants (Åhman 2010). The advantage of this process is that it can be used with a variety of feed stocks. Biomethane is a high-quality energy carrier that can be used in varieties of applications and it is highly efficient. Furthermore, biomethane can be used for transportation, which can therefore be integrated into the existing natural gas distribution network (Åhman 2010).

8.3 Biogas-Biodiesel Fuel Mix for SI Engines

Small co-generation systems are almost always attributed to the use of spark ignition engines that run directly on biogas (with impurities like water, carbon dioxide, or hydrogen sulphide). The conversion of SI engines to make them compatible with gaseous fuels, is simple, however, it only requires the installation of a simple gas-fuel adaptor and probably toughened valves and valve seats (Yang and Zeng 2018; Lebedevas et al. 2019). The articles published for biogas-powered spark ignition engines primarily include works on the engines' overall emission and performance relative to when the system is fueled with gaseous and liquid fuels. More on this will be discussed in the later sections. Furthermore, quite a number of studies with and/ or without computer simulation have been discovered in the literature for gaseous fuel systems. Almost all of the investigators used simulated biogas instead of raw biogas, with typical composition of methane (~60%) and carbon dioxide (~40%) in their studies (Glaude et al. 2010). Other studies have also shown that, biogas being a dilute gaseous fuel may have substantial effects on the engine's combustion, emission and performance characteristics (Lee et al. 2001; Adelt et al. 2011; Matthias et al. 2012). Table 8.1 shows the properties of single fuel systems that can be blended as fuels for SI engines.

Table 8.1 Properties of different single gaseous fuels

Parameter	Unit	Natural gas properties (* 10^{-1})	Town gas (typical properties) (* 10^{-1})	Biogas properties with: CH_4 (60%), CO_2, (38%, other gases (2%) (* 10^{-1})
Theoretical air requirement	m^3 air/m^3 gas	95.3	38.3	57.1
Calorific value (Lower)	MJ/m^3	361.4	161	214.8
Max. ignition velocity	m/s	3.9	7.0	2.5
Dew point	°C	590	600	600–1600
Density	kg/m^3	8200	5100	12,100

Adapted from Huyskens et al. (2011)

8.3.1 *Potential Single Fuel Systems that Can Be Blended and Their Characteristics*

See Table 8.1.

8.3.2 *Dual Fuel Blending Techniques: Methods of Preparation, Homogenization and Their Selection Criteria*

ICEs powered by CNG are used in industries. Compressed natural gas engines have many advantages over diesel and gasoline combustion engines. In the lean operating region, CNG engine performance can be improved (Salazar and Kaiser 2011). Conversely, very low flame increases the CNG velocities and severely limit the lean operating limits of compressed natural gas engines. H_2 has a fast flame speed and a wide operating range that includes lean conditions. The addition of hydrogen to a compressed natural gas engine makes it a feasible and cost-effective method of extending the lean operating limit, improving performance and lowering emissions (Scarcelli et al. 2011). However, H_2 has some shortcomings as fuel source such as low power density as a result of low-heating value per unit volume when compared to compressed natural gas, as well as gas susceptibility to pre-ignition and engine knock due to low minimum ignition energy and wide flammability limits (Sandalcı and Karagöz 2014). Thus, combining H_2 and compressed natural gas brakes the limitations of each fuel-type.

Researchers have conducted experiments on alternative fuels that can be blended with gasoline fuels. ICEs also use alternative fuels such as natural gas, H_2, ethanol,

Table 8.2 Fuel properties of acetylene, hydrogen, CNG, gasoline and ethanol

Properties	Acetylene $*10^1$	H_2 $*10^1$	CNG $*10^1$	Ethanol $*10^1$	Gasoline $*10^1$
Formula	C_2H_2	H_2	CH_4	C_2H_5OH	$C_4 - C_{12}$
Density (kg/m^3)	0.11	0.0081	0.07	81.0	70–75
Ignition temperature (°C)	30.5	57.2	54.0	36.3	25.7
Motor octane number (MON)	4.5 -5.0	13.0	10.5	90.0	9.7
Flammability limit in air (vol%)	0.25–8.1	0.4–7.4	0.53–1.5	0.3–1.9	0.14–0.76
Minimum quenching diameter (mm)	0.09	0.09	0.35	0.30	0.30
Maximum flame speed (m/s)	0.15	0.35	0.04	0.06	0.05
Lower heating value (kJ/kg)	4825.5	12,000	5000	2700	4300
Adiabatic flame temperature (K)	250	240	233	209.3	231.0

Source Adapted from Karim and Wierzba (1992) (no need for copyright, values were modified)

biofuels and acetylene (Oni et al. 2021; Saravanan and Nagarajan 2008). Hydrogen is approximately fourteen times lighter than air in the gas phase. Moreover, it is the cleanest fuel on the planet. Due to its high ignition limit (5–76%) and low ignition energy, it requires a special design to be used in its pure form in ICEs. It has been demonstrated that adding 20% hydrogen to fuels improves the fuel's emission, performance, and combustion characteristics. Natural gas is primarily composed of CH_4 (85–96%) which can be used as fuel in ICEs. In ICEs, pure ethanol can be admixed with other fuels or used as fuel alone. NG, hydrogen, NG + H_2 (HCNG), ethanol, ethanol + gasoline, ethanol + H_2, acetylene, acetylene + gasoline mixtures have all been studied for their effects on engine emission and performance characteristics (Oni et al. 2021; Kaiser and White 2009).

To reduce emissions from ICEs, the properties of acetylene, H_2, ethanol, and NG which are alternative fuels that can be used without changes in SI engine modification, and their effects on the emission and performance of the engine, have been studied. Table 8.2 shows the chemical and physical properties.

8.3.3 Conditions for Maximizing the Combustion Potentials of Dual Fuels in ICEs

As shown in Table 8.2, H_2 has distinct physicochemical properties when compared to diesel, CNG, and gasoline commonly used as transportation, fuels. The engine performances involving these fuels in various engine-modes are comparable to that of H_2 for the same engine types. One benefit of using H_2 in internal combustion engine as a clean alternative fuel is that it contains no carbon (Murphy and Power 2009). This means that carbon-based emissions, primarily carbon dioxide, carbon monoxide and soot can be completely removed leaving NOx gases. Because of its high energy density, H_2 can provide almost 3-times as much energy by mass as diesel, CNG, and gasoline fuels as evidenced by its lower heating value. In lieu of the aforementioned, there are lots of limitations associated with H_2 owing to its extremely low density (Karim and Wierzba 1992). Due to the extreme low molecular weight of H_2, its temperature and density at atmospheric pressure are in an order of magnitude that is less than that of NG. Because of its low boiling point, compressed hydrogen is likely to be the most popular storage option. Due to limited vehicle space, this poses a challenge for the implementation of H_2 in ICE on-road applications. In order to increase hydrogen density as well as its volumetric energy content, the storage pressure of the gas must be increased (Mansor et al. 2017). For example, at 370 bar and 273 K, the density of hydrogen can increase to 32 kg/m^3 and its volumetric energy content can be as high as 3800 MJ/m^3. H_2 is expected to be highly resistant to knocking as a result of its high auto-ignition temperature and RON, 130 in comparison to other common fuels. However, it has a lower minimum ignition energy in air of stoichiometric requirement than hydrocarbon fuels, thus indicating that H_2 can be ignited easily by hot spots in the chamber (Chu et al. 2018). This can result in fuel pre-ignition, which is defined as combustion during the compression stroke prior to ignition. This causes knock, loss of combustion phasing control and, failure in engine. However, the global effects of knock and pre-ignition cannot be distinguishable (Ehsan et al. 2003), owing to the fact that pre-ignition leads to knock. Previous works have shown that MON of H_2 is much less than its RON, compared to a normal 8.0–10.0—point reduction for gasoline; although the exact value of H_2 MON has not been clearly understood, nonetheless, MON is confirmed to be a more accurate knock resistance metric in H_2 engine designs (Dimitriou and Tsujimura 2017). This explains why knock is frequently reported in H_2 engine applications. In comparison to conventional hydrocarbon fuels, hydrogen has a short quenching distance. As a result, higher temperature-gradients close to the walls of the chamber are to be expected which may result in higher heat losses. When H_2 is applied in engines, the short quenching distance combined with the high flame speed in air, indicates a greater proclivity for flame backfiring into the intake manifold. This problem can be solved by changing the engine geometry, and completely eliminating abnormal discharge in the ignition system.

Hydrogen's unique physical and thermochemical properties can aid in the design of highly efficient ICEs. This is because, the dispersion of H_2 is 4-times faster than compressed natural gas, as evidenced by comparing their diffusion coefficients in air (Baldwin 2008). In ICEs, this can promote air mixing and in-cylinder fuel. The volume fraction of the required stoichiometric hydrogen is 29.43 vol%. Nonetheless, the wide flammability limit of 4–76 vol% H_2 in air, combined with the high flame speed, implies that the use of H_2 in internal combustion engines can give rise to significantly lean fuels and enhanced thermal efficiencies (Adelt et al. 2011). In stoichiometric amounts, the adiabatic flame temperature of hydrogen is relatively high, which promotes NOx formation. Nevertheless, due to the wide flammability limits or lean operations, a high level of EGR can be used to reduce NOx emissions (Hänggi et al. 2019). The use of H_2 in internal combustion engines may raise additional safety concerns regarding the on-board fuel storage and delivery system, especially when high pressure hydrogen is used (Ogunkunle and Ahmed 2019).

8.3.4 Factors Effecting Dual Fuel Characteristics in SI Engines

8.3.4.1 Effect of In-Cylinder Pressure

The presence of the water/carbon dioxide that is inherent in biogas results in a reduction in the in-cylinder pressure (Fig. 8.3a) and the rate of pressure rise. When the diluent concentration increases, the delay in the position of maximum combustion pressure occurs (Szwaja and Grab-Rogalinski 2009). This can possibly be clarified in terms of low burning velocity, lean mixture, low energy input per cycle, and variations in mixture strength. However, higher compression ratios (CRs) can significantly improve the cylinder pressures. According to the works of Kalsi and Subramanian (2017), an optimal CR of 13:1 for a biogas mixture is achieved for biogas containing 40% carbon dioxide but can be reduced above 40% concentration thus causing reduction in the engine efficiency.

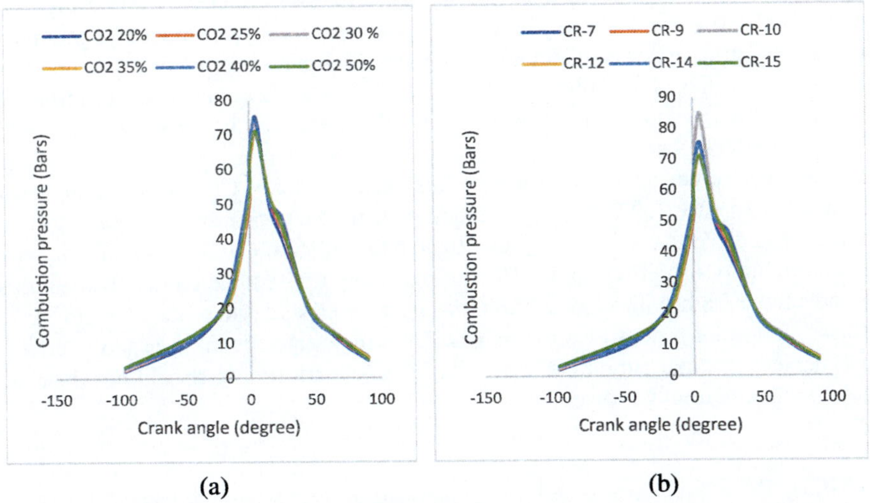

Fig. 8.3 **a** Cylinder pressure versus crank angle for different CO_2 fractions in biogas, speed 2000 rpm, 15:1 and A/F = 0.98 (the lowest peak for the highest CO_2 content); Fig. 8.3 **b** Cylinder pressure versus crank angle for different Compression ratios at 2500 rpm, A/F = 0.97 and CO_2 fraction in biogas = 37.5% (the lowest peak for the lowest)

8.3.4.2 Effect of Air–Fuel Ratio

The occurrence of carbon dioxide in biogas reduces the level of flammability of the air/fuel ratio. Air–fuel ratio decreases as the CO_2 level in fuel rises (Helmolt and Eberle 2007). The engine operating point shifts away from the stoichiometric mixture and towards the weak-mixture flammability limit with high CO_2 content, and the combustible mixture range becomes more sensitive to the mixture's velocity level and turbulent characteristics of the mixture to be burnt (Stürmer 2017). Ambarita (2017) calculated the highly combustible air- fuel ratio range for various fuel proportions. The authors stated that as the carbon dioxide concentration in biogas improved, the carburetor itself made the mixture weaker. Because CO_2 is less dense than CH_4, thus lowers the volumetric flow rate of fuel upon induction. The occurrence of diluent CO_2 in biogas has a serious influence on the spreading of flames into the flowing streams of fuel/air mixtures in SI engines. Because the specific heat of CO_2 tends to increase at a faster rate as temperature rises; also, its share of the fraction of energy released by fuel oxidation rises at elevated temperatures (Roubaud and Favrat 2005). At such high temperatures, CO_2 will undergo high endothermic dissociation reactions. As a result of the presence of CO_2 in biogas, the net effect is a significant reduction in the adiabatic flame temperature of the biogas (Bortel et al. 2019). Because of the reduction in diffusivity, rate of reaction, flame temperature, and transport phenomena of the mixing properties, both laminar and turbulent burning velocities decrease significantly. Yadvika et al. (2004) calculated the mass fraction burnt using a 2-zone combustion model. The model was modified to fit the combustion geometry of diesel engine under consideration thus making it easy to compute the turbulent burning velocities. The authors discovered a 50% significant decrease in turbulent burning velocity with a biogas mixture of 70% CH_4 and 30% CO_2 as compared with 100% CH_4. Lower flame propagation rates result in a significant increase in the length of the period of combustion, which increases as the CO_2 in biogas increases (Stockhausen et al. 2002). In biogas combustion, the MBT timings are typically more sophisticated. Verma et al. (2018) investigated the effect of CO_2 dilution in biogas on the burn durations/periods and MBT timings using a SI engine model. The authors discovered a highly developed MBT timing as well as higher (i.e., 10 and 90%) mass-fraction burn durations for biogas. To compensate for these negative effects and achieve optimal engine performance, biogas spark timing should be advanced compared to natural gas or gasoline.

8.3.4.3 Effect of Power and Fuel Consumption

In comparison to natural gas or methane, the diluent CO_2 in biogas lowers the calorific value of the fuel which in turn reduces the brake power output (BPO) and increases the Brake specific energy consumption. The published engine power data for biogas fueling shows a wide range of variations. The large differences in power output are caused by differences in ignition timing, engine specifications, and operating air–fuel ratios. Midkiff et al. (2001) found that biogas (60% CH_4: 40% CO_2) reduced the

BPO by 14% at ER = 1 when compared to the baseline fuel natural gas. The BSEC increased as the carbon dioxide content of the fuel increases, thus signifying that dilution has no effect on brake specific energy consumption with ER or φ ranging from 0.7 to 0.9. This could be because, at lower equivalence ratios, air becomes a more dominant diluent than CO_2 in biogas. Depending on the operating conditions and engine configurations, the reduction in power output compared to methane or natural gas ranges from 3 to 22%. According to Anwar et al. (2019), such a large reduction in power output in the case of biogas fueling could be due to incorrect optimization of the ignition timing and air/fuel ratio. For example, Lebedevas et al. (2019) did not optimize biogas ignition timing, and the CH_4 flow rate remained unchanged when CO_2 was added. This led to a higher ER and a 20% decrease in output power. Suarez-Bertoa et al. (2019) tested stoichiometric air–fuel ratios with a fixed ignition timing (12° BTDC). They discovered a 2% decrease in output for every percentage point increase of CO_2 in the biogas. However, Wimmer et al. (2005) discovered a 9% decrease in power output when switching from CH_4 to biogas in stoichiometric amounts. This was done with a predetermined equivalence ratio (ER). This reduction was discovered to be primarily due to a decrease in entrained energy on the inside of the cylinder, as well as a partial increase in combustion duration (White et al. 2006).

8.3.4.4 Effects of Diluent

Biogas is mostly methane, which is an exceptional fuel for spark ignition engines due to its high resistance to knock and auto-ignition. This means that the engine can function at a broader range of compression ratios and that the biogas will not cause unexpected knocking. The amount of CO_2 tends to lower the mixture temperature which in turn reduces the burning velocity, however, this does not increase the knocking tendency for most engines and operating conditions (Scarcelli et al. 2011; Yadvika et al. 2004). Salazar and Kaiser (Niculescu et al. 2017) calculated the effect of diluents in a fuel/CH_4 on incidence of knock. Knock would occur, according to the model, when the energy generated by the end gas's pre-ignition reaction activity ahead of the spreading flame gets strong enough to cause auto-ignition (Mohammadi et al. 2007). The results revealed that the increased diluent in biogas inhibited the pre-ignition reaction of the end gas region, thus suggesting low knock relative to the pure fuel CH_4 mixture under the same process conditions.

8.3.4.5 Effects of Engine Emissions

Nox

Nitrogen oxide (NOx) formation is strongly influenced by the engine's combustion temperature and O_2 concentration. The combustion temperature is lower in leaner mixtures, but the O_2 concentration is higher. On the other hand, as the mixtures become more concentrated stoichiometrically, the combustion temperature rises to its maximum, but the O_2 concentration falls (Meijden et al. 2011). As a result of the following competing effects, most HC fuels emit their highest NOx gases at just below the critical stoichiometric amount of air required for the operation (Maggio et al. 2019). NOx emissions are reduced via any method that lowers the combustion temperature. However, diluents in CH_4 lowers the combustion temperature, biogas combustion emits less NOx than NG or gasoline. Various researchers have obtained similar results. Midkiff et al. (2001) discovered a decrease in NOx for all biogas mixtures when compared to the NG they used. Peak NOx emissions from biogas, between the range of 0.85–0.9, showed 50% reduction. Anwar et al. (2019) conducted an analytical study on the formation of NOx emissions from biogas and discovered that they are lower than CH_4 emissions. The researchers explained the cause to be the existence of lower burnt gas temperatures caused by inert gas dilution in the biogas. The burnt gas temperature was found to be reduced by 130 K when compared to CH_4, which was enough to cutdown the NO level in the exhaust by half. The peak value of the equilibrium predicted NO was found to be at par with the actual value received at the exhaust according to the results of the simulation (Shrestha and Karim 2001). However, the peak value of equilibrium predicted NO was found to be higher than the initial kinetically predicted value, and the kinetically predicted concentration of NOx value was found to be higher than the equilibrium towards the end of combustion process. According to Kalsi, (Kalsi and Subramanian 2017), higher compression ratios increased the NOx emissions in contrast to the perceived lower compression ratios. But the peak NOx for biogas was still lower than that obtained for Natural gas (NG) of the highest composition (35%) relative to those of 10 and 25% NG respectively.

CO and CO_2 Emissions

CO emissions from biogas are found to be low in lean mixtures and do not vary with the CO_2 fraction in the mixture. CO emissions for a biogas mixture (60% CH_4, 40% CO_2) are found to be approximately 10% higher than NG for the leaner mixtures ($\varphi = 0.65$–0.90). Because O_2 becomes scarce as the mixtures are enriched, CO emissions rise rapidly with the CO_2 fraction due to incomplete combustion (Bhattachariya 2004). Bari (1996) also suggested that high CO emissions from biogas appeared to be linked to lower exhaust temperatures rather than with fuel CO_2 dissociation. According to the Bhattacharya et al. (Bhattachariya 2004) investigation, the higher the CO_2 fraction in biogas, the lower the exhaust temperature, and the end result is higher CO emissions in the exhaust. Increase in carbon dioxide emissions is directly proportional to the amount of CO_2 in biogas. Feroskhan et al. (2018), on the other hand, reported that with mixtures richer than the required stoichiometric amounts of air and fuel, the increase in CO_2 emissions can be modified by dissociation reactions which lead to an increase in CO emissions. Also, according to reports, the results from the emissions of CO_2 can be up to about 80% more than that of natural gas for the biogas (Adelt et al. 2011).

Unburnt HC

Unburned HC emissions have been reported to be higher in biogas compared to natural gas, and they increase as the fractions of CO_2 in the biogas mixtures increase (Chang et al. 2011). The dominant parameters causing increased HCs in leaner mixtures are quenching effects due to lower cylinder temperature and incomplete combustion due to slower flame speed. Incomplete combustion occurs in richer mixtures due to a lack of high O_2 concentration, resulting in increased hydrocarbon emissions (Caton and Pruitt 2009) (Table 8.3).

8.3.5 Dual Fuel Systems and Engine Life

Only a few research articles have been published that investigated the impact of biogas fueling on engine life. Barik and Murugan (2016) looked into the severe corrosion of biogas-powered engines that failed after 1000–3000 h of operation. The failure of the small end and camshaft bearing (especially copper-based) was nearly always found to be the cause of failure due to H_2S attack. In the study, the formation of blowby, its composition and flowrate were measured, modeled, and simulated using a simple computer program. The simulation revealed that the complex nature of the quenching process in the combustion chamber affected the fuel content of blowby. According to Di Iorio et al. (2017), the H_2S attack could only be caused by unburnt fuel entering the crankcase and the rate of ingress increased as the operating hours of the engine increased. It was also suggested that copper-based bearings should be avoided or protected by a positive flow of engine oil (Wallner et al. 2009). During their investigations into biogas-powered vehicles, other researchers made similar findings as listed below:

- SI engines tolerate emission level of hydrogen sulphide up to 1000 ppm.
- between 1000 and 2000 ppm concentrations of H_2S the acidification problem can be controlled (Tanoue et al. 2000).

Tanoue et al. (2000) studied the effect of adding a minute quantity of H_2 to the lean CH_4 mixture. The burning velocity significantly increased as it were for the misfire limits, which were the main disadvantages of using biogas fuel. Verhelst (2014) used unscavenged combustion pre-chamber instead of a direct ignition type to achieve a cleaner and more rational biogas while incurring no significant economic costs. When compared to NG operation, CO and total hydrocarbon (THC) emissions were reduced by 15% and 8%, respectively, for the same rated power output and NOx emissions. Alternatively, efficiencies greater than 36% were achieved while the combustion process remained essentially unchanged (Wallner et al. 2013).

Table 8.3 Summary of some biofuel types, their engine emissions, combustion characteristics and performance

Fuel/experimental details	BTE	NOx	UHC	CO	CO_2	Smoke	Ignition delay	Rate of heat release	References
D90B10; D80B20; D50B50; D95E5; D90E10; E15D85—Cottonseed Biodiesel	Slightly higher	NA	Decreases	Decreases for all blends Increases with the percentage of biofuel in the blend	Decreased	Decreases substantially with increase in bioethanol	Increases with increase in ethanol	Very high at the start of combustion and lower later on the expansion stroke for diesel/Biodiesel blends	Tutak et al. (2016)
Diesel–Methanol dual-fuel—Separate injections	Increases with methanol content in the blend	Decreases significantly	NA	Decreases	NA	Decreases	NA	NA	Bharadwaz et al. (2016)
B10E10D80—Biodiesel obtained from microalgae oil Variable compression ratio diesel engine at different speeds	A maximum BTE of 28.2% was observed	Decreases	Decreases	Decreases	Decreases	NA	NA	NA	Niculescu et al. (2017)

(continued)

Table 8.3 (continued)

Fuel/experimental details	BTE	NOx	UHC	CO	CO_2	Smoke	Ignition delay	Rate of heat release	References
D90M10; D80M20; D70M30	NA	Decreases with an increase in the percentage of methanol in diesel blends	NA	NA	Decreases with the increasing percent of methanol	Increases with an increase in the percentage of methanol in diesel blends	Higher at all loads	The maximum value increases with increasing the methanol fraction in the blends	Bae and Kim (2017)
Ethanol/Isobutanol as additive in the biodiesel/diesel blends – Fish oil Biodiesel	Higher	Increased	Decreased	Decreased	NA	Decreases	NA	NA	Paredes-Sánchez et al. (2019)

D diesel; *B* biodiesel; *E* ethanol, *M* methanol

8.3.6 Current Trends in the Use of Biofuels as High-Performance Engine Fuels

Since the current body of literature is largely flooded with works related to CI engines relative to ICEs (very scarce), it is common knowledge that same fuel systems can be modified with compatible properties for ICEs. Also, most of the fuel proportions adopted for CI engines are not usually desirous below or above the optimum composition, however, it is then recommended that some of the fuel misfits can be tried for their compatibilities with ICEs.

In dual mode, CH_4 and diesel combustion occurs in 3-modes: flame progression in methane–air mixture, pilot diesel combustion, and CH_4 combustion around the diesel spray. Exceptionally poor blends, increase the flame zones, and incomplete combustion of the fuel for all three sections at minimum loads which can lead to an increase in hydrocarbon and carbon monoxide emissions (Prakash et al. 1999).

Homogeneous charge compression ignition engines may be the ultimate solution to lowering NOx, since they emit almost no NOx. They have the advantages of diesel and Otto engines. Homogeneous charge compression ignition engines have many benefits, such as high thermal efficiencies, low emission, and high heat release rates, due to the premixed and lean mixtures, self-ignition at high compression rates (Mohammadi et al. 2007). It has some disadvantages such as difficulty in controlling the starting point of ignition over variations of speed. The combustion phase has been controlled with variable valve timing and EGR to overcome this HCCI challenge. The most effective way to control combustion, however, is to inject pilot diesel.

Meijden et al. (2011) transformed a heavy-duty compression ignition engine to dual-fuel by injecting a diesel and NG mixture into the compression ignition engine and injecting small amounts of diesel near TDC. In that study, a heavy-duty engine is transformed to dual fuel operation.

Baldwin (2008) investigated the effects of dissolved CH_4 in diesel fuel on engine emission and performance in a single-cylinder direct-injection diesel compression ignition engine. They discovered that as methane concentration increased, the maximum heat release rate decreased while the ignition delay increased. Diesel fuel containing dissolved CH_4 emits less NOx, and the amount of smoke produced is proportional to methane concentration.

Chong and Hochgreb (2011) injected 60% methane and 40% carbon dioxide by volume into the intake manifold of an engine. Their findings show that particulate matter (PM), smoke, and NOx, levels decreased, however, the combustion stability was compromised, and then the resulting carbon dioxide and HC emissions were seen to have increased. As a result, they introduced syngas to the liquid in attempt to optimize combustion stability emissions.

Oni et al. (2021) investigated the combustion, performance, and emission characteristics of moringa biodiesel as an additive in (HCNG). According to the authors, five hybrid HCNG—Moringa biodiesel (MB) oil samples were analyzed in a (Petter PH1W diesel engine). At higher engine loads, there was improvement in the engine performance between 19 and 33.9% for all the fuel blends which confirms the suitability of MB as an additive for improving the BTE of the engine with reductions in HC, CO_2, CO, and NOx emissions.

8.4 Future Prospects of Dual-Fuel System as an Alternative Fuel

The transition to a 100% renewable energy system is a complex process which may be saddled/fraught with technical and economic difficulties. To achieve these goals, several steps should be taken concurrently, including increasing the energy efficiency, reducing primary energy consumption, and deploying variable energy sources (Wallner et al. 2009). This transition by the energy sector channeled towards decarbonization necessitates the use of new technologies and energy sources (Suresh et al. 2018). Increased use of renewable energy sources necessitates the installation of energy storage systems to store excess electricity and improve system efficiencies (Rimkus et al. 2018). These electricity surpluses, which will become more common in the future energy systems, could be effectively used to produce alternative fuels. The majority of the alternative fuels being considered for future applications are already known chemicals or products that are currently used for other purposes (Lee et al. 2001). Another significant benefit of some alternative fuels is their ability to serve as an energy-carrier. This feature may be critical when discussing their potential for use and future development. Fuels that can be used for both power generation and as an energy carrier will play a larger role in the future and are likely to be used on a larger scale (Mansor et al. 2017). Renewable energy sources, on the other hand, such as biomass, are already widely used, and their roles in future systems are unquestionable despite the fact that a significant increase in biomass consumption raises serious concerns about its sustainability and necessitates the development of new approaches (Huyskens et al. 2011). Figure 8.4 shows some renewable energy sources as potential sources for sustainable biofuel production.

8.5 Sustainable Technologies for Alternative Fuels and Future Challenges

8.5.1 Sustainable Technologies

8.5.1.1 Production of H_2 for CNG Fuel-Systems

Production of H_2 from fossil fuels is a common process that uses coal or natural gas as feedstock (Stančin et al. 2020). H_2 is most commonly produced via steam reforming

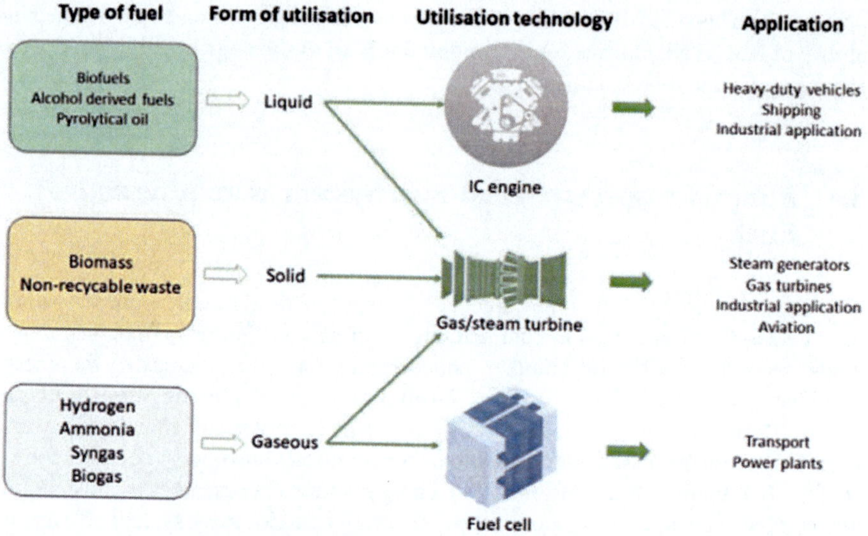

Fig. 8.4 Technology utilization perspectives for alternative fuels. Reprinted from Stančin et al. (2020) with copyright permission obtained from Elsevier

of CH_4, but it can also be produced through partial or autothermal oxidation or gasification (Salazar and Kaiser 2011). Nonetheless, production from fossil fuels will be impossible in a future decarbonized energy system, hence, progressive efforts must shift towards sustainable solutions. Biomass pyrolysis or gasification is required for its production using renewable energy sources (Glaude et al. 2010). Moreover, H_2 can be produced directly through solar, nuclear, or industrial waste heat utilization since higher temperatures are required if hydrogen is produced directly from solar energy, because hydrogen production from concentrated solar power (CSP) appears to be the best solution (Baldwin 2008). Nuclear energy production necessitates the incorporation of waste heat for high-temperature electrolysis, which necessitates additional research (Boisen 2008). Blending H_2 with natural gas is a major breakthrough and a promising mixture fuel for CI engines, hence, it is suggested that higher percentages can be blended with natural gas with compositions lighter than those of diesel engines as test-fuel mixtures for ICEs.

8.5.1.2 Methanol Synthesis

The use of light to drive or assist chemical reactions and processes to produce clean methanol is an innovative technology that is becoming more visible in the scientific literature (Manoharan et al. 2019). The prospect of synthesizing methanol using solar energy, CO_2, and water could lead to an economically viable technology capable of replacing fossil fuel with a renewable alternative (Das et al. 2020). There are several methods for producing light-assisted chemical products, including direct conversion of carbon dioxide and H_2O via photochemistry, photo-electrochemistry or solar thermochemistry. Another alternative possible explanation is to gasify biomass with the aim of producing hydrogen and syngas (Glaude et al. 2010). Solar concentrators increase the amount of sunlight that strikes the biomass gasifier. The temperatures reached should be high enough (850 °C) to impact biomass gasification without additional heating (Wallner et al. 2009). The methanol produced via this technology can be made available for blending biofuels for use in ICEs.

8.5.1.3 Biomethane from Anaerobic Digestion (AD)

AD is a 4-step bio-metabolism process that converts wastes which are biodegradable into valuable biogas, primarily CH_4 (Lee et al. 2001). The AD process is ideal for treating a biodegradable fraction of MSW, agricultural waste (harvest remains, energy crops, algal biomass, animal manures), food industry waste (starch, beverage processing, dairy), or sludge from sewage (Wallner et al. 2009). The four-step process can be applied to single-stage or multi-stage AD systems, with the last step necessitating additional modifications. Complex relationships between various operating parameters, growth factors, system design, and reactor type determine the overall process efficiency (Verhelst 2014). The feedstock type is critical for the selection of system design and reactor type, as well as for growth factors and operational parameters (Hänggi et al. 2019). During the process, pH is strictly controlled because it affects the efficiency of the bacteria and, as a result, the process's success rate. Nowadays, the majority of AD systems operate in a continuous single-stage mode while processing a variety of biodegradable waste (Salazar and Kaiser 2011). The biomethane produced can be used to atomize biodiesel such that the hybrid-fuel exhibits enhanced spray characteristics when used in ICEs.

8.5.2 Current Challenges and Future Trends

Presently, there are several limitations for the efficient use of alternative fuels. First, fossil fuel availability makes it very difficult for alternative fuels to meet cost-competitive production costs (Glaude et al. 2010). Considering waste fuel and biofuels, attaining standard quality is the main concern. Also, thermal stability, higher acidity, lower heating value, and similar limits, provide wider deployment of current commercially available biofuels. However, in this field, research is still ongoing with the constant improvement on the fuels produced, indicating that the role of these fuels is not uncertain in the future (Salazar and Kaiser 2011). Alternatively, ammonia, alcohols, and hydrogen derived fuels have familiar production procedures, which are mostly synthesized for industrial use, thus indicating that higher costs of production are not a concern for such processes, however, further reduction is to be expected if the purpose is for it to be used as fuel. Additionally, the use of new fuels involves modification on the existing technologies (Wellinger and Lindberg 2000). The effective use of alcohol and biofuel in ICEs is very much possible with little or improved engine modifications of the existing ICEs, however, in the case of H_2 and NH_3 significant modification or possibly development of new technology is often required (Bordelanne et al. 2011). The development of fuel cells, for the utilization of H_2, demonstrates exceptional perspective to be deployed of both portable and stationary applications, even if further modifications are expected to increase efficiency as well as optimize the operating conditions. Lastly, another challenge of alternative fuels is in the production process, which is expected to shift towards sustainable and clean solutions (Salazar et al. 2009). In biofuels, for example, this implies the utilization of industrial biomass residues and agricultural wastes that will yield high-quality clean fuels. Concurrently, in achieving carbon neutrality, synthetic fuel production should tend towards new solutions that do not require the use of fossil fuels as precursor/feedstock.

Direct utilization of solar energy for fuel synthesis is a trending research, which is still ongoing. One advantage of this process is that it does not require an external source of energy (Antunes et al. 2008). Nonetheless, the low conversion efficiency of this type of energy is influenced by some factors such as overall process efficiency, thus making solar production unprofitable and cost effective. Furthermore, efforts are being made by researchers to bring innovative technologies that can be operated in flexible mode especially on a larger scale (Scarcelli et al. 2011). This is important for carbon capture and electrolysis technologies which are used to produce feedstock (hydrogen and carbon dioxide) for alternative fuel synthesis. Concerted efforts are being made by researchers to bring about thermochemical conversion methods for alternative fuel production on a larger scale basis (You et al. 2020). Gasification and pyrolysis are very important processes since they can process varieties of waste materials, thus converting them into chemicals and/or fuels. Currently, research focus is shifted to enhancing the properties of biofuels through co-gasification or co-pyrolysis with high calorific waste materials which will also help in waste management (Yadvika et al. 2004).

8.6 Conclusion

Dual fuel systems are promising fuels which hold future prospects as transport fuels in ICEs since a considerable number of options exist with respect to CI engines. The importance of dual fuel can be ascribed to improving thermal efficiencies, and combustion properties with reduced emissions when compared to fossil fuel. Some dual fuel types were discussed in this book which include; biogas-biomethane blends, biomethane-CNG hybrid fuels, among others. The performance, emission and combustion characteristics of dual fuels were discussed as well as other properties which could affects the fuel quality such as air -fuel ratio, power and fuel consumption, effect of dual fuel on engines lifespan and the effects of diluents on the fuel. Based on reports, it is clear that besides the use of NG and alcohol blends as fuels for ICEs, other fuel options explored for CI engines including fractions comprising of far lower ratios of biofuels relative to gasoline can be considered as these will render the fuels as light as gasoline without clearly altering the properties of gasoline such that they are fit for application in such ICEs. There is needs to install a simple gas-fuel adaptor with toughened valves and valve seats which slightly modifies the fuel adaptor in the regular engine configuration for SI engines. Furthermore, with the recent prospects of hydrogen as transport fuel, issues related to sustainable production and storage are currently being overcome which then increases the chances of the fuel becoming a replacement/an alternative fuel for gasoline such that hybrid cars that can run on either hydrogen/NG and gasoline are designed for use as modern-day vehicles. With the current advancement in today's research, there is an inkling that the road to the envisaged future of hybrid gasoline-based cars running on hydrogen/NG and NG will soon become a reality.

References

Adelt M, Wolf D, Vogel A (2011) LCA of biomethane. J Nat Gas Sci Eng 3(5):646–650. https://doi.org/10.1016/j.jngse.2011.07.003

Åhman M (2010) Biomethane in the transport sector—an appraisal of the forgotten option. Energy Policy 38(1):208–217. https://doi.org/10.1016/j.enpol.2009.09.007

Ambarita H (2017) Performance and emission characteristics of a small diesel engine run in dual-fuel (diesel-biogas) mode. Case Stud Therm Eng 10:179–191

Antunes JG, Mikalsen R, Roskilly AP (2008) An investigation of hydrogen-fuelled HCCI engine performance and operation. Int J Hydrogen Energy 33:5823–5828

Anwar M, Rasul MG, Ashwath N (2019) The efficacy of multiple-criteria design matrix for biodiesel feedstock selection. Energy Convers Manag 198:111790

Arnoersson H (2011) A feasibility study of using biomethane as an alternative fuel for taxis in the Reykjavik Capital area. University of Iceland and University of Akureyri

Bae C, Kim J (2017) Alternative fuels for internal combustion engines. Proc Combust Inst 36:3389–3413

Baldwin J (2008) The case for biomethane fueled vehicles. CNG services ltd.

Banerjee S, Musa MN, Jaafar AB (2017) Economic assessment and prospect of hydrogen generated by OTEC as future fuel. Int J Hydrogen Energy 42:26–37

Bari S (1996) Effect of carbon dioxide on the performance of biogas/diesel dual-fuel engine. Renew Energy 9:1007–1010

Barik D, Murugan S (2016) Experimental investigation on the behavior of a DI diesel engine fueled with raw biogas—diesel dual fuel at different injection timing. J Energy Inst 89:373–388

Bharadwaz YD, Rao BG, Rao VD, Anusha C (2016) Improvement of biodiesel methanol blends performance in a variable compression ratio engine using response surface methodology. Alexandria Eng J 55:1201–1209

Bhattachariya SC (2004) Fuel for thought—the status of biomass energy in developing countries. Renew Energy World 7:122–130

Boesel J (2009) Taking advantage of the lowest carbon fuel and improving the environmental performance of NGVs. Natural bio Gas Vehicle Association (NGVA) Summit

Boisen P (2008) The role biomethane plays for the growing use of NG/biomethane as a vehicle fuel. Natural & bio Gas Vehicle Association (NGVA)

Bordelanne O, Montero M, Bravin F, Prieur-Vernat A, Oliveti-Selmi O, Pierre H, Papadopoulo M (2011) Biomethane CNG hybrid: a reduction by more than 80% of the greenhouse gases emissions compared to gasoline. J Nat Gas Sci Eng 3(5):617–624. Elsevier B.V. https://doi.org/10.1016/j.jngse.2011.07.007

Bortel I, Vávra J, Takáts M (2019) Effect of HVO fuel mixtures on emissions and performance of a passenger car size diesel engine. Renew Energy 140:680–691

Caton PA, Pruitt JT (2009) Homogeneous charge compression ignition of hydrogen in a single-cylinder diesel engine. Int J Engine Res 10:45–63

Chang IS, Zhao J, Yin X, Wu J, Jia Z, Wang L (2011) Comprehensive utilizations of biogas in Inner Mongolia, China. Renew Sustain Energy Rev 15(3):1442–1453. https://doi.org/10.1016/j.rser.2010.11.013

Chintala V, Subramanian KA (2017) A comprehensive review on utilization of hydrogen in a compression ignition engine under dual fuel mode. Renew Sustain Energy Rev 70:472–491

Chong CT, Hochgreb S (2011) Measurements of laminar flame speeds of liquid fuels: Jet-A1, diesel, palm methyl esters and blends using particle imaging velocimetry (PIV). Proc Combust Inst 33:979–986

Chu S, Lee J, Kang J, Lee Y, Min K (2018) High load expansion with low emissions and the pressure rise rate by dual-fuel combustion. Appl Therm Eng 144:437–443

Das S, Kashyap D, Kalita P, Kulkarni V, Itaya Y (2020) Clean gaseous fuel application in diesel engine: a sustainable option for rural electrification in India. Renew Sustain Energy Rev 117:109485

Di Iorio S, Magno A, Mancaruso E, Vaglieco BM (2017) Analysis of the effects of diesel/methane dual fuel combustion on nitrogen oxides and particle formation through optical investigation in a real engine. Fuel Process Technol 159:200–210

Dimitriou P, Tsujimura T (2017) A review of hydrogen as a compression ignition engine fuel. Int J Hydrogen Energy 42:24470–24486

Ehsan M, Bhuiyan MSA, Naznin N (2003) Multi-fuel performance of a petrol engine for small scale power generation. SAE Technical Paper No. 2003-32-0063, Small Engine Technology Conference & Exhibition, Madison, Wisconsin, USA, September 15–18

Elfasakhany A (2018) Exhaust emissions and performance of ternary iso-butanolebio-methanolegasoline and n-butanolebio-ethanolegasoline fuel blendsin spark-ignition engines: assessment and comparison. Energy 158(2018):830–844

Elfasakhany A (2021) State of art of using biofuels in spark ignition engines. Energies 14:779. https://doi.org/10.3390/en14030779

Feroskhan M, Ismail S, Reddy MG, Sai Teja A (2018) Effects of charge preheating on the performance of a biogas-diesel dual fuel CI engine. Eng Sci Technol Int J 21:330–337

Glaude PA, Fournet R, Bounaceur R, Molière M (2010) Adiabatic flame temperature from biofuels and fossil fuels and derived effect on NO_x emissions. Fuel Process Technol 91:229–235

Hänggi S, Elbert P, Bütler T, Cabalzar U, Teske S, Bach C, Onder C (2019) A review of synthetic fuels for passenger vehicles. Energy Rep 5:555–569

Heffel JW (2003) NO_x emission and performance data for a hydrogen fueled internal combustion engine at 1500 rpm using exhaust gas recirculation. Int J Hydrogen Energy 28:901–908

Helmolt RV, Eberle U (2007) Fuel cell vehicles: status 2007. J Power Sources 165:833–843

Huyskens P, Van Oost S, Goemaere PJ, Bertels K, Pecqueur M (2011) The technical implementation of a retrofit hydrogen PFI system on a passenger car; SAE Paper 2011-01-2004. SAE International, Warrendale, PA, USA

Juknelevičius R, Rimkus A, Pukalskas S, Matijošius J (2019) Research of performance and emission indicators of the compression-ignition engine powered by hydrogen—diesel mixtures. Int J Hydrogen Energy 44:10129–10138

Kaiser S, White C (2009) PIV and PLIF to evaluate mixture formation in a direct-injection hydrogen-fuelled engine. SAE Int J Engines 1:657–668

Kalsi SS, Subramanian KA (2017) Effect of simulated biogas on performance, combustion and emissions characteristics of a bio-diesel fueled diesel engine. Renew Energy 106:78–90

Karim GA, Wierzba I (1992) Methane-carbon dioxide mixtures as a fuel. SAE Technical Paper No. 921557

Kukharonak H, Ivashko V, Pukalskas S, Rimkus A, Matijošius J (2017) Operation of a spark-ignition engine on mixtures of petrol and N-butanol. Procedia Eng 187:588–598

Lebedevas S, Pukalskas S, Daukšys V, Rimkus A, Melaika M, Jonika L (2019) Research on fuel efficiency and emissions of converted diesel engine with conventional fuel injection system for operation on natural gas. Energies 12:2413

Lee JT, Kim YY, Lee CW, Caton JA (2001) An investigation of a cause of backfire and its control due to crevice volumes in a hydrogen fueled engine. J Eng Gas Turbines Power 123:204–210

Lee KJ, Kim YR, Byun CH, Lee JT (2013) Feasibility of compression ignition for hydrogen fueled engine with neat hydrogen-air pre-mixture by using high compression. Int J Hydrogen Energy 38:255–264

Liu J, Dumitrescu CE (2019) Methodology to separate the two burn stages of natural-gas lean premixed-combustion inside a diesel geometry. Energy Convers Manag 195:21–31

Lyng K-A, Stensgård AE, Hanssen OJ, Modahl IS (2018) Relation between greenhouse gas emissions and economic profit for different configurations of biogas value chains: a case study on different levels of sector integration. J Clean Prod 182:737–745

Maggio G, Nicita A, Squadrito G (2019) How the hydrogen production from RES could change energy and fuel markets: a review of recent literature. Int J Hydrogen Energy 44:11371–11384

Manoharan Y, Hosseini SE, Butler B, Alzhahrani H, Senior BTF, Ashuri T, Krohn J (2019) Hydrogen fuel cell vehicles; Current status and future prospect. Appl Sci 9:2296

Mansor MRA, Abbood MM, Mohamad TI (2017) The influence of varying hydrogen-methane-diesel mixture ratio on the combustion characteristics and emissions of a direct injection diesel engine. Fuel 190:281–291

Martins J, Brito FP (2020) Alternative fuels for internal combustion engines. Energies 13:4086. https://doi.org/10.3390/en13164086

Martins J (2020) Motores de Combustão Intern, 6th edn. Quantica Editora, Porto, Portugal. ISBN 9789898927842

Matthias NS, Wallner T, Scarcelli R (2012) A hydrogen direct injection engine concept that exceeds U.S. DOE light-duty efficiency targets. SAE Int J Engines 5:838–849

Meijden CMVD, Rabou LPLM, van der Drift A, Vreugdenhil BJ, Smit R (2011) Large scale production of bio methane from wood. International Gas Union Research Conference IGRC, Seoul, South Korea. Retrieved from http://www.ecn.nl/docs/library/report/2011/m11098.pdf

Midkiff KC, Bell SR, Rathnam S, Bhargava S (2001) Fuel composition effects on emissions from a spark-ignited engine operated on simulated biogases. Natural Gas Engines, ICE Vol. 123, Transactions of the ASME, pp 132–138

Mohammadi A, Shioji M, Nakai Y, Ishikura W, Tabo E (2007) Performance and combustion characteristics of a direct injection SI hydrogen engine. Int J Hydrogen Energy 32:296–304

Murphy JD, Power N (2009) Technical and economic analysis of biogas production in Ireland utilising three different crop rotations. Appl Energy 86(1):25–36. https://doi.org/10.1016/j.apenergy.2008.03.015

Niculescu R, Iosub I, Clenci A, Zaharia C, Iorga-Siman V (2017) Development of a test method for distillation ˇ of diesel-biodiesel-alcohols mixtures at reduced pressure. IOP Conf Ser Mater Sci Eng

Obermair H, Scarcelli R, Wallner T (2010) Efficiency improved combustion system for hydrogen direct injection operation; SAE Paper 2010–01-2170. SAE International, Warrendale PA USA

Ogunkunle O, Ahmed NA (2019) A review of global current scenario of biodiesel adoption and combustion in vehicular diesel engines. Energy Rep 5:1560–1579

Oni BA, Sanni SE, Ibegbu AJ, Aduojo AA (2021) Experimental optimization of engine performance of a dual-fuel compression-ignition engine operating on hydrogen-compressed natural gas and Moringa biodiesel. Energy Rep 7(2021):607–619

Ozcanli M, Bas O, Akar MA, Yildizhan S, Serin H (2018) Recent studies on hydrogen usage in Wankel SI engine. Int J Hydrogen Energy 43:18037–18045

Paredes-Sánchez JP, López-Ochoa LM, López-González LM, Las-Heras-Casas J, Xiberta-Bernat J (2019) Evolution and perspectives of the bioenergy applications in Spain. J Clean Prod 213:553–568

Prakash G, Ramesh A, Shaik AB (1999) An approach for estimation of ignition delay in a dual fuel engine. SAE Paper No. 1999-01-0232

Rimkus A, Matijošius J, Bogdevičius M, Bereczky A, Török Á (2018) An investigation of the efficiency of using O_2 and H_2 (hydrooxile gas –HHO) gas additives in a ci engine operating on diesel fuel and biodiesel. Energy 152:640–651

Roubaud A, Favrat D (2005) Improving performances of a lean burn cogeneration biogas engine equipped with combustion prechambers. Fuel 84:2001–2007

Salazar VM, Kaiser SA, Halter F (2009) Optimizing precision and accuracy of quantitative PLIF of acetone as a Tracer for hydrogen fuel. SAE Int J Fuels Lubr 2:737–761

Salazar V, Kaiser S (2011) Interaction of intake-induced flow and injection jet in a direct-injection hydrogen-fueled engine measured by PIV; SAE Paper 2011-01-0673. SAE International, Warrendale, PA, USA

Sandalcı T, Karagöz Y (2014) Experimental investigation of the combustion characteristics, emissions and performance of hydrogen port fuel injection in a diesel engine. Int J Hydrogen Energy 39:18480–18489

Sanni SE, Alaba PA, Okoro E, Emetere M, Oni B, Agboola O, Ndubuisi AO. Strategic examination of the classical catalysis of formic acid decomposition for intermittent hydrogen production, storage and supply: a review. Sustain Energy Technol Assess 45:101078

Santoso WB, Bakar RA, Nur A (2013) Combustion characteristics of diesel-hydrogen dual fuel engine at low load. Energy Procedia 32:3–10

Saravanan N, Nagarajan G (2008) An experimental investigation of hydrogen-enriched air induction in a diesel engine system. Int J Hydrogen Energy 33:1769–1775

Scarcelli R, Wallner T, Matthias N, Salazar V, Kaiser S (2011) Mixture formation in direct injection hydrogen engines: CFD and optical analysis of single- and multi-hole nozzles. SAE Int J Engines 2:2361–2375

Shrestha SO, Karim GA (2001) Prediction the effects of the presence of diluents with methane on spark ignition engine performance. Appl Therm Eng 21:331–342

Stančin H, Mikulčić H, Wang X, Duić N (2020) A review on alternative fuels in future energy system, vol 128, Aug 2020, 109927

Stenlåås O, Christensen M, Egnell R, Johansson B, Mauss F (2004) Hydrogen as homogeneous charge compression ignition engine fuel. J Fuels Lubr 113:1317–1326

Stockhausen WF, Natkin RJ, Kabat DM, Reams L, Tang X, Hashemi S, Szwabowski SJ, Zanardelli VP (2002) Ford P2000 hydrogen engine design and vehicle development program; SAE Paper 2002–01-0240. SAE International, Warrendale, PA USA

Stürmer B (2017) Biogas—part of Austria's future energy supply or political experiment? Renew Sustain Energy Rev 79:525–532

Suarez-Bertoa R, Kousoulidou M, Clairotte M, Giechaskiel B, Nuottimäki J, Sarjovaara T, Lonza L (2019) Impact of HVO blends on modern diesel passenger cars emissions during real world operation. Fuel 235:1427–1435

Suresh M, Jawahar C, Richard A (2018) A review on biodiesel production, combustion, performance, and emission characteristics of non-edible oils in variable compression ratio diesel engine using biodiesel and its blends. Renew Sustain Energy Rev 92:38–49

Szwaja S, Grab-Rogalinski K (2009) Hydrogen combustion in a compression ignition diesel engine. Int J Hydrogen Energy 34:4413–4421

Tanoue K, Kido H, Hamatake T, Shimada F (2000) Improving the turbulent combustion performance of lean methane mixture by hydrogen addition. In: Proceeding of FISITA World Automotive Congress, Seoul, Korea, F2000A115

Tutak W, Jamrozik A, Pyrc M, Sobiepański M (2016) Investigation on combustion process and emissions characteristic in direct injection diesel engine powered by wet ethanol using blend mode. Fuel Process Technol 149:86–95 [CrossRef]

Verhelst S (2014) Recent progress in the use of hydrogen as a fuel for internal combustion engines. Int J Hydrogen Energy 39:1071–1085
Verma S, Das LM, Kaushik SC, Tyagi SK (2018) An experimental investigation of exergetic performance and emission characteristics of hydrogen supplemented biogas-diesel dual fuel engine. Int J Hydrogen Energy 43:2452–2468
Wallner T, Scarcelli R, Nande AM, Naber JD (2009) Assessment of multiple injection strategies in a direct-injection hydrogen research engine. SAE Int J Engines 2:1701–1709
Wallner T, Matthias NS, Scarcelli R, Kwon JC (2013) A Evaluation of the efficiency and the drive cycle emissions for a hydrogen direct-injection engine. J Automob Eng 227:99–109
Wellinger A, Lindberg A (2000) Biogas upgrading and utilization—IEA bioenergy task 24: energy from biological conversion of organic waste. International Energy Association, Paris, France, 20 p
White CM, Steeper RR, Lutz AE (2006) The hydrogen-fueled internal combustion engine: a technical review. Int J Hydrogen Energy 31:1292–1305
Wimmer A, Wallner T, Ringler J, Gerbig F (2005) H2-direct injection—a highly promising combustion concept; SAE Paper 2005–01-0108. SAE International, Warrendale, PA USA
Wu Z, Han Z (2020) Micro-GA optimization analysis of the effect of diesel injection strategy on natural gas-diesel dual-fuel combustion. Fuel 259:116288
Xue S, Song J, Wang X, Shang Z, Sheng C, Li C, Zhu Y, Liu J (2020) A systematic comparison of biogas development and related policies between China and Europe and corresponding insights. Renew Sustain Energy Rev 117:109474
Yadvika S, Sreekrishnam TR, Kohli S, Rana V (2004) Enhancement of biogas production from solid substrates using different techniques—a review. Biores Technol 95:1–10
Yang B, Zeng K (2018) Effects of natural gas injection timing and split pilot fuel injection strategy on the combustion performance and emissions in a dual-fuel engine fueled with diesel and natural gas. Energy Convers Manag 168:162–169
You J, Liu Z, Wang Z, Wang D, Xu Y (2020) Impact of natural gas injection strategies on combustion and emissions of a dual fuel natural gas engine ignited with diesel at low loads. Fuel 260:116414
Zhang W, Chang S, Wu W, Dong L, Chen Z, Chen G (2019a) A diesel/natural gas dual fuel mechanism constructed to reveal combustion and emission characteristics. Energy 179:59–75
Zhang Y, Fu J, Shu J, Xie M, Zhou F, Liu J, Zeng D (2019b) A numerical investigation of the effect of natural gas substitution ratio (NGSR) on the in-cylinder chemical reaction and emissions formation process in natural gas (NG)-diesel dual fuel engine. J Taiwan Inst Chem Eng 105:85–95
Zheng J, Liu X, Xu P, Liu P, Zhao Y, Yang J (2012) Development of high pressure gaseous hydrogen storage technologies. Int J Hydrogen Energy 37:1048–1057

Chapter 9
Compressed Natural Gas Utilization in Dual-Fuel Internal Combustion Engines

Sulav Kafle, Ankur Kalwar, Hardikk Valera, and Avinash Kumar Agarwal

Abstract Unrestrained consumption of petroleum products and harmful emissions generated from them are of great concern globally. Because of stringent emission legislation, researchers are now focusing on controlling emissions and performance improvement in the existing vehicles. Dual-fuel operation using CNG is an option for emission compliance while improving fuel economy. The dual-fuel mode can be used in both compression ignition (CI) and spark ignition (SI) engines, with gaseous fuel (compressed natural gas: CNG) being the primary fuel and a small proportion of liquid fuel being injected as pilot fuel (Secondary fuel). Due to vast availability, cleaner combustion characteristics, and superior economics, CNG has emerged as an attractive fuel for light-duty and heavy-duty engines. Advanced injection timings, use of oxygenated pilot fuels, EGR, piston bowl geometry optimization, and injector nozzles optimization are the strategies that could be adopted in conventional engines for improving their performance and reducing the emissions in dual-fuel mode. HC, CO, and smoke reduction were observed in dual-fuel engines with advanced injection timings. Combustion simulations and modelling of CNG-diesel dual-engine using the G-equation model are also discussed.

Keywords CNG · Dual-fuel · Pilot diesel · Gasoline direct injection · Efficient combustion · Emissions

9.1 Introduction

Rapid depletion of fossil fuels due to excessive usage and increasing petroleum prices has forced automotive manufacturers to focus on alternative fuels, combustion concepts, and powertrains. Alternative environment-friendly fuels are essential for sustainable transport. Growing climate change due to GHG emissions and health issues due to pollutants in urban areas, and stricter emission regulations are the factors

S. Kafle · A. Kalwar · H. Valera · A. K. Agarwal (✉)
Engine Research Laboratory, Department of Mechanical Engineering, Indian Institute of Technology Kanpur, Kanpur 208016, India
e-mail: akag@iitk.ac.in

A. K. Agarwal et al. (eds.), *Advanced Combustion for Sustainable Transport*, Energy, Environment, and Sustainability, https://doi.org/10.1007/978-981-16-8418-0_9

pushing for the adaption of alternate fuels. Stringent transport sector emission regulations enforced worldwide are moving towards near-zero-emissions. Though diesel engines have higher efficiency, they emit higher NO_x, soot, and polycyclic aromatic hydrocarbons (PAHs). NO_x reacts with water vapour present in the atmosphere to cause acid rain, harming biotic and abiotic components of the ecosystem. Soot emissions cause cardiovascular problems, affect the development of lungs in small children, and lead to numerous other chronic health issues (Manisalidis et al. 2020). Significant advances have been made for emission control using exhaust gas after-treatment devices such as selective catalytic reduction (SCR), lean NO_x traps (LNT), diesel particulate filter (DPF), etc. However, there is scope for further improvement in performance, combustion, and emissions of the current generation automotive engines using suitable alternative fuels and implementing innovative strategies for controlling NO_x and unburned hydrocarbons to meet the prevailing and upcoming emission legislations. Dual-fuel engines with CNG as primary fuel can be an option to provide higher efficiency and higher diesel substitution with lower emissions. Dual-fuel engines exhibited noticeable improvement in particulate emissions with a higher proportion of diesel substitution by CNG (Cordiner et al. 2007). These engines also reported lower NO_x and PM emissions than baseline diesel-fueled engines (Mustafi et al. 2013). Two important challenges must be addressed for making this technology feasible and market-friendly: (i) availability of an after-treatment system suitable for low exhaust gas temperature and (ii) finding the primary cause of pre-ignition and addressing them. The combustion chamber should be designed to extend the limits of lean-burn combustion, and the control system should also limit the maximum air–fuel ratio at lower loads. The properties of natural gas, its application in dual-fuel configuration, and the possibilities and challenges of using it in the transport sector are discussed extensively in this chapter. The dual-fuel technology can be implemented in both CI and SI combustion engines. Therefore, this dual-fuel mode combustion can be a possible future direction for obtaining sustainable transport. Different types of CNG dual-fuel engines such as CNG-diesel, CNG-biodiesel, CNG-diesel + alcohol, CNG-gasoline, HCNG etc., are discussed along with their pros and cons in this chapter.

9.2 Natural Gas

Compressed natural gas (CNG) is produced by compressing the natural gas, mainly containing methane (CH_4). CNG is used as a cooking/heating fuel in domestic households. CNG was used as a fuel for transport as early as 1930 in Italy (Yeh 2007). During the aftermath of oil scarcity in the 1970s, CNG gained considerable traction as a promising alternative fuel for transport. However, its uses exponentially increased after the early 2000s, when oil prices increased sharply, and natural gas remained cheaper and cleaner. Natural gas is expected to be the fastest-growing primary energy source soon, and its usage is expected to double between 2020 and 2040 (Khan et al. 2015). There are three types of natural gas vehicles based on fuel supply systems: (i) dedicated CNG engines, (ii) retrofitted bi-fuel gasoline engines, and (iii) dual-fuel engines. The dedicated CNG engines are spark-ignition engines that run only on natural gas. These vehicles are designed and manufactured by accounting for the combustion characteristics of natural gas. They are optimized to exploit the higher octane number of natural gas. These are cleaner engines and produce lesser harmful gases. Bi-fuel engines are gasoline engines that can operate either on gasoline or CNG. The normal gasoline engines are retrofitted by advancing the spark timing to use natural gas (as CNG), which has a higher ignition delay and longer combustion duration. Dual-fuel vehicle technology has become popular in CI engines also. These engines run on two fuels, where the primary fuel is a gaseous fuel such as CNG, which is ignited at the end of the compression stroke using a pilot liquid fuel such as diesel. These vehicles are highly efficient vehicles with superior fuel economy and lower operating costs.

The physicochemical properties of CNG are comparable to conventional fuels such as Diesel and Gasoline (Table 9.1). Policymakers and automotive companies view CNG as a fuel with great potential due to its huge availability and favourable physicochemical properties. Therefore many OEMs are converting their existing engines to CNG.

Natural Gas is a widely available environment-friendly fuel that is lighter than air, hence a safer fuel. It does not contain toxic constituents. Its sulfur content is very low. Thus the sulfate emissions are negligible. Since the carbon–carbon molecular bond is absent, the formation of carcinogens such as polycyclic aromatic hydrocarbons and soot is minimal. Natural gas has a higher octane number of ~130, due to which it possesses strong anti-knocking behaviour. This allows the use of higher compression ratios. Even under lean conditions, natural gas has stable combustion due to its wider flammability limits. CO_2 and NO_x emissions are lower because of the lowest carbon

Table 9.1 Physicochemical properties of CNG and Conventional fuels (Khan et al. 2015)

Properties	CNG	Gasoline	Diesel
Octane number	120–130	85–95	–
Cetane number	–	–	45–55
Molar mass (kg/mol)	17.3	109	204
Stoichiometric A/F	17.2	14.7	14.6
Stoichiometric mixture density (kg/m^3)	1.25	1.42	1.46
LHV (MJ/kg)	47.5	43.5	42.7
LHV of stoichiometric mixture (MJ/kg)	2.62	2.85	2.75
Volumetric energy density (gas) (MJ/m^3)	24.6	42.7	36
Flammability limit in air (% v/v in air)	4.3–15.2	1.4–7.6	1–6
Flame speed (m/s)	0.41	0.5	–
Adiabatic flame temperature (°C)	1890	2150	2054
Auto-ignition temperature (°C)	540	258	316
Wobbe index (MJ/m^3)	51–58	–	–

to hydrogen ratio amongst hydrocarbon fuels and lower adiabatic flame temperature. Natural gas-fuelled engines are quieter than diesel engines. Hence they are more attractive in urban areas. The higher efficiency of CNG vehicles is due to the higher mixing and diffusion of gaseous and lighter fuel in the air, which aids in combustion. CNG fueled engines have lower maintenance costs since they have a low dilution of lubricants. The high octane number of natural gas allows higher compression ratios in the engines, increasing the thermal efficiency (Bielaczyc et al. 2015; Khan et al. 2015).

The engine power output and torque of CNG engines are affected by lower flame speed, lower volumetric efficiency, and the absence of fuel evaporation. Lower flame speed stretches the combustion duration, increasing the possibility of a misfire, and unburned hydrocarbons may exit in the exhaust, especially in SI engines. This problem can be addressed by mixing the CNG with higher flame speed fuels such as hydrogen, which improves the lean-burn capability of CNG engines (Ma et al. 2007). The use of a bowl-in-piston combustion chamber increases turbulence, which increases the flame speed. In CI engines with a higher compression ratio, flame speeds are higher. Natural gas is a potential candidate for CI engines for displacing diesel in terms of carbon content reduction. Due to the higher auto-ignition temperature of natural gas, it isn't easy to fully replace diesel. However, with appropriate modifications in the engine design and using diesel as pilot fuel, this can be made technically feasible. There is a reduction in volumetric efficiency of the engine with CNG since it is gaseous and displaces a large volume of the intake air. This can be addressed by changing the design of the intake manifold, adjusting valve timing and lift profile, and using turbocharging (Kalam et al. 2011; Kubesh et al. 1995).

The purchase cost of CNG dedicated vehicles is initially higher than gasoline-fueled engines. Though natural gas can be stored in compressed form at a pressure of 200 bar, it still occupies 3–4.5 times more volume than gasoline in liquid form for the same energy content. Therefore, natural gas storage and transport are challenging. Refuelling infrastructures are limited, and the time for refuelling is also more than the refuelling of liquid petroleum fuels. Exhaust emissions contain methane in higher proportions. Therefore catalytic converters with higher loading of catalysts, especially palladium, are required. Natural gas is hydrophilic, i.e., it absorbs moisture, which may freeze in cold conditions and potentially block the CNG pipeline. The composition of natural gas varies with the geography of its source, affecting its stoichiometric air/fuel ratio (Bielaczyc et al. 2015; McTaggart-Cowan et al. 2006; Korakianitis et al. 2011a).

Worldwide, demand for CNG vehicles is increasing significantly, as shown in Fig. 9.1. One of the reasons for increasing demand is public interest and the interventions of fuel suppliers.

Among various segments of the transport sector, owners of 3-Wheelers and 4-Wheelers are attracted towards CNG and are converting existing vehicles to standalone CNG-operated vehicles or bi-fuel vehicles. A bi-fuel system means that the engine can operate on both CNG and conventional fuels, but one fuel at a time. The driver decides which fuel will burn for powering the engine before switching on the ignition system. Nowadays, automotive OEMs have started offering this option

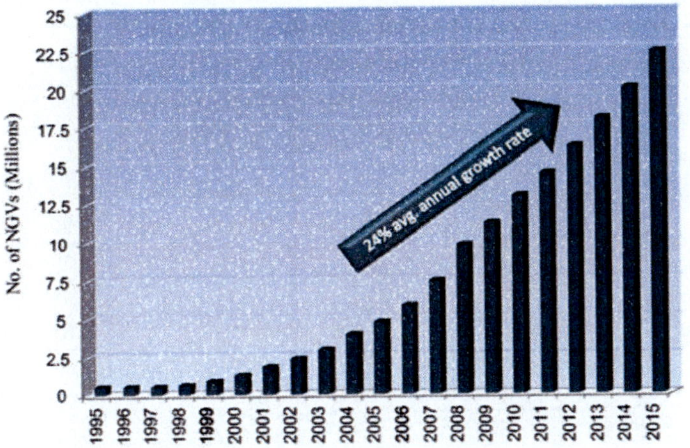

Fig. 9.1 Growth of CNG fuelled vehicle globally (Khan et al. 2015)

in the vehicle due to very high demand. It is an understanding that if automotive companies provide a bi-fuel system, more customers will be attracted. Otherwise, customers will purchase a vehicle with a single fuel system and then retrofit a bi-fuel system. However, only a few countries have an abundance of CNG. Therefore, some automotive OEMs from other countries are working on the development of dual-fuel vehicles. Dual-fuel vehicle means either vehicle runs on a conventional fuel only or mixtures of CNG and conventional fuel (such as CNG ignited with pilot diesel). Diesel is directly injected into the combustion chamber, and CNG is introduced in port using a carburettor or a port fuel injector.

Fifteen countries have a great experience of on-road CNG fueled vehicles (Table 9.2). CNG price is ~50% lower than conventional fuels, namely diesel, and gasoline.

Lower fuel price is the main reason for deeper penetration of CNG fueled vehicles in the transport sector of countries in the Asia–Pacific region. The economics of CNG fueled vehicles vis-a-vis gasoline and diesel-fueled vehicles were studied for the FY 2011–12 (Table 9.3).

A systematic procedure for evaluating different automotive technologies from all possible points associated with their life cycle is called 'Life Cycle Assessment.' The life cycle can be assessed for two main aspects: (i) vehicle cycle and (ii) fuel cycle. From different research studies conducted for the CNG life-cycle, the greenhouse gas (GHG) emissions are mainly from the methane leakage during the production stage of CNG. There are several merits of CNG over gasoline and diesel, which were highlighted by several Well-to-Wheels GHGs studies conducted by researchers world-over (Ou et al. 2012, 2013; Weiss et al. 2000). From the life-cycle study conducted by Argonne National Laboratory (ANL) for light-duty vehicles, CNG vehicles produced 6–11% fewer GHG emissions than gasoline vehicles (Wang and Huang 2000). Rose et al. real-time engine operation experiment (Rose et al. 2013) showed a ~24% reduction in GHG emissions when vehicles switched from diesel to

Table 9.2 Retail fuel prices in 15 CNG user countries in US$ (Khan et al. 2015)

	Country	Fuel Prices in US$			
		Gasoline	Diesel	CNG per liter gasoline equivalent	CNG per liter diesel equivalent
1	Iran	0.42	0.17	0.30	0.34
2	Pakistan	1.02	0.79	0.72	0.80
3	Argentina	1.44	1.44	0.33	0.39
4	Brazil	1.72	1.11	0.92	1.05
5	China	1.05	0.98	0.56	0.63
6	India	1.38	0.85	0.60	0.69
7	Italy	2.03	1.85	0.85	0.95
8	Colombia	1.31	0.96	0.80	0.92
9	Uzbekistan	1.03	0.98	0.30	0.34
10	Thailand	1.25	1.06	0.27	0.32
11	Bolivia	0.83	0.66	0.30	0.29
12	USA	1.02	1.12	0.60	0.68
13	Armenia	1.31	1.19	0.49	0.56
14	Bangladesh	0.79	0.56	0.27	0.29
15	Egypt	0.33	0.20	0.07	0.09
Average		1.13	0.93	0.49	0.56

Table 9.3 Cost comparison of CNG versus gasoline and diesel-fueled vehicles (Khan et al. 2015)

Items	CNG	Gasoline	Diesel
Vehicle type	Bus	Bus	Bus
km travelled per annum per vehicle	80,000	80,000	80,000
Total annual consumption of fuel in liters (consider unit of 'N m^3' in case of CNG)	36,184	39,400	32,000
Retail fuel price per litre in US$ (consider the unit of 'N m^3' in case of CNG)	0.52	1.02	0.92
Annual fuel cost (US$)	18,816	40,188	29,440
% Fuel cost saving CNG versus gasoline	113		
% Fuel cost saving CNG versus diesel	57		

CNG. A study conducted in Beijing, China, by Karman (2006) concluded a significant reduction in CO_2 emissions when vehicles were switched to CNG. The major exhaust product of concern from CNG vehicles is methane, which can be converted to CO_2 and water using a suitable catalyst. Therefore, the global warming potential of CNG vehicles is lower than that of other fuels.

From a safety aspect point of view, natural gas is far safer than gasoline. CNG has a narrow flammability range bandwidth (4.3–15.2% v/v) in air. It means that below 4.3% and above 15.2% concentrations in air, natural gas will not burn even in the presence of an ignition source such as a spark. Also, it has a higher auto-ignition temperature than gasoline and diesel, which makes it an inherently safer fuel. CNG is lighter than air, which means it will disperse rapidly, avoiding the chances of an accidental fire in the event of a leakage. CNG cylinders are generally designed using a high factor of safety. Hence CNG is as safe to use as diesel and gasoline.

9.3 Dual-Fuel Engines

The dual-fuel engines are a hybrid of diesel and petrol engines from a combustion perspective. They have both CI and SI engine characteristics along with their uniqueness. It isn't easy to characterize its combustion process. This is a conventional CI and SI engine technology with minor modifications to retain similar power performance. Dual-fuel engines have the advantage of the use of two fuels along with improvements in emissions and efficiency. Operating the engine only in diesel or petrol is possible when CNG or other fuels are unavailable. Therefore, it is a more flexible concept of modern engine technology. The seamless transition between dual-fuel mode and diesel or petrol mode seems to be a much more sustainable concept in the long run. Though the initial cost may be slightly higher for this technology, the operating cost of dual-fuel engines is significantly lesser due to its superior fuel economy.

In terms of fuel injection, there are two types of dual-fuel engines (i) direct-injected dual-fuel engines and (ii) port-injected dual-fuel engines. The direct-injected configuration uses a special injector for direct injection of both diesel/petrol and natural gas. For a CI engine, in the beginning, a small quantity of pilot diesel was injected ahead of the main injection for raising the temperature, creating a hot environment for the combustion of natural gas. Then, natural gas is injected, which combusts similar to diesel, in a diffusion flame. This strategy enables unthrottled operation, but the formation of particulates and NO_x is an issue. It is a complicated and expensive strategy, and it is not possible to operate only on diesel; hence the flexibility of switching fuel is also not available. The natural gas and air are premixed in the port-injected dual-fuel engines since the natural gas is injected into the intake port. In this type, the diesel pilot initiates the combustion of the CNG-air mixture. Port injected dual-fuel engine has the benefit of fuel flexibility. Therefore, this technology can be adopted in the market quite easily (Dronniou et al. 2014; Guzman et al. 2013).

9.4 CNG-Diesel Dual-Fuel Engines

Figure 9.2 shows the schematic of a dual-fuel CNG-diesel turbocharged engine with

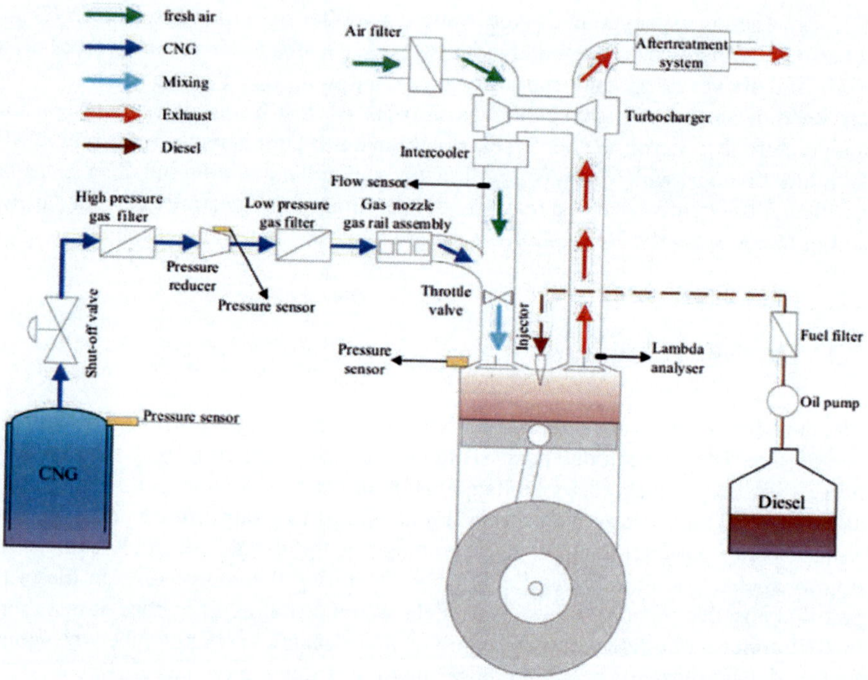

Fig. 9.2 Experimental layout of CNG-diesel dual-fuel engine (Zhang et al. 2019)

a gas nozzle gas rail assembly (Zhang et al. 2019). Initially, the engine operates on pilot diesel during starting and then switches to dual-fuel mode.

In this setup, ECU was retrofitted to cater to different engine operating conditions with fuel flexibility such as idling, low load, medium load, and high load. Throttle valve adjustment was provided in the inlet manifold. In this dual-fuel configuration, combustion was affected by the pulse width of pilot diesel injection, start of injection, and fuel rail pressure. CNG was injected into the intake port using a CNG supply system. Before injection, CNG passed through high pressure and low-pressure stages to make it suitable for injection. The pressure was controlled with the help of a shut-off valve and a pressure reducer. In most dual-fuel configurations, flame trap and pressure control valves were used to remove the possibility of backfire. A Coriolis mass flow meter was used to measure the CNG flow rate. Pressure losses during these stages were measured with pressure gauges. A typical example of a CNG injection control circuit is shown in Fig. 9.3 (Paul et al. 2015).

In the CNG cylinder, the pressure was ~196 bar, whereas the working pressure for the solenoid injector was very low at ~1.2 bar (Fig. 9.3). The injector was placed at a suitable position at around 1.5 D (D = Outer diameter of the manifold) from the inlet valve to enable the formation of a homogenous charge. The injector was synchronized with an angle encoder for timely induction of CNG at the corresponding crank angle to control the fuel injection duration. There are different CNG injection strategies for

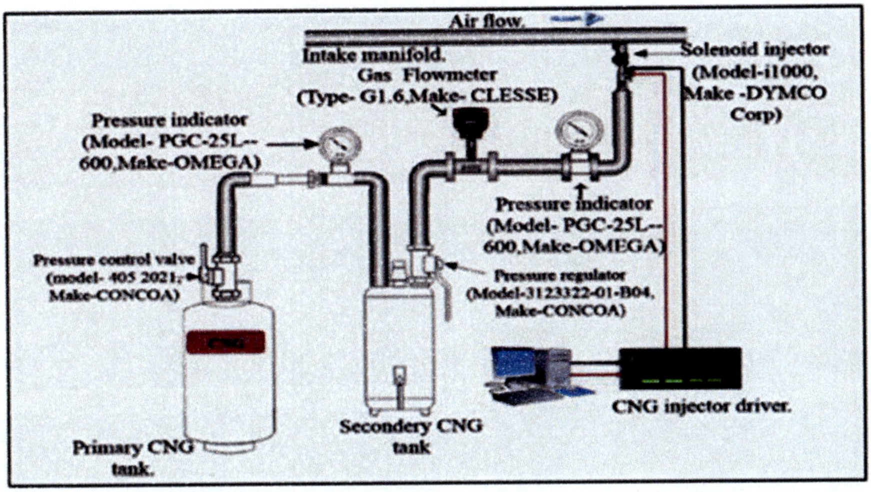

Fig. 9.3 Typical CNG injection circuit (Paul et al. 2015)

controlling the CNG flow rate at different engine speeds. CNG induction depends upon the valve timing diagram of the engine. Injection strategies are formulated for an effective intake valve opening duration.

The combustion characteristics of a CNG-diesel dual-fuel engine are similar to CI and SI combustion. There are mainly four regimes (Azimov et al. 2011). The first part is the combustion of pilot fuel. The premixed combustion of natural gas follows this. The next regime is the propagation of flames of the natural gas-air mixture. And the last regime is the bulk ignition of end gas. The natural gas substitution ratio affects the combustion characteristics and dictates how it fits between diesel and gasoline engine combustion.

Diesel-syngas combustion images are shown in Fig. 9.4 to throw light on the combustion phenomena involved in dual-fuel engines (Tomita 2007). A small amount of diesel was injected using a diesel injector having four holes. The images show that combustion initiates in those locations where the diesel spray is injected. The combustion then progresses throughout the combustion chamber because of flame propagation. The glowing particles seen in these images are the soot particles originating from the combustion of pilot diesel.

The following observations were noted by measuring different emissions of a dual-fuel engine using a chassis dynamometer (Thipse et al. 2015). When the quantity of gaseous fuel increased at the part-load operation, it led to a sharp rise in CO concentration. It was due to the incomplete combustion of gaseous fuel. Overall, CO emissions under dual-fuel mode were slightly higher than diesel. An increased natural gas flow rate resulted in a reduction in NO_x emissions compared to baseline diesel. There was a dramatic reduction of NO_x emissions at higher loads compared to the part-load operations. It was because of a sharp reduction in in-cylinder temperature due to an increased heat capacity and slower combustion due to lower oxygen

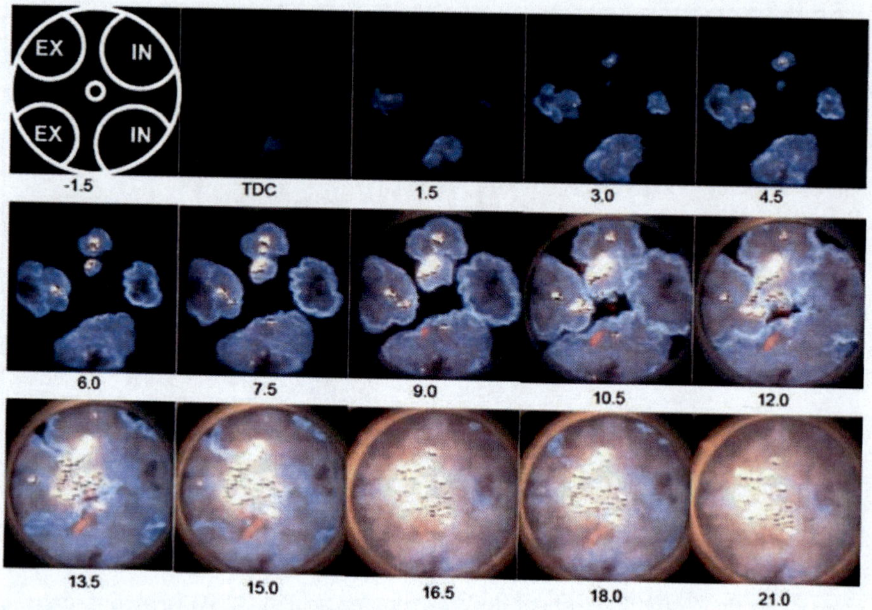

Fig. 9.4 Images of dual-fuel diesel-syngas combustion engine (Tomita 2007)

availability at a higher natural gas ratio. HC emissions were also higher at lower load due to incomplete combustion. At higher loads, HC emissions increased. Still, after a certain limit, they decreased due to the rise in combustion temperature, promoting better oxidation.

Researchers have attempted several studies related to strategies and parametric optimization to improve CNG-diesel dual-fuel combustion engine performance. In one such study by Verma et al. (2017), low load engine performance and emission characteristics were improved by advancing the injection timing of pilot diesel. Experiments were performed by varying the injection timings from 20° to 32° bTDC for a maximum diesel substitution of ~88%. However, diesel substitution was not affected significantly by advancing the injection timings. It was observed that HC, CO, and smoke emissions reduced, and NO_x increased with advanced injection timings. Although, the optimum injection timing varied per operating load and RPM. In another study (Shu et al. 2019a), the effect of injector spray angle was analyzed using computational tools. Increasing the spray cone angle from 60° to 140° increased the in-cylinder pressure. However, a further increase to 160° reduced the in-cylinder pressure. The start of combustion and overall combustion duration decreased upon increasing the spray cone angle from 60° to 140°. Hence, faster combustion and higher in-cylinder temperature resulted in higher NO_x emissions. CO emissions were found to be reduced for spray angles between 120° and 160°.

Unburnt methane emissions were not affected much by the spray cone angle variations. The maximum power output of dual-fuel injection was achieved at a spray angle of 140°. Work was also done to improve the design of piston bowl shape for better combustion and lower emissions of CNG-diesel dual-fuel combustion engines (Donateo et al. 2013). However, multi-level optimization was required considering the effect of speed and CNG-diesel energy proportions. Hosmath et al. (2016) performed response surface methodology analysis. They found that a higher compression ratio and advanced injection timing with a lower CNG flow rate were beneficial in augmenting the BTE and reducing CO, HC, and smoke. The natural gas substitution ratio is defined as the amount of energy from the natural gas to the total energy supplied from all fuels. It alters the reaction and emissions formation pathways in CNG-diesel dual-fuel engines. For the chemical kinetics, N-heptane was used to represent diesel, and methane represents the natural gas in the reduced chemical mechanism. A closed homogenous reactor model was assumed. The typical chemical pathway of natural gas-diesel dual-fuel is shown in Fig. 9.5 for different natural gas substitution ratios (NGSR).

The chemical reaction pathways initiate with hydrogen abstraction, oxygen addition, isomerization reactions, the second oxygen addition, and abstraction of OH radical, respectively. The intermediate compounds undergo a decomposition reaction. The reactions represented in the pathway and conversion of fragment intermediates to C_1 molecules lead to a low-temperature combustion process. The OH radicals formed from the decomposition of intermediates further promote the low-temperature combustion phenomenon.

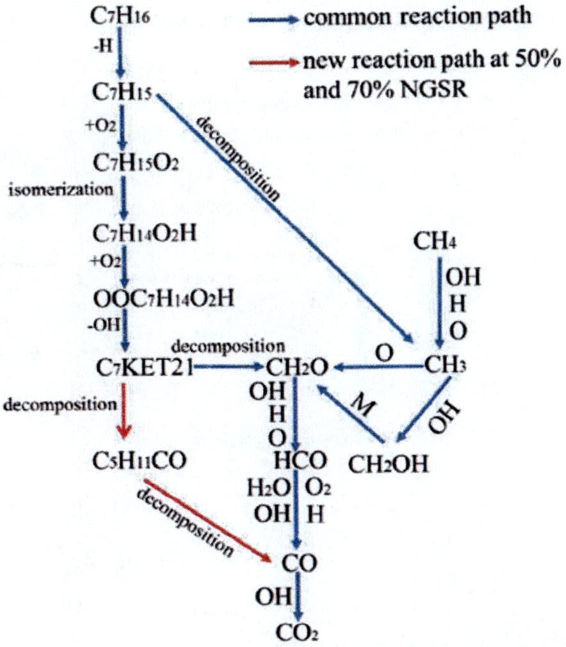

Fig. 9.5 Dual-fuel NG-diesel chemical reaction pathways at different NGSR (Zhang et al. 2019)

At the end of the low-temperature reaction pathways, small C_1–C_2 molecules are formed. In a high-temperature reaction pathway, these small C_1 molecules accumulated react with OH, releasing a large amount of heat. The decomposition of H_2O_2 and other chain branching reactions release OH free radicals, which are important in the high-temperature combustion stage. The chemical reaction of CO with OH is the most important since this exothermic reaction releases the maximum fraction of fuel heat.

The above shows two phases of combustion reactions: (i) low-temperature pathway and (ii) high-temperature pathway (Pitz and Mueller 2011). If the NGSR is 50% and 70% in the low-temperature process, there is a new reaction pathway where intermediates decompose to form CO. Therefore, CO concentration rises sharply for low NGSR. With an increase in diesel proportion, this decomposition reaction of intermediate species increases, which enhances a sharp rise in CO formation further. This sharp surge in CO concentration profile increases the temperature rapidly, which reduces the ignition delay. It is also worth mentioning that with a reduction in NGSR, the equivalence ratio decreases, and therefore oxygen concentration increases, leading to an overall reduction in CO emissions. Hydrocarbon combustion involves chemical kinetics of H_2 and CO oxidation. Figures 9.5 and 9.6 show that the oxidation of two hydrocarbons (C_7H_{16} and CH_4) mixture is the same as that of a single hydrocarbon (C_7H_{16} or CH_4) with a lower effect of NGSR on CO formation mechanism in NG-diesel dual-engine (Pitz and Mueller 2011; Zhang et al. 2019; Zheng and Yao 2006).

Higher temperature and increased oxygen concentration are important conditions for NO_x formation (Shu et al. 2019b). NO_x emissions decrease with an increasing

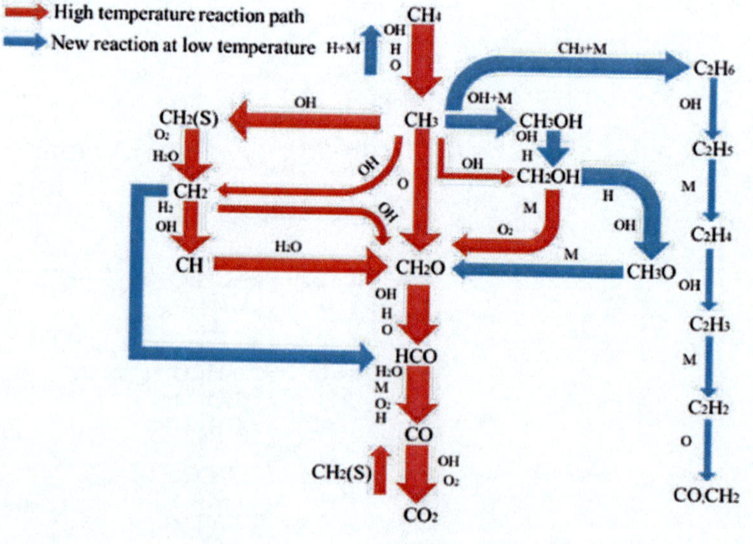

Fig. 9.6 Chemical reaction pathway of methane at high and low temperature (Zhang et al. 2019)

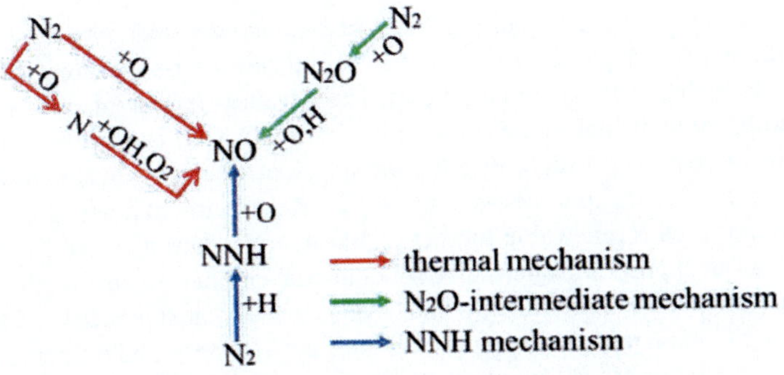

Fig. 9.7 Chemical reaction pathways for NO formation (Zhang et al. 2019)

NSGR, though it has a lower effect on the formation mechanisms. Temperature is not the only factor responsible for NO_x emissions, as the final temperature in all NSGR cases is the same. Equivalence ratio also plays an important role in NO_x emission formation in dual-fuel engines as a leaner mixture generates higher NO_x emissions. Since there are two stages in dual-fuel NG-diesel oxidation (high temperature and low temperature), the NNH mechanism is a major process that forms most NO emission under the low-temperature stage (< 1200 K), N_2O intermediate mechanism occurs during low equivalence ratio (< 0.8) and lower temperature conditions, while the thermal mechanism is a dominant mechanism for higher temperature conditions (Fig. 9.7).

Some studies used ethanol–diesel blends as a pilot, with the ethanol proportion between 5 and 10% v/v. CNG-diesel dual-fuel engines reported a reduced CO and HC emissions (Paul et al. 2013). A rise in the ethanol percentage in the pilot fuel blend increased the brake thermal efficiency and decreased the brake-specific energy consumption. The CNG-diesel dual-engine produced higher hydrocarbon emissions but decreased with the increasing ethanol content in the pilot fuel blend since hydrocarbons reacted with inherent oxygen present in the ethanol. NO_x emissions also decreased with increasing ethanol fraction in the pilot fuel blend due to extra fuel oxygen provided by ethanol.

However, there are few shortcomings in using ethanol in a pilot fuel blend. The miscibility of ethanol in diesel is extremely limited. Different common additives such as alkanols, alkanol amides, and deca-glycerol mono-oleate increase ethanol's miscibility in diesel (Dzulkefly et al. 2002). The most common co-solvent is biodiesel for increasing the miscibility of ethanol in diesel (Kwanchareon et al. 2007; Sureshkumar et al. 2008). Biodiesel also increases the cetane number and oxygen content of the blend, which minimizes hydrocarbon emissions. Namasivayam et al. (2009, 2010) and Korakiantis et al. (2011b) tested various biodiesels and obtained higher efficiency and reduced HC and CO emissions with the natural gas-biodiesel dual-fuel engine operation. A trade-off study by Paul et al. (Paul et al. 2015) exhibited that the optimum natural gas injection up to 9000 μs was suitable for superior engine performance and lower NO_x emissions. It also concluded that a higher CNG injection duration was required with a higher proportion of oxygenated fuels in pilot fuel blends such as ethanol and biodiesel. The advancement in injection timing with CNG-biodiesel dual-fuel mode resulted in lower HC, CO, and smoke, though NO_x emissions increased considerably (Basavarajappa et al. 2018). NO_x emissions could be controlled by using appropriate exhaust gas after-treatment devices. Clean natural gas used along with renewable biodiesel (like Honge, Jatropha, etc.) blended with alcohol was an appropriate alternative for dual-fuel engines for sustainable transport.

As the demand for alternate fuels like CNG is increasing, it is necessary to improve the existing CNG injectors for improving their performance and reducing their cost. There are consistent efforts for improving the injector designs. The enhanced design injectors can be used with conventional systems by ensuring ECU software and hardware compatibility with the injectors. If the new injectors do not match ECU hardware, the injector driver circuit should be modified. ECU software compatibility should also be taken into account. ECU contains maps to adjust the CNG fuelling at different loads. If there is information such as flow rate, invalid injection time, etc., for both injectors at different conditions, it is easier to calibrate and save time. Misfiring and unstable combustion should be avoided during ECU calibration. CNG combustion is more stable in the leaner region. There are conventional and modern methods for carrying out calibration. The conventional approach involves a sequence of steps like a design study, ECU calibration approach, On-vehicle calibration, and On-vehicle validation. On-vehicle calibration consumes half of the development time and is the most expensive part of this conventional calibration method. The previously calibrated information from a large number of testings is of no significance while developing a model with old injectors using the conventional approach. Therefore, to overcome the limitation of the conventional method, a new method was developed by implementing the strategy to offload the testing in the lab by incorporating mathematical calculations, decreasing the time and the cost. In this method, on-desk calibration replaces the on-vehicle calibration. The difference in the design of new injectors should be analyzed. Hardware specifications of the new injectors should be compatible with ECU. Still, there may be some structural modifications, such as in the collar and welded structures, as the size of new injectors is smaller. Then, the ECU calibration is carried out for the new design injectors using pulse width calculation. It is the sum of fuelling time from the maps with temperature and pressure compensation factors and invalid injection time.

$$\text{Drive time for injector} = \text{Fuelling calculation from the base map} * C_p * C_t + t$$

where t is Invalid injection time; C_p is pressure compensation for fuelling; C_t is temperature compensation for fuelling.

Different calibrations should be performed for the calibration of new CNG injectors. These include (i) time of invalid injection, (ii) base fuelling, (iii) gas-temperature compensation, (iv) warm-up compensation, (v) transient compensation, and (vi) startability. Invalid injection time is based on the map of fuel temperature versus supply voltage. It is necessary to recalibrate it as the electric coil is changed in the new injector designs for CNG operation. Base fuelling depends on fuelling quantity, and the map is engine load versus rpm. Therefore, it should be recalibrated for the new injectors. Temperature compensation calibration is performed to compensate for the fuel injection temperature, which depends on gas temperature and the load applied to the engine. Warm-up compensation calibration is used to maintain the air–fuel ratio during the engine's warm-up or starting phase. Transient compensation is calibrated to compensate for fuel flow effects during acceleration and deceleration. The startability calibration is used for extremely hot and cold operating conditions.

After completing these calibration steps for different zones (i.e., idling, part load, and full load), a mathematical model was developed based on the flow rate of injectors and invalid injection time at different fuel temperatures and voltage supplied to the injector. After on-desk calibration, on-vehicle validation is done. Base-fuel validation is checked for lambda deviation of ±2% for the air–fuel ratio of CNG, which was 17.2. Gas temperature was validated by choosing different gas temperatures and their corresponding deviation in lambda at a particular rpm. Lean combustion was more stable for CNG. Therefore, the lambda deviations up to 5% were considered on the leaner side, while lambda deviations up to 2% were considered on the richer side. Finally, since all parameters in the ECU were calibrated and validated, emissions were also assumed to be similar to the old injectors and were confirmed in different test cycles. Therefore, new CNG injectors were calibrated using this approach (Dhillon et al. 2013).

A common rail injection system in the CI engine is the most appropriate technology for the dual-fuel mode operation. It provides flexibility in controlling injection timing and fuel quantity while optimizing diesel substitution ratio with a lower effect on engine efficiency. The amount of fuel that can be injected into the combustion chamber depends on the fuel injection pressure. Therefore, by reducing the common rail pressure, diesel fuel injection quantity could be reduced. Few studies were undertaken for replacing the diesel ECU with dual-fuel ECU. However, studies for controlling the common rail engines in duel-fuel engines are rather scarce in the open literature. In the control setup for dual-fuel engines, the electronic gas control unit (EGCU) and diesel ECU take signals from various sensors such as crank angle sensor, accelerator pedal position sensor, common rail sensor. EGCU does not change signals from other sensors, excluding rail pressure sensors. This gas control unit matches the rail pressure sensor signal and controls the CNG injector's diesel injection timing and quantity. Initially, the calibration of rail pressure should be done for initializing the EGCU operation. For this, voltage signals at 0% and 100% position of accelerator pedal are found first. After that, the calibration of engine speed is done. The voltage signal corresponding to rail pressure at idling and a certain speed (say 3000 rpm) is obtained, and the relationship between the signal is established for matching the pressure signal. After this, for each cycle of dual-mode operation, diesel and CNG injection values were set by knowing accelerator pedal position and engine speed. During this dual-fuel configuration, the objective was to reduce diesel injection quantity and use CNG in maximum proportion. Therefore, the rail pressure signal was not supplied to diesel ECU but was given to EGCU, which used the diesel look-up table for corresponding dual-fuel utilization. This injection signal was then given to the diesel ECU. There are separate look-up tables for CNG and diesel for different accelerator pedal positions and engine speed. The setup values in the look-up table are absolute and are multiplier of the actual pressure signal value that should be fed into ECU. For achieving equivalent performance, lambda values of dual-fuel were matched with diesel. Therefore, by this method, dual-fuel control could be obtained at various operating points. The details of the discussed methodology are shown in Fig. 9.8 (Ismail et al. 2018).

3D CFD simulation software can provide information that cannot be captured by experimental observations and existing data acquisition systems. These software packages use complicated phenomenological sub-models with solution-adaptive mesh refinement and coarsening along with robust chemistry solvers (McNenly et al. 2015; Verma et al. 2016). Engine CFD simulation packages such as Converge CFD, ANSYS Forte, etc., contain the models that can be used to simulate a CNG-diesel dual-fuel engine. The entire combustion chamber (360°) with an appropriate injector can be modelled in this software. To model turbulence, the RNG k-Ɛ model can be used. This two-equation model is preferred because it possesses numerical stability. The n-heptane reduced chemical mechanism developed by Lawrence Livermore National Laboratory (LLNL) can be used for combustion in a CNG-diesel dual-fuel engine. This mechanism consists of 1282 reversible chemical reactions and 159

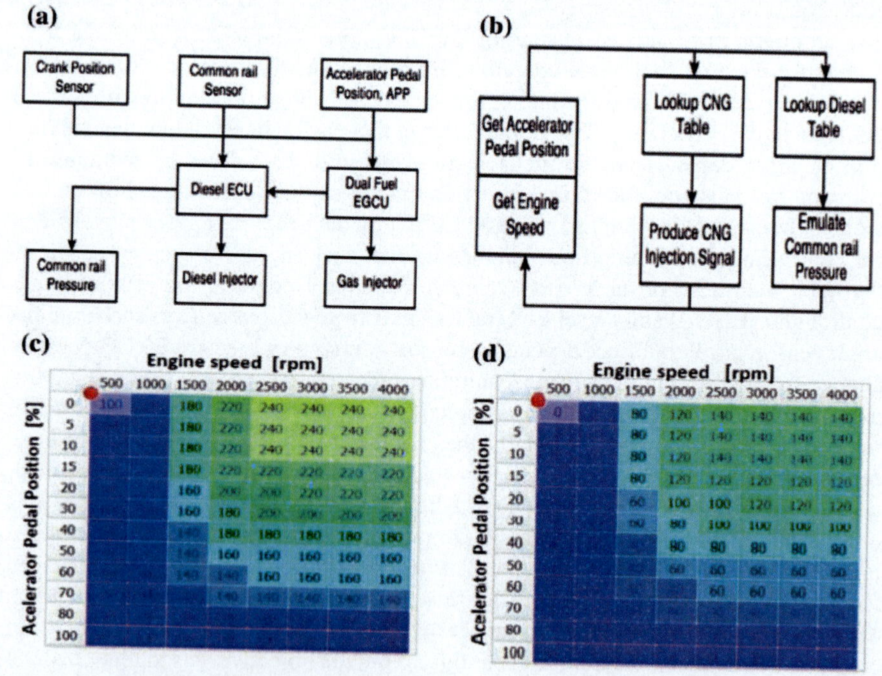

Fig. 9.8 **a** Dual-fuel control setup, **b** the cyclic nature of dual-fuel control, **c** Diesel look-up table, and **d** CNG look-up table (Ismail et al. 2018)

chemical species. A homogeneous CNG-air mixture can be assumed to focus on the effects of pilot fuel injection timing (Liu et al. 2018; Mehl et al. 2009). For a dual-fuel engine, the initiation of flame travel due to auto-ignition of pilot diesel can be modelled using a detailed chemical kinetics model. After ignition, a G-equation model along with a detailed chemistry model can be implemented. The G-equation model is a scalar transport model, which determines turbulent flames inside the cylinder without taking any source term. This G term measures the scalar distance between the instantaneous and the average flame-front. This G-equation uses several tunable parameters to calibrate sub-models and improve the accuracy of the simulations. It is because accuracy depends not only on engine geometry but also on the combustible mixture and flame properties when a reduced chemistry mechanism is used. It is up to the user to choose the tuning values, depending on how close their simulation matches with the experimental observations. After fixing these constants for several operating conditions, simulation results can predict the in-cylinder pressure and heat release rate. Combustion characteristics of the dual-fuel engine are very sensitive to pilot injection timing and pilot fuel quantity. This model should be turned on before pilot injection to reduce ignition delay incorporating late injection events (Liang et al. 2006; Talekar et al. 2016). In this model, both burned and unburned regions are characterized by positive and negative values of G scalar, respectively. The highest temperature achieved by the auto-ignition of diesel was noted in the first iteration of the simulation. Then, this obtained temperature triggered the G-equation. This G-equation could be solved using a level set numerical approach for finding the turbulent flame location (Ewald and Peters 2005; Tan and Reitz 2003).

$$\frac{\partial \rho G}{\partial t} + \frac{\partial \rho G u_i}{\partial x_i} = -D_t K \left| \frac{\partial G}{\partial x_i} \right| + \rho_u S_t \left| \frac{\partial G}{\partial x_i} \right| \tag{9.1}$$

where

ρ_u = density of the unburned region
G = non-reacting scalar
D_t = turbulent diffusion
K = curvature of flame
S_t = turbulent flame speed and is calculated using (9.2)

$$S_t = S_l + u_f \left\{ \left[\left(\frac{a_4 b_3^2 Da}{2b_1} \right)^2 + a_4 b_3^2 Da \right]^{\frac{1}{2}} \right\} \tag{9.2}$$

where

u_f = root mean square of turbulent fluctuating velocity.
$a_4 = 0.78$, $b_1 = 2.0$, $b_3 = 1.0$ (Mustafi et al. 2013; Turns 1996).
Da = Damkohler number = $S_l l_t / u l_f$.

L_f = laminar flame thickness.

The laminar flame speed can be determined using Gulder's coefficients for methane. Turbulent flame speed can be computed using the reference value of laminar flame speed. The length-scale of turbulent flow is useful in finding out the turbulent thickness of flame, given by the Eq. (9.3).

$$l_t = \frac{C_\mu k^{\frac{3}{2}}}{\varepsilon} \quad (9.3)$$

where,

$C_\mu = 0.47$.
k = Turbulent kinetic energy.
ε = Turbulent dissipation.

k and ε can be determined from the RNG k-ε equation model.

The Reynolds-Averaged Navier Stokes (RANS) equations, RNG k-ε model, and G-equation model complete the closed set of equations that can be used to determine the turbulent flame speed, which can be used to find out heat release and emissions formation (Cordiner et al. 2007). When diesel substitution ratio is very high, i.e., ~96–99% at low load in dual-fuel configuration, it is very difficult to predict major combustion phases with only an auto-ignition model. Therefore, the G-equation flame propagation model and detailed chemistry are significantly effective in finding end-gas ignition in RANS and LES modelling. When used along with detailed chemistry, this G-equation model is computationally efficient compared to the conventional detailed chemistry model. The G-equation model heavily relies on laminar flame speed. It is easy to find laminar flame speed for single fuel such as methane and is well established in the open literature. However, the dual-fuel mode is affected by the presence of heavy molecules in the combustible mixture. Therefore they need to be taken into account while making corrections to the laminar flame speed. Gulder's correlation for flame speed calculations can be used at a fixed range of equivalence ratios. It tends to over-predict the laminar flame speed for stoichiometric combustible mixtures, and therefore, it should be taken into account for determining the in-cylinder combustion (Mustafi et al. 2013).

9.5 CNG-Gasoline Dual-Fuel Engine

The dual-fuel combustion of CNG-gasoline allows improvement in methane combustion characteristics which otherwise suffer from lower flame speeds. Detailed combustion and emission characteristics were investigated by Di Iorio et al. (2014a) to study this aspect. Experiments were conducted on an optically accessible 250 cc motorcycle engine. The dual-fuel injection was achieved by direct methane injection and gasoline port injection. This configuration was selected to reduce the issue of intake air displacement and enhance volumetric efficiency. A high-speed camera with a high spatial and temporal resolution was used to measure the natural flame imaging in the visible range to investigate the combustion phenomena optically. Emission Spectroscopy in UV–visible wavelength allowed the detection of free radicals such as OH* and CH* as markers of combustion reactions.

The engine had a centrally located spark plug and a side-mounted direct injector. The methane injection pressure was maintained at 8 bar via a pressure regulator. A Synerject strata injector was employed for directly injecting methane in the vapour phase at 60° bTDC to cause sufficient charge stratification. The tests were performed at 2000 RPM, WOT condition at varying dual-fuel ratios (DF80 and DF50), and were compared with only gasoline and methane fueling conditions. DF80 and DF50 refer to energy displacement by methane to 80% and 50%, respectively. Only gasoline combustion showed bright spots at various locations. These indicate diffusive flames due to the combustion of fuel droplets and deposits on the chamber walls. The homogeneous mixture of gasoline resulted in radial flame propagation.

On the opposite side, only methane and dual-fuel combustion resulted in asymmetric and wrinkled flame behaviour. It was mainly a result of in-cylinder turbulence and air–fuel ratio distribution. Direct injection of methane was responsible for inducing turbulence into the flow, causing charge stratification. Progress in the centroid of luminosity data revealed the rapid flame travel at the bottom in the direction of the injector for cases involving direct methane injection. For a homogeneous gasoline-air mixture, the location was nearly in the center of the cylinder. An increase in methane content in the cylinder enhanced the degree of stratification and thereby distortion of flame while moving towards the intake wall. For global combustion prediction, the integral luminous intensity was analyzed for ten consecutive cycles. Differences in the flame speed of gasoline and methane led to significant variations in the luminosity rate. It was observed that DF50 and DF80 resulted in faster combustion, induced by stratification of the charge compared to only-methane, which was evident by their high luminosity rate.

However, gasoline combustion reported the highest luminosity during the combustion. In the late combustion phase, i.e., after 50° CA, dual-fuel combustion resulted in a lower luminosity rate than methane combustion. Evaluation of emission intensity markers, namely OH* ($\lambda = 309$ nm) and CH* ($\lambda = 431$ nm) showed higher value for gasoline and decreasing values with an increasing methane proportion during combustion. These radicals are indicative of AFR and the temperature (Gupta et al. 2011). The breakup of fuel into simpler hydrocarbons shows the presence of the CH* marker. Its higher intensity is an indicator of partial vaporization of the liquid fuel (Higgins et al. 2001). The exhaust gas measurement data showed a significant reduction in CO and CO_2 emissions for dual-fuel combustion cases compared to gasoline due to a lower percentage of carbon content in methane. The dual-fuel engine showed higher methane in the hydrocarbons emitted. However, the overall hydrocarbons, including non-methane hydrocarbons, were the highest for gasoline. Reduced gasoline content and faster flame propagation rate played an important role in the oxidation of unburnt hydrocarbons. NO_x emissions were comparable for dual-fuel and gasoline combustion, while it significantly reduced for only methane combustion. This exhibited that NO_x emissions depended on the liquid fuel since gasoline combustion resulted in higher in-cylinder temperatures. Methane-gasoline dual-fuel combustion showed quite low smoke emissions, though.

It is preferred to add hydrogen to CNG for improving combustion. However, hydrogen usage is limited due to its explosive nature and challenges in handling hydrogen. The combination of hydrogen and CNG provides a superior way to use hydrogen in the transport sector. The laminar burning velocity of hydrogen is 290 cm/s, significantly higher than the burning velocity of methane, which is 48 cm/s (Korb et al. 2016). Due to higher burning velocity, hydrogen enhances the flame speed to improve the thermal efficiency of CNG fueled engines. This type of engine is known as the HCNG engine. Hydrogen addition to CNG increases the peak heat release rate (HRR_{max}), extends the lean-burn operation, and reduces the combustion duration and carbon monoxide emissions (Kalsi and Subramanian 2017; Di Iorio et al. 2014b). Simulations revealed that adding hydrogen in CNG increased H and OH radical concentrations, improving flame propagation speed and combustion efficiency (He et al. 2016). In the hydrogen-CNG fueled engine, the CO_2 emission reduction depends on the extent of CNG substitution by hydrogen. HC and CO emissions also reduce from the HCNG engines, but NO_x emissions increase due to higher in-cylinder temperatures.

CNG-fueled vehicles are getting attention in several countries; however, its global market is struggling to penetrate the transport sector market in large numbers. The struggle of these CNG fueled vehicles is attributed to the few challenges listed in Table 9.4.

Table 9.4 Parameters and barriers for CNG fueled vehicles

Parameters	Reason
Driving range	1 m^3 of CNG at NTP = 1.0 L of gasoline and 1.1 L of diesel. Due to the lower energy density of CNG, it consumes more space (3–4 times) than diesel and petrol, even in highly compressed form, drastically reducing the vehicle's driving range
Storage/cargo space	CNG cylinders are quite large and occupy a lot of space in the car. Generally, they are located in the boot space, reducing the space available for cargo and storage in a car
Refuelling time	Refuelling time for the CNG fuelled vehicles is relatively longer than a conventional vehicle. This creates long queues at refuelling station and sometimes cause traffic congestion on the road
Refuelling infrastructure	Few countries have invested in the infrastructure of CNG refuelling stations. On the other hand, many other countries lack infrastructure, which is not equivalent to gasoline or diesel refuelling
Accident	Any accident in the CNG pipeline may create a shut down of the supply line of the entire city, crippling transportation in the city

9.6 Summary and Way-Forward

The utilization of CNG in IC engines has always attracted researchers. Due to its cleaner combustive properties, CNG is considered one of the most promising and clean fuels, which can meet the stringent emission norms and help reduce pollutants. Its application in dual-fuel mode offers dual advantages of avoiding the drawbacks of only CNG combustion and reducing dependence on conventional fuels. Adopting dual-fuel technology offers the advantage of both fuels because CNG limits the emissions and conventional fuels (gasoline/diesel) maintain the engine performance. In the CI engines, CNG is generally port injected and ignited with the help of pilot diesel. In contrast, CNG is either port-injected or directly injected in the SI engines, and accordingly, gasoline can be either directly or port-injected. Late direct injection of CNG in SI engines and gasoline port injection has shown similar engine performance characteristics as a gasoline PFI engine, without losing power and volumetric efficiency. In both SI and CI engines, CNG addition reduces carbonaceous and NO_x emissions. Studies have also investigated strategies for optimizing the control parameters such as fuel injection timing, piston shape, EGR, injector parameters, etc., for improving the low load engine performance of CNG fueled engines. Specifically, CNG dual-fuel direct injection can be an effective technology that can embrace the future in the automobile sector in the coming decades. An approach for ECU calibration and common rail pressure control is also discussed for implementing dual-fuel technology in existing engines. It can be concluded that CNG is a safer and cleaner fuel that can rapidly penetrate the market via dual-fuel technology.

References

Azimov U, Tomita E, Kawahara N, Harada Y (2011) Premixed mixture ignition in the end-gas region (PREMIER) combustion in a natural gas dual-fuel engine: operating range and exhaust emissions. Int J Engine Res 12(5):484–497

Basavarajappa YH, Banapurmath NR, Pradeep G, Vinod R, Yaliwal V (2018) Performance and emission characteristics of a CNG-Biodiesel dual-fuel operation of a single-cylinder four-stroke CI engine. IOP Conf Ser: Mater Sci Eng 376(1):012028)

Bielaczyc P, Szczotka A, Woodburn J (2015) Regulated and unregulated exhaust emissions from CNG fueled vehicles in light of Euro 6 regulations and the new WLTP/GTR 15 test procedure. SAE Int J Engines 8(3):1300–1312

Cordiner S, Rocco V, Scarcelli R, Gambino M, Iannaccone S (2007) Experiments and multi-dimensional simulation of dual-fuel diesel/natural gas engines (No. 2007–24–0124). SAE Technical Paper

Dhillon SPS, Sappal S, Singh M, Vashisth AK (2013) ECU calibration approach to implement improved CNG injectors in existing CNG vehicles (No. 2013-01-2752). SAE Technical Paper

Di Iorio S, Sementa P, Vaglieco BM, Catapano F (2014a) An experimental investigation on combustion and engine performance and emissions of a methane-gasoline dual-fuel optical engine (No. 2014-01-1329). SAE Technical Paper

Di Iorio S, Sementa P, Vaglieco BM (2014b) Experimental investigation on the combustion process in a spark-ignition optically accessible engine fueled with methane/hydrogen blends. Int J Hydrogen Energy 39(18):9809–9823

Donateo T, Strafella L, Laforgia D (2013) Effect of the shape of the combustion chamber on dual-fuel combustion (No. 2013-24-0115). SAE Technical Paper
Dronniou N, Kashdan J, Lecointe B, Sauve K, Soleri D (2014) Optical investigation of dual-fuel CNG/diesel combustion strategies to reduce CO_2 emissions. SAE Int J Engines 7(2):873–887
Dzulkefly K, Lim WH, Hamdan S, Norlizah S (2002) Effects of alkanols on the solubilization of methanol and ethanol in red palm oil for biofuel. J Oleo Sci 51(4):237–242
Ewald J, Peters N (2005) A level set based flamelet model for the prediction of combustion in spark-ignition engines. In: 15th international multidimensional engine modeling user's group meeting, Detroit, MI
Gupta SB, Bihari BP, Biruduganti MS, Sekar RR, Zigan J (2011) On use of CO_2* chemiluminescence for combustion metrics in natural gas-fired reciprocating engines. Proc Combust Inst 33(2):3131–3139
Guzman PMB, Ribas X, García Sr JM, Pita Sr M (2013) PM and CO_2 reduction in a dual-fuel heavy-duty diesel engine during the freeway part of transient worldwide emission tests (No. 2013–01–2759). SAE Technical Paper
He Z, Gao Z, Zhu L, Li S, Li A, Zhang W, Huang Z (2016) Effects of H_2 and CO enrichment on the combustion, emission and performance characteristics of spark ignition natural gas engine. Fuel 183:230–237
Higgins B, McQuay MQ, Lacas F, Candel S (2001) An experimental study on the effect of pressure and strain rate on CH chemiluminescence of premixed fuel-lean methane/air flames. Fuel 80(11):1583–1591
Hosmath RS, Banapurmath NR, Khandal SV, Gaitonde VN, Basavarajappa YH, Yaliwal VS (2016) Effect of compression ratio, CNG flow rate and injection timing on the performance of dual-fuel engine operated on honge oil methyl ester (HOME) and compressed natural gas (CNG). Renew Energy 93:579–590
Ismail MM, Ali MFM, Osman SA (2018) CNG-diesel dual-fuel controlling concept for common rail diesel. Int J Integr Eng 10(3)
Kalam MA, Kazi SN, Jayed MH (2011) Power boosting of a modified natural gas engine. Int J Phys Sci 6(28):6548–6557
Kalsi SS, Subramanian KA (2017) Experimental investigations of effects of hydrogen blended CNG on performance, combustion and emissions characteristics of a biodiesel fueled reactivity controlled compression ignition engine (RCCI). Int J Hydrogen Energy 42(7):4548–4560
Karman D (2006) Life-cycle analysis of GHG emissions for CNG and diesel buses in Beijing. In: 2006 IEEE EIC climate change conference. IEEE, pp 1–6
Khan MI, Yasmin T, Shakoor A (2015) Technical overview of compressed natural gas (CNG) as a transportation fuel. Renew Sustain Energy Rev 51:785–797
Korakianitis T, Namasivayam AM, Crookes RJ (2011a) Natural-gas fueled spark-ignition (SI) and compression-ignition (CI) engine performance and emissions. Prog Energy Combust Sci 37(1):89–112
Korakianitis T, Namasivayam AM, Crookes RJ (2011b) Diesel and rapeseed methyl ester (RME) pilot fuels for hydrogen and natural gas dual-fuel combustion in compression–ignition engines. Fuel 90(7):2384–2395
Korb B, Kawauchi S, Wachtmeister G (2016) Influence of hydrogen addition on the operating range, emissions and efficiency in lean-burn natural gas engines at high specific loads. Fuel 164:410–418
Kubesh JT, Podnar DJ, Guglielmo KH, McCaw D (1995) Development of an electronically-controlled natural gas-fueled John Deere PowerTech 8.1 L engine. SAE Trans, 788–795
Kwanchareon P, Luengnaruemitchai A, Jai-In S (2007) Solubility of a diesel–biodiesel–ethanol blend, its fuel properties, and its emission characteristics from diesel engine. Fuel 86(7–8):1053–1061
Liang L, Reitz RD, Yi J, Iyer CO (2006) A G-equation combustion model incorporating detailed chemical kinetics for PFI/DI SI Engine simulations. In: SAE Congress, April, vol 2
Liu J, Szybist J, Dumitrescu C (2018) Choice of tuning parameters on 3D IC engine simulations using G-equation (No. 2018-01-0183). SAE Technical Paper

Ma F, Wang Y, Liu H, Li Y, Wang J, Zhao S (2007) Experimental study on thermal efficiency and emission characteristics of a lean-burn hydrogen-enriched natural gas engine. Int J Hydrogen Energy 32(18):5067–5075

Manisalidis I, Stavropoulou E, Stavropoulos A, Bezirtzoglou E (2020) Environmental and health impacts of air pollution: a review. Front Public Health 8:14

McNenly MJ, Whitesides RA, Flowers DL (2015) Faster solvers for large kinetic mechanisms using adaptive preconditioners. Proc Combust Inst 35(1):581–587

McTaggart-Cowan GP, Reynolds CCO, Bushe WK (2006) Natural gas fuelling for heavy-duty on-road use: current trends and future direction. Int J Environ Stud 63(4):421–440

Mehl M, Pitz WJ, Sjöberg M, Dec JE (2009) Detailed kinetic modelling of low-temperature heat release for PRF fuels in an HCCI engine (No. 2009-01-18). SAE Technical Paper.

Mustafi NN, Raine RR, Verhelst S (2013) Combustion and emissions characteristics of a dual-fuel engine operated on alternative gaseous fuels. Fuel 109:669–678

Namasivayam AM, Crookes RJ, Korakianitis T, Olsen J (2009) Assessment of combustion in natural gas dual-fuelled compression ignition engines with dimethyl ether and rapeseed methyl ester pilot ignition. Int J Engine Res 10(3):165–174

Namasivayam AM, Korakianitis T, Crookes RJ, Bob-Manuel KDH, Olsen J (2010) Biodiesel, emulsified biodiesel and dimethyl ether as pilot fuels for natural gas-fuelled engines. Appl Energy 87(3):769–778

Ou X, Yan X, Zhang X, Liu Z (2012) Life-cycle analysis on energy consumption and GHG emission intensities of alternative vehicle fuels in China. Appl Energy 90(1):218–224

Ou X, Zhang X, Zhang X, Zhang Q (2013) Life cycle GHG of NG-based fuel and electric vehicle in China. Energies 6(5):2644–2662

Paul A, Bose PK, Panua RS, Banerjee R (2013) An experimental investigation of performance-emission trade-off of a CI engine fueled by diesel–compressed natural gas (CNG) combination and diesel–ethanol blends with CNG enrichment. Energy 55:787–802

Paul A, Panua RS, Debroy D, Bose PK (2015) An experimental study of the performance, combustion and emission characteristics of a CI engine under dual-fuel mode using CNG and oxygenated pilot fuel blends. Energy 86:560–573

Pitz WJ, Mueller CJ (2011) Recent progress in the development of diesel surrogate fuels. Prog Energy Combust Sci 37(3):330–350

Rose L, Hussain M, Ahmed S, Malek K, Costanzo R, Kjeang E (2013) A comparative life cycle assessment of diesel and compressed natural gas-powered refuse collection vehicles in a Canadian city. Energy Policy 52:453–461

Shu J, Fu J, Liu J, Ma Y, Wang S, Deng B, Zeng D (2019a) Effects of injector spray angle on combustion and emissions characteristics of a natural gas (NG)-diesel dual-fuel engine based on CFD coupled with reduced chemical kinetic model. Appl Energy 233:182–195

Shu J, Fu J, Liu J, Wang S, Yin Y, Deng B, Becker SM (2019b) Influences of excess air coefficient on combustion and emission performance of diesel pilot ignition natural gas engine by coupling computational fluid dynamics with reduced chemical kinetic model. Energy Convers Manage 187:283–296

Sureshkumar K, Velraj R, Ganesan R (2008) Performance and exhaust emission characteristics of a CI engine fueled with Pongamia pinnata methyl ester (PPME) and its blends with diesel. Renew Energy 33(10):2294–2302

Talekar AP, Lai MC, Zeng K, Yang B, Jansons M (2016) Simulation of dual-fuel-CI and single-fuel-SI engine combustion fueled with CNG (No. 2016-01-0789). SAE Technical Paper

Tan Z, Reitz RD (2003) Modelling ignition and combustion in spark-ignition engines using a level set method. SAE Trans, 1028–1040

Thipse S, Kulkarni AV, Vispute SJ, Rairikar SD, Sonawane SB, Sagare VS, Dev S, Kavathekar KK, Mengaji P, Karle US, Marathe NV, Sinha K (2015) Development of dual-fuel (Diesel-CNG) engine for SUV application in India. SAE Int J Engines 8(1):341–349

Tomita E (2007) Combustion characteristics and performance of supercharged pyrolysis gas engine with micro-pilot ignition. In: Proceedings of 25th CIMAC world congress on combustion engine technology (CIMAC 2007) (No. PaperNo. 178, pp. 1–10)
Turns SR (1996) Introduction to combustion, vol 287). McGraw-Hill Companies, New York, NY, USA, p 569
Verma S, Das LM, Bhatti SS, Kaushik SC (2017) A comparative exergetic performance and emission analysis of pilot diesel dual-fuel engine with biogas, CNG and hydrogen as main fuels. Energy Convers Manage 151:764–777
Verma I, Bish E, Kuntz M, Meeks E, Puduppakkam K, Naik C, Liang L (2016) CFD modeling of spark ignited gasoline engines-Part 1: modeling the engine under motored and premixed-charge combustion mode (No. 2016-01-0591). SAE Technical Paper
Wang MQ, Huang HS (2000) A full fuel-cycle analysis of energy and emissions impacts of transportation fuels produced from natural gas (No. ANL/ESD-40). Argonne National Lab., IL (US)
Weiss MA, Heywood JB, Drake EM, Schafer A, AuYeung FF (2000) On the road in 2020—a life-cycle analysis of new automobile technologies
Yeh S (2007) An empirical analysis on the adoption of alternative fuel vehicles: the case of natural gas vehicles. Energy Policy 35(11):5865–5875
Zhang Y, Fu J, Shu J, Xie M, Zhou F, Liu J, Zeng D (2019) A numerical investigation of the effect of natural gas substitution ratio (NGSR) on the in-cylinder chemical reaction and emissions formation process in natural gas (NG)-diesel dual-fuel engine. J Taiwan Inst Chem Eng 105:85–95
Zheng Z, Yao M (2006) Numerical study on the chemical reaction kinetics of n-heptane for HCCI combustion process. Fuel 85(17–18):2605–2615

Part V
Miscellaneous

Chapter 10
Analysis of the Potential Metal Hydrides for Hydrogen Storage in Automobile Applications

Dosawada Pavan Sai, Niraj Kumar, and Vishal Saxena

Abstract Considering the increase in air pollution and depletion of the available fossil fuels there is a need for the implementation of other energy sources for industrial use and automobiles. Hydrogen fuel is considered as the possible solution to act as a new energy source. It is the ideal means for energy storage for automobiles as it leads to zero emissions in case it is used in fuel cells. Unlike electrical vehicles the hydrogen fuel cells offer advantage of fast recharging and at present there are lot of technologies available which can make the hydrogen production easy and economical in future. However, the wide spread of hydrogen is restricted due its storage as the boiling point of hydrogen is 20.4 K at 1 atm. Thus, the storage of hydrogen in liquid form requires higher pressures and cryogenic conditions which consumes external energy. In addition, storing hydrogen in gaseous form is not feasible as it requires higher volumes. Storing hydrogen in the form of metal hydrides requires hydrogen to react with other metals and it can be released whenever required by breaking bond. Thus, metal hydrides have potential to overcome the problems associated with storage and has many advantages such as thermal stability, automobile safety, no fuel losses and long-term storage, etc. However, lower gravimetric storage density is a major issue beside it requires external source for hydrogen release. According to present automobile manufacturing requirement, the ultimate hydrogen storage material should have high gravimetric reversibility preferably greater than 10wt % of hydrogen, high reversibility which should be greater than 1500 cycles. Many metal hydrides with the suitable catalyst or nanoparticles have shown tendency towards satisfying the above conditions. In the present study, thermodynamic and kinetics for every such metal hydride are mathematically analyzed using the equations based on pore size and surface area. By knowing the thermodynamic and kinetic requirements of metal hydride reactions the suitable material for the automobile can be predicted and the technologies required for the proper release of the hydrogen can be suggested.

Keywords Metal hydride · Hydrogen · Thermodynamic analysis

D. P. Sai · N. Kumar (✉)
Department of Mechanical Engineering, ASET, Amity University, Noida, India

V. Saxena
Department of Fuel (FMME) Engineering, I.I.T (ISM), Dhanbad, Jharkhand, India

A. K. Agarwal et al. (eds.), *Advanced Combustion for Sustainable Transport*, Energy, Environment, and Sustainability, https://doi.org/10.1007/978-981-16-8418-0_10

10.1 Introduction

The decrease in availability of fossil fuels and the undesirable environmental impacts caused by them led to the initiative of using the most promising alternative fuels. In the search of such alternative fuels, one of the most advantageous and eco-friendly fuel found to be hydrogen (H_2) which has higher energy density compared to gasoline and unlike hydrocarbon fuels it won't release carbon emissions out. Hydrogen has 3 times higher energy content than gasoline. Lower heating values are 33.3 kWh/kg for hydrogen and 11.1 kWh/kg for gasoline. Hydrogen is not freely available on the earth surface but it can be generated through many methods such as thermochemical process which use energy in sources like natural gas, coal, or biomass to release hydrogen from their molecular structures, electrolytic processes which use electricity to split water into hydrogen and oxygen, direct solar water splitting where the radiant light energy is used to split water into hydrogen and oxygen and biological processes where bacteria and microalgae can produce hydrogen through biological reactions using sunlight and organic matter. All these methods are in research phase and there is lot of advancements required to make hydrogen available at low cost and without any scarcity (Kumar 2019).

The main challenge in utilization of hydrogen as fuel is the difficulty in its storage. Hydrogen has a boiling point of −252.9 °C at 1 atm. In-order to store the hydrogen in liquid form we need to provide the cryogenic temperatures or the higher pressures to increase its boiling point. Storing of hydrogen in gaseous form requires higher volume and pressure containers which will also post a major challenge in automobile application. The various targets set for hydrogen storage for the year 2020 and 2025 are shown in Table 10.1. To overcome the limitations of liquid and gaseous hydrogen storage, researchers are focusing on chemical storage of hydrogen within the suitable materials. Hydrogen storage material includes complex metal hydrides, metal hydrides, chemical hydrides, carbon nanotubes, inorganic nano tubes and sorbents such as metal organic frameworks.

Table 10.1 DOE Targets for hydrogen storage systems for on-board vehicular applications (Schneemann et al. 2018)

Storage parameters	Units	2020	2025	Ultimate
System gravimetric capacity:	$kWh\ kg^{-1}$ (wt%)	1.5 (4.5)	1.8 (5.5)	2.5 (6.5)
Volumetric capacity: usable energy density from H_2	$kWh\ L^{-1}$ ($kg\ H_2\ L^{-1}$)	1.0 (0.03)	1.3 (0.04)	1.7 (0.05)
Min/max delivery temperature	°C	−40/85	−40/85	−40/85
Cyclelife (uptake/release cycles)	Cycles	1500	1500	1500
Delivery pressure	bar (abs)	5	5	5
System fill time	Min	3–5	3–5	3–5
Fuel purity (H_2 from storage)	% H_2	99.97	99.97	99.97

According to the bond formation, the hydrogen storage materials are classified in three categories such as physisorption materials, off board regenerable hydrides and on-board reversible hydrides. Physisorption is the adsorption process in which the forces involved are interatomic forces. These materials include metal–organic frameworks, organic polymers, porous carbons and zeolites. H_2 is physiosorbed on the surface of the pores where the storage capacity majorly depends on the surface area and pours volume. The use of physisorption materials in hydrogen storage systems is limited due to its weak Van-der wall forces. The main advantage of these materials in their fast desorption and adsorption kinetics.

Chemisorption is the adsorption that is related to the strong chemical bond formed between the adsorbate and adsorbent molecules. Off-board regenerable hydrides and On-board reversible hydrides comes under this category. Materials like covalently bound metal hydrides, metal hydrides, borohydrides and metal amides come under the category of on-board reversible hydrides. In these materials the hydrogen is released by breaking the chemical bond by endothermic reactions which generally takes place in reasonable temperature and pressure.

Off-board regenerable hydrides store hydrogen in the form of covalently bound hydrogen materials which release hydrogen exothermally. It involves complex chemical process which limited its use in onboard vehicle hydrogen storage. Some of the materials are ammonia borane and hydrocarbons.

In the present study, the metal hydrides with potential hydrogen storage properties were briefly mentioned. The report gives the details about types of metal hydrides available and their gravimetric capacity, kinetic and thermal properties. Various methods are suggested to improve the performance of metal hydrides. The mathematical equations related to the thermodynamics of dehydration and hydration process in the metal hydrides is also shown.

10.2 Physisorption-Based Hydrogen Storage

The hydrogen storage is possible through two methods such as physisorption and chemisorption. In physisorption process the materials involved are metal organic frameworks, organic polymers, porous carbons and zeolites.

10.2.1 Metal Organic Frameworks (MOFs)

Metal organic frameworks (MOFs) are one of the promising materials for hydrogen storage. MOFs are originally crystalline materials which consists of metal ions linked together by various organic ligands which generate micro pores (2 nm) and channels. The frame topology, surface area and pore sizes can by readily modified by selecting the molecular building blocks. MOFs generally have permanent porosity, high specific surface area and defined structures. The highest excess hydrogen storage capacity which was reported so far for MOFs is 99.5 mg g^{-1}at 56 bar and 77 K. The MOF-5 which is prepared from the materials, benzene-1, 4-dicarboxylate (BDC) and Zn (II) salt have shown 4.5 wt% at 77 K and 1.0 atm and 1.0 wt% at room temperature. The preparation and handling conditions have also affected the hydrogen uptake preparation and handling under nitrogen (N_2) atmosphere led to H_2 uptake of excess 7.1 wt% at 77 K and 40 bar. MOFs for hydrogen storage which were based on physisorption should have a combination of high specific surface area and small pores but these two will not be proportional. Materials with small pores often leads to a reduced specific surface area. To meet the U.S. DOE targets the highly stable MOFs with pore volume and surface area which has many open metal sites must be investigated (Suh et al. 2012).

10.2.2 Porous Carbons

For achieving the high-capacity hydrogen storage, porous carbon should possess low bulk density, highly accessible adsorption surface and strong hydrogen adsorption enthalpy. Various carbon materials such as carbon aerogels, activated carbon, carbon nano tubes, templated carbon, and graphene contain high surface area pores. The properties such as total pore volume, specific surface area, pore size, micropore surface area and micropore volume, can be tailored by relevant experimental parameters and proper synthesis methods (Saxena et al. 2021). The thermodynamics of porous carbon material H_2 binding can be improved by two strategies. The first approach is to increase the specific surface area of carbon with suitable pore width which may allow multiple adsorption layers to be formed. The second approach is doping with the light weight elements which helps to functionalize the porous carbon (Xia et al. 2013).

10.2.2.1 Carbon Nanotubes

Carbon nanotubes are basically cylindrical molecules which consists of rolled up sheets of single layered carbon atoms. Carbon nanotubes have lower surface area and pour volume than that of porous carbon. Therefore, the hydrogen storage capacities of carbon nano tubes are lower than 1 wt% at ambient temperature and 100 bars pressure. However, at cryogenic temperatures the storage capacity can be increased. H_2 uptake of 2.4 wt% at 77 K for purified single walled carbon nanotubes has been observed. In addition, single walled carbon nanotubes exhibit increased storage capacity from 1 to 1.7 wt% at 77 K and 1 bar after subjected to acid treatment. It has been reported that the single walled and multi walled carbon nano tubes can normally reversibly adsorb H_2 of 1.5 wt% per 1000 m^2g^{-1}under ambient conditions. According to the above observations we can conclude that the carbon nano tubes are not potential material for hydrogen storage at ambient conditions (Xia et al. 2013).

10.2.2.2 Activated Carbons

Activated carbon is a form of carbon processed to have small and low volume pores that increase the surface area available. The usually available specific surface area is up to 3000 m^2g^{-1} and with abundance of micropore size less than 1 nm. The experimental investigations on activated carbons demonstrated that H_2 uptake about 2.5 wt% at low pressure (1–40 bar) (Yamaura et al. 2002) and 5.5 wt% at high pressures up to 60 bar and 77 k were achievable (Dehouche et al. 1999). The highest gravimetric capacity for commercially available activated carbon is found to be 1 wt% at room temperature and at 700 bar pressure (Xia et al. 2013).

10.2.2.3 Carbon Aerogels

Carbon aerogels are the new class of activated carbons which has a specific surface area of 3200 m^2g^{-1}. They exhibit a maximum adsorption capacity of 5.3 wt% at 77 k and pressure of 20–30 bar. The hydrogen adsorption enthalpy of aerogels is found to be 7.0 kJ mol^{-1}. Hydrothermal carbonization of organic material followed by chemical activation with KOH can generate carbon materials with specific surface area of 2700 m^2g^{-1} and H_2uptake of 6.4 wt% at 77 k and 20 bar (Xia et al. 2013).

10.2.3 Zeolites

Adsorption of gases in Zeolites is a well-known phenomenon. The main principle is that the guest molecules are forced under elevated temperatures and pressures into the cavities of molecular host. Upon cooling to room temperatures, the hydrogen gets entrapped inside the cavities of zeolites. It can be released again by raising the temperatures. One of the major advantages of using zeolite as hydrogen storage material is their thermal and chemical stability. The use of different exchangeable cations will offer a different thermodynamic heat of adsorption which allows zeolites to be tuned to provide improved adsorption at elevated temperatures. The hydrogen uptake capacity of the zeolite like carbon material is among the highest ever reported for carbon or any other porous material. Zeolite templated porous carbon have shown a hydrogen storage capacity of 4.5 wt% at 20 bar and 77 K (Xia et al. 2013).

10.3 Chemisorption-Based Hydrogen Storage

10.3.1 *Complex Metal Hydrides*

Complex metal hydrides are the components composed of metal cations such as group 1 and 2 elements and hydrogen containing complex anions such as amides, borohydrides and alanates. Complex metal hydrides are of interest because of their light weight and high hydrogen storage capacities. They release hydrogen by heating or chemical reactions. Some of the complex hydrides have reported to release hydrogen even in presence of water or aqueous solution. Some of the intermetallic hydride combinations are shown in Table 10.2.

10.3.1.1 Metal Amides

The amide hydride composite hydrogen storage system has a general form of $M(NH_2)_xM'H_y$ where M refers to the metal and x and y refers to the valency of M. $LiNH_2$–LiH (lithium amide and lithium hydride composite) was the first amide identified which shown the possibility of storing the hydrogen at large quantities. In the chemical reaction between $LiNH_2$ (lithium amide) and LiH (lithium hydride), firstly one equivalent.H_2 and Li_2NH are formed (Wiberg and Amberg 1971). Then the both products will further react with LiH to form two equivalent H_2 and form Li_3N. In total 10.5 wt% of H_2 can be adsorbed/desorbed in the complete reaction (Gizer et al. 2020).

$$LiNH_2 + 2LiH \Leftrightarrow Li_2NH + LiH + H_2 \Leftrightarrow Li_3N + 2H_2 \quad (10.1)$$

LiH is a highly stable hydride which has a formation enthalpy of −91 kJ/mol, desorbs hydrogen at the temperature higher than 600 °C. The decomposition of $LiNH_2$ takes place at low temperature of 300 °C. Therefore, the composite will release hydrogens at very low temperature which is less than 180 °C.

Table 10.2 Examples of intermetallic hydrides (Dornheim 2011)

S. No.	Composition	A	B	Compounds
1	A_2B	Mg, Zr	Ni, Co, Fe	Zr_2Fe, Mg_2Ni, Mg_2Co,
2	AB	Ti, Zr	Ni, Fe	TiFe, TiNi,, ZrNi
3	AB_2	Zr, Ti, Y, La	V, Cr, Mn, Fe, Li	YNi_2, $LaNi_2$, ZrV_2, YMn_2, $ZrCr_2$, $ZrMn_2$, $TiMn_2$
4	AB_3	La, Y, Mg	Ni, Co	$LaCo_3$, YNi_3, $LaMg_2NI_9$
5	AB_5	Ca, La, rare earth	Ni, Cu, Co, Pt, Fe	$LaNi_5$, $CeNi_5$, $CaNi_5$, $LaPt_5$, $LaFe_5$

H_2 release depends on the chemical interaction between the reactants. $H^{\partial+}$ and $H^{\partial-}$ have a high chemical potential which can be considered as one of the dehydrogenations driving forces. Many variations of amides are possible by matching the materials of $H^{\partial+}$ such as $LiNH_2$, with the materials of $H^{\partial-}$ such as LiH (lithium hydride), MgH_2 (magnesium hydride), $Mg(BH_4)_2$ (magnesium borohydride), $LiBH_4$ (lithium borohydride), $LiAlH_4$ (lithium aluminium hydride) which gives the amide composites. $Mg(NH_2)_2$–2LiH (magnesium amide and lithium hydride composite) have shown a gravimetric capacity of 5.6 wt% with high reversibility and hydrogen desorption at 1 bar pressure and temperature of 90 °C (Gizer et al. 2020).

Amide hydride hydrogen storage have many kinetic barriers which will be associated with the hydrogen dissociation, interface reactions, nucleation and diffusion process. Intensive ball milling of species such as $LiNH_2$–LiH and $Mg(NH_2)_2$–2LiH have shown a remarkable improvement in hydrogen storage due to the decrease in the particle size and increase in the homogenous distribution of reaction mixture. In the continuous rehydration and de-hydration, the small particles are prone to get aggregated into larger once which eventually slows down the reaction kinetics. Some of the methods such as introduction of proper dispersants which confine with scaffold materials can reduce the aggregate growth of reactant particles to some degrees. Some of the catalytic additives are identified which can improve the kinetic performance of amides. Catalytic effects of transition metals such as Fe, Co, Ni and Pt were significantly seen in amides. Introduction of multifunctional additives is required to improve the storage capacity and kinetics of amide hydrogen storage (Wang et al. 2013).

10.3.1.2 Borohydrides

Metal borohydride $M(BH_4)_n$ has a cation M^{n+} and an anion BH^{4-} in which B is covalently bonded with 4 hydrogen atoms. Borohydrides such as $LiBH_4$ (lithium tetrahydroborate), $Mg(BH_4)_2$ (magnesium borohydride) and $Ca(BH_4)_2$ (calcium borohydride) have shown the gravimetric capacity higher than 10 wt%.

Many metal borohydrides are found to be irreversible which limited their use in on-board applications. Research carried out on exploring the reversibility of lithium calcium and magnesium borohydrides. $LiBH_4$ is found to be reversible at higher temperatures of 600 °C. Ca and Mg borohydrides have shown reversibility at temperatures and pressures lower than lithium but they have some pre-assumed end products. Decreasing the dehydrogenation enthalpy is the major point of research in using the Li (lithium), Ca (calcium) and Mg (magnesium) borohydrides as the on-board hydrogen storage materials. Combining $M(BH_4)_n$ with other metals, metal hydrides and complex hydrides is the major approach to reduce the dehydrogenation enthalpy. $2LiBH_4$–MgH_2 was the mostly studied element in this context (Lu et al. 2019). Pure $LiBH_4$ releases 13.8 mass% of hydrogen according to the reaction shown below.

$$LiBH_4 \leftrightarrow LiH + B + 3/2H_2 \tag{10.2}$$

In this reaction the ΔH_{des} (change in enthalpy for dehydrogenation) is found to found to be 67 kJ/mol. The same reaction when carried out again with MgH_2 has shown the ΔH_{des} of 46 kJ/mol according to the following reaction.

$$2LiBH_4 + MgH_2 \leftrightarrow 2LiH + MgB_2 + 4H_2. \quad (10.3)$$

This combination has shown the decrease in T_d (dehydrogenation temperature) along with the improvement in the kinetic performance of re-hydrogenation which improves the reversibility. But the pressure higher than 0.3 MPa is essential to supress the formation of $Li_2B_{12}H_{12}$ (Orimo et al. 2007).

Many additives such as metals, oxides, halides, and carbon-based materials when mixed with borohydrides have shown good performance in encountering the sluggish kinetics of metal borohydrides. In this regard adding $0.2MgCl_2$–$0.1TiCl_3$ (magnesium chloride and titanium chloride-based catalyst) into $LiBH_4$ have shown a super catalytic effect, releasing 5 mass% of hydrogen from 60 to 400 °C. Addition of $ZrCl_4$ into the $2LiBH_4$–MgH_2 has influenced its dehydrogenation kinetics and significantly reduced the incubation period of first dehydrogenation. Nanoconfinement is an approach which can improve the thermodynamics and kinetics of hydrogen storage materials. $LiBH_4$ which are nanoconfined to carbon scaffolds with a pore size of 13-nm has shown 50 times increase in dehydrogenation rates at a temperature of 300 °C and the capacity loss for 3 cycles is decreased from 72 to 40% compared to the bulk $LiBH_4$. Nano porous carbon with a smaller pore size of 2 nm can supress the release of B_2H_6 (Wang et al. 2013).

10.3.1.3 Metal Alanates

Metal alanates are also known as metal aluminium hydrides which contain metal cations and aluminium hydride anions. Alanates are moisture and air sensitive. The gravimetric capacities of alanates such as $LiAlH_4$ and $NaAlH_4$ are reported to be 10.6 wt% and 7.4 wt% respectively. One of the major metal alanate was found to be $LiAlH_4$ (lithium aluminium hydride). The equation which involves the dehydrogenation of $LiAlH_4$is given as follows.

$$LiAlH_4 \Leftrightarrow 1/3Li_3AlH_6 + 2/3\ Al + H_2 \quad (10.4)$$

The $LiAlH_4$ undergo the phase transition at a temperature of 160–177 °C before it undergoes the above dehydrogenation reaction which occurs at 187–218 °C. This first dehydrogenation reaction was exothermic with ΔH (change in enthalpy) of –10 kJ/(mol of H_2). Then the second dehydrogenation reaction of Li_3AlH_6 takes place endothermically at 228–282 °C with ΔH of 25 kJ/(mol of H_2).

$$Li_3AlH_6 \Rightarrow 3LiH + Al + 3/2\ H_2 \quad (10.5)$$

Then the LiH undergo dehydrogenation at 370–483 °C with ΔH of 140 kJ/mol.

Table 10.3 Storage data for low and high storage hydrides

S. No.	Material	Volume (l)	Weight (kg)	H_2 quantity stored (kg)	References
1	TiFe	50	200	4	Reilly and Wiswall (1974)
2	Mg_2Ni	50	100	4	Kohno et al. (1996)
3	MgH_2	50	100	8	Buchner (1978)

Doping of the $LiAlH_4$ with Ti (titanium) has enhanced the dehydrogenation kinetics. It was found that with 3 mol% addition of $TiCl_4$ (titanium tetrachloride) and 5 min of mechanical milling, the first dehydrogenation of $LiAlH_4$ into Li_3AlH_6 and aluminium will take place at room temperature. With 2 mol% addition of $TiCl_3$ in Li_3AlH_6, the decomposition into LiH and Al was found to be occurring at temperature as low as 100 °C (Orimo et al. 2007).

10.3.2 Metal Hydrides

Many metals form metal hydrides by reacting reversibly with hydrogen. The equation involved in formation of metal hydride is as follows:

$$Me + x/2H_2 \Leftrightarrow MeH_x + Q \quad (10.6)$$

Here, x = cH [H/Me] and Q is the heat of reaction. The enthalpy and entropy of the metal hydrides formation and hydrogen release determines the favourable temperature and pressures of the hydrogen storage materials. Most of the metal hydrides have higher thermodynamic stabilities which require the temperatures greater than 300°C at pressure of 1 bar. Some of the examples of metal hydrides are of alkaline earth metals, alkaline metals, rare earth metals, as well as transition metals of Sc (scandium)-, Ti (titanium)- and V (vanadium)-group (Dornheim 2011). The storage data for high and low temperature hydrides is given in Table 10.3.

10.4 Metal Hydride Properties

10.4.1 Metal Hydrides Available

The light weight metals such as Li (lithium), Be (beryllium), Na (sodium), Mg (magnesium), B (boron), and Al (aluminium) form a variety of metal hydrides which are majorly known for their light weight and number of hydrogen atom per metal atom (Bobet et al. 2000).

10.4.1.1 Mg Based Metal Hydride

Mg and Mg based alloys have shown high gravimetric hydrogen storage densities. Mg is the eight most frequent elements in earth which makes it relatively inexpensive. The Mg–H bond are highly stable with high enthalpy of hydride formation. Mg based hydrogen storage can be classified into three categories which are, Pure Mg, Mg-based alloys, Mg-based composites. In case of pure hydrogen the hydrogen storage capacity is extremely high with a reported gravimetric capacities of 7.6 wt% but the dissociation of compound will require higher temperatures and the reaction kinetics are very slow, it takes 50 h (50 h) for the complete formation of MgH_2 (magnesium hydride) at 350 °C and the enthalpy of desorption is $\Delta H = -74$ kJ/mol of H_2 which limits the use of pure Mg based hydrogen in practical applications. Moreover, the stability of MgH_2 made it to be useful in many suitable applications (Ouyang et al. 2020).

To improve the performance those pure Mg hydrides, some catalysts are added as additives such as transition metals or metal oxide elements. These additives are responsible for the change of surface properties, microstructures and grain growth with minimum affect to the storage capacity. To further improve the performance of MgH_2 some of the sample preparation methods such as smelting, diffusion, powder sintering, heat treatment and surface treatment methods can be adopted. Among them the most commonly used technique is reactive mechanical milling which will vary the microstructures of the metal to modify their properties. The main advantages of the method are the generation of reactive and fresh surfaces, the ease of formation of multiphase nano composites and creation of micro strains in crustal lattice. Several parameters of milling can be varied considering the end product of ball milling, such as milling speed, ball to powder weight ratio, milling time, milling atmosphere and temperature of the milling (Zhao and Zhang 2014).

If we change the entropy and enthalpy and of a hydrogenation reaction in an effective way to decrease the desorption temperature of MgH_2 systems. To modify those properties some of the methods such as alloying, nano structuring, changing the reaction paths and metastable phase formation are identified (Song et al. 2020).

Alloying

In case of alloying the Mg based alloys are usually made with a hydrogen non forming element which forms a less stable hydride decreasing the hydrogen reaction enthalpy. Materials such as counter rare earth (eg: Ce (cerium), La (lanthanum)), non-transition metals (eg: Al (aluminium), In (indium)) and transition metals (eg: Ni (nickel), Co (cobalt), Fe (iron), Cu (copper), Ag (silver), Sc (scandium), Y (yttrium), etc., have been identified to form a suitable-alloys for performance improvement of Mg based hydrides (Cui et al. 1999). Based on the bond formation three-categorizations of Mg based alloys are found such as solid solution compounds, intermetallic compounds, and other compounds (Ouyang et al. 2020).

Intermetallic Compound

Some of the intermetallic compounds such as Mg_2NiH_4 (magnesium nickel hydride), Mg_2FeH_6 (magnesium iron hexahydride), Mg_2CoH_5 (dimagnesium cobalt pentahydrate) have shown potential change in thermodynamic and kinetic properties (Cracco and Perchergon-Guegon 1998). In those materials if we consider Mg_2NiH_4 compound, Mg_2Ni is a potential hydrogen storage material with gravimetric capacity of 3.6 wt%, Mg_2NiH_4 has of $\Delta H = -64.5$ kJ mol^{-1} H_2 which gives the desorption temperature of 523 k with plateau pressure of 3 bar at room temperature (Ouyang et al. 2020).

Similar to Mg–Ni system, the Mg_2Cu alloy can form from the Mg and Cu (Sun et al. 1999). When Mg_2Cu is applied with 30 MPa hydrogen pressure, it decomposes to $MgCu_2$ and MgH_2 according to the following reaction:

$$Mg_2Cu + H_2 \Rightarrow MgH_2 + MgCu_2 \tag{10.7}$$

The hydride enthalpy of formation is of -73.5 kJ(mol^{-1}H_2), hydrogen storage capacity is of 2.6 wt% and equilibrium temperature for 1 bar of hydrogen pressure is reduced to about 513 K but unfortunately this reaction is irreversible (Ouyang et al. 2020). In case of Mg_2FeH_6 which is formed by the mechanical milling of Mg and Fe in presence of hydrogen has hydrogen formation enthalpy of 98 kJ (mol^{-1} H_2) (Liang et al. 1998). It has shown a volumetric hydrogen density of 150 kg m^{-3} and gravimetric capacity of 5.6 wt% the which is more than that of liquid hydrogen at 20.2 K and moderate pressure up to 20 bar (Raman et al. 2002). Similar to the Mg–Fe system, ball milling of Mg and Co in the presence of hydrogen led to formation of Mg_2CoH_5 material with gravimetric capacity of 4.5 wt% and ΔH of -82 kJ mol^{-1} H_2 and desorption temperature of 513 K at 1 bar pressure (Wang et al. 2006).

Mg Based Solid Solution

Formation of Mg based solid solution is another strategy for destabilization which applies a minor modulation to the composition and structure (Zhu et al. 1999). Cadmium exhibits unlimited solubility in Mg. Mg^{-1} in the presence of Cd alloy can absorb 5 wt% hydrogen at 673 K in 24 h (Zhang et al. 2019). Mg_3Cd alloy using high energy milling have exhibited a gravimetric capacity of 2.8 wt% at 573 k in 120 s and ΔH is -65.5 kJ mol^{-1} (Skripnyuk and Rabkin 2012).

In the same way Mg–In system has high solubility of 10 at% indium at wide range of temperatures. The $Mg_{0.95}In_{0.05}$ solid solution can react with hydrogen reversibly at 573 k according to the following reaction (Cui et al. 2018):

$$Mg_{95}In_5 + 0.9H_{22} \Leftrightarrow 0.9MgH_2 + 0.05MgIn \tag{10.8}$$

$Mg_{0.95}In_{0.05}$ have shown a gravimetric capacity of 5.3 wt% with the poor kinetics of desorption/absorption reaction. Therefore, ternary alloys such as Mg (In, Al), Mg (In, Cd) and Mg (In, Y) were investigated (Wang et al. 2014; Zhong et al. 2011).

In case of $Mg_{90}In_5Cd_5$the reaction with H_2 goes as follows:

$$Mg_{90}In_5Cd_5 + 70\ H_2 \Leftrightarrow 70MgH_2 + 5MgIn + 5Mg_3Cd \quad (10.9)$$

Here the dehydriding reaction enthalpy is found to be 86 kJ (mol^{-1} H_2). It has shown gravimetric capacity of 4.3 wt% (Xu et al. 2019). It also exhibited improved hydrogen kinetics and reduced hydrogenation activation energy of 61 kJ mol^{-1} but its dehydrogenation rate is slow because of long range diffusion of Cd and In and in MgH_2 (Lu et al. 2015).

Nano-structuring

In nano structuring the dimensionality of the architecture is the main feature for example: (i) zero dimensional (quantum dots, nanoparticles, and hollow spheres), (ii) one dimensional (nanorods, nanowires, nanofibers, and nanobelts) and (iii) three dimensional (frameworks, nanoflowers, and dendric structures) (Jepsen et al. 2017). Nanomaterials offer numerous advantages for hydrogen storage systems as the decrease in size of the material, its surface to volume ratios increases apparently. The atoms at the surface have dangling bonds which are extremely active and form strong bonds with surrounding molecules (Ouyang et al. 2020).

Nano particles or nano crystalline materials

Numerous methods can be used in the preparation of MgH_2 nano particles which includes melt spinning, ball milling and hybrid combustion. The surface energy density of magnesium is smaller than that of MgH_2, and some of the heat of formation can be stored as excess surface energy which reduces the heat release in the process of hydrogen release and hydride formation. Wagemans et al. (2005). The nanostructured uniform mixture of MgH_2–$0.1TiH_2$ powder has a cyclic hydrogen storage capacity 6 wt% and the ΔH_{des} is found to be 68 kJ (mol^{-1} H_2) (Zhang et al. 2019). Asano et al. concluded that embedding nano sized clusters of Mg in an immiscible metal matrix is found to beh the powerful strategy to destabilize MgH_2 which results in lower working temperatures (Asano et al. 2015). Carbon based nano materials has shown the potential in improving the gravimetric performance of MgH_2 (Zhang et al. 2017). Nanocomposite $(Mg\text{-}V)_{nano}/G$ has shown a reversible gravimetric capacity of 6.03 wt% with enthalpy of formation of −68 kJ mol^{-1} of H_2 (Schneemann et al. 2018).

Nanowires and Nanofiber

Mg nanowires which have thin diameter has the advantage of easy activation for desorption/absorption of hydrogen. Changes in the thermodynamic properties are reported when the diameters of nanowires are thinner than 30 nm (Zaluska et al.

2001). Energy stability of Mg and MgH_2 nanowires were calculated using first principles density functional theory. The ΔH_{des} was calculated to be −20.64, 34.54, and 61.86 kJ(mol^{-1} H_2) for nano wires of A1_MgH_2(⌀0.68 nm), A2_MgH_2(⌀0.85 nm) and A3_MgH_2(⌀1. 24 nm) (where ⌀ indicates diameter of the nano wire) by which we can conclude that the decrease in diameter of the nano wire leads to energy instability in MgH_2 and Mg nanowires (Li et al. 2009).

Nanofilms

Thin films are largely applied in Mg materials. Mg nanostructured films were synthesized through various techniques which involve electron beam evaporation, Magnetron sputtering deposition, direct current and radio frequency, and plasma spatter. Thin film fabrication is an accurate method to control the nano structure, interfacial properties and composition of hydrogen storage materials (Schneemann et al. 2018).

10.4.1.2 Na Based Metal Hydrides

Sodium based metal hydrides can store the hydrogen in different forms. One method of storing the hydrogen is in the form of plastic-encapsulated NaH (sodium hydride) pellets. These pellets are cut into small slices and immersed in water where the NaH surface will react with water and produces sodium hydroxide and water. This method has shown a hydrogen storage density of 4.3 wt%. Sodium hydride and water reaction is strongly exothermic, so the reaction should take place in controlled environment.

$$NaH + H_2O \Rightarrow NaOH + H_2 \quad \Delta H = -35\,\text{kcal/mol} \tag{10.10}$$

The sodium hydroxide can again be converted into NaH by following certain steps such as (1) drying the sodium hydroxide, (2) raising the temperature of the system, (3) regenerating the hydroxide. The primary conversion involves reaction of sodium hydroxide with methane gas as shown in the below reaction:

$$NaOH + CH_4 \Rightarrow Na + CO + 2.5H_2 \tag{10.11}$$

$$Na + 1/2H_2 \Rightarrow NaH \tag{10.12}$$

This method of hydrogen storage is not feasible for on-board applications as the conversion of NaOH to NaH is a high energy consuming process (DiPietro and Skolnik 2000). In all the alloys of sodium-based metal hydrides, sodium alanates are extensively investigated. One of the compounds which drawn attention is $NaAlH_4$.The atomic structure of sodium alanate consists of isolated $[AlH_4]^-$ tetrahedra structure where the sodium atoms are surrounded by the eight $[AlH_4]^-$ tetrahedra structures in a distorted square antiprismatic geometry. The reversible release and uptake of hydrogen of the selective additives dopped $NaAlH_4$ can take place in two steps as described by the Eqs. (10.1 and 10.2).

$$3NaAlH_4 \Leftrightarrow Na_3AlH_6 + 2Al + 3H_2\left(3.7\,wt\%\ H_2,\quad \Delta H = 37\,KJ\,mol^{-1}H_2\right) \tag{10.13}$$

$$Na_3AlH_4 \Leftrightarrow 3NaH + Al + 1.5H_2,\ \left(1.8\,wt\%\ H_2\quad \Delta H = 47\,KJ\,mol^{-1}H_2\right) \tag{10.14}$$

The first reaction releases about 3.7 wt% of H_2 upon heating to a temperature of 190 °C and second reaction releases 1.8 wt% of H_2 upon heating to a temperature of 225 °C. The hydrogenation condition were considered to be severe as they have the operating temperature of 200–400 °C and pressures of 10–40 MPa. In-order to optimize the operating temperatures and pressures for practical applications, doping of $NaAlH_4$ with other suitable materials is thoroughly investigated (Li et al. 2013).

Doping of TiB_2 in $NaAlH_4$ after milling has shown a gravimetric capacity of 5.21 wt% at 190 °C (Li et al. 2012). Chen et al. (2009) stated that when $CeCl_3$ (cerium chloride) is doped with $NaAlH_4$ the activation energies for first and second dehydrogenation were calculated to be 80.76 and 97.27 kJ mol^{-1}. Bitter et al .(2006) investigated the carbon nanofiber-supported $NaAlH_4$ with different size ranges of particles such as 1–10 μm, 2–10 nm and 19–30 nm prepared by wet chemical synthesis. The results showed that the maximum desorption rate temperature (T_{max}) was found to be 70 °C for 2–10 nm $NaAlH_4$, 164 °C for 19–30 nm $NaAlH_4$ and 186 °C for 1–10 μm $NaAlH_4$. The dehydrogenation of Ti-doped $NaAlH_4$ composite has occurred at appreciable rates at a temperature of about 100 °C, which shows their potential to be used as on-board hydrogen storage system. After adding the $CeAl_2$, $NaAlH_4$ has shown rapid reaction kinetics with reload of 4.9 wt% hydrogen in 20 min and 70% of hydrogen can be released in just first 6 min (Fan et a 2011).

Another approach to enhance the kinetic performance is to decrease the $NaAlH_4$ particle size. The micro meter sized Al and NaH which are formed by agglomeration during hydrogenation/dehydrogenation reaction will be prevented by adding nano-sized sodium alanate. The hydrogenation/dehydrogenation rates were enhanced and hydrogenation/dehydrogenation temperatures were decreased. After impregnation, $NaAlH_4$ nano particles were amorphous, desorption temperature of hydrogen got reduced significantly and reaction kinetics improved (Li et al. 2013).

$NaAlH_4$ Doping Methods

$NaAlH_4$ doping methods

In metal infiltration, $NaAlH_4$ is loaded in a autoclave along with the inert scaffold material, where the hydrogen gas is pressurized to 150–200 bar and heated to a temperature of 170–200 °C (Bogdanović et al. 2000). Nielsen et al. (2011) reported that melt infiltrated $NaAlH_4$ into nano porous carbon aerogels could release hydrogen in the room temperature itself ($T_{onset} = 33$ °C) and at 125 °C hydrogen release rates reached a maximum value when functionalized with 3 wt % $TiCl_3$ catalyst.

Solvent mediated infiltration

In solvent mediated infiltration method, hydrogen storage material is dissolved in solvent to form a homogenous solution. The main advantage of this method is that the process can be performed under milder conditions when compared to melt infiltration. Sun et al. (2008) proposed a space-confined nano materials obtained by $NaAlH_4$confining into ordered mesoporous silica. The space-confined $NaAlH_4$have shown faster kinetics, lower dehydrogenation temperature, higher reversibility and higher loading of 20% was achieved (Li et al. 2013).

High energy ball milling

The main advantages of using high energy ball milling are increase in surface area, formation of nano/micro structures and lattice defects creation on the interior and surface of metal hydride. Beattie et al. studied the structural difference between un-milled and milled $NaAlH_4$ and observed that milled $NaAlH_4$ performed better than the un-milled $NaAlH_4$ as a reversible hydrogen storage material (Ashby et al. 1963).

10.4.1.3 Boron Based Metal Hydrides

Boron based hydrogen storage materials which are generally represented as boranes and borohydrides are extensively investigated for potential hydrogen storage capability.

Boron based metal hydrides chemically store $4H^{\delta-}$ in the boron-hydride bonds. The main challenge involved is to recover 100% $H^{\delta-}$ in the form of H_2 molecule in mild conditions. The dehydrogenation can be carried out in two ways. The first one is hydrolysis of BBMs. In the presence of water, one $H^{\delta-}$ of BBM will react with one $H^{\delta+}$ of the H_2O molecule which liberates the H_2 molecule. The second method is thermolysis which consists of (i) intermolecular reaction and (ii) intramolecular reaction for the hydrogen release (Moussa et al. 2013).

Boron Based Hydrides for Hydrolysis

From all the borohydrides available so far $LiBH_4$, $NaBH_4$ and KBH_4 are considered under hydrolysis process.

$$NaBH_4 + 4H_2O \Leftrightarrow NaBO_2 \cdot 2H_2O + 4H_2 \quad (10.15)$$

$$LiBH_4 + 4H_2O \Leftrightarrow NaBO_2 \cdot 2H_2O + 4H_2 \quad (10.16)$$

$$KBH_4 + 4H_2O \Leftrightarrow KBO_2 \cdot 2H_2 \quad (10.17)$$

Among them only $NaBH_4$ is investigated excessively because of problems associated with other two such as hygroscopicity and spontaneous H_2 evolution, low gravimetric hydrogen storage capacity, slow H_2 release kinetics and clogging (Zhu et al. 2006). $NaBH_4$ also faces certain issues relating to low net hydrogen storage capacity, catalyst durability and positioning. The gravimetric hydrogen storage capacity of $NaBH_4$–H_2O is 7.3 wt% when there are 4 mol of H_2O for one mole of $NaBH_4$ for the hydrolysis reaction. The net gravimetric hydrogen storage capacity reported is 6–7 wt% in ambient conditions. The studies on $NaBH_4$were more focused on prototyping, design of reactor, scalability, and coupling with reactor prototype. $NaBH_4$ was found to have more potential as a hydrogen carrier for mobile and portable applications (Moussa et al. 2013).

Same as $NaBH_4$, NH_3BH_3, hydrolysis in the presence of suitable catalyst the absolute net gravimetric capacity of the couple NH_3BH_3–$2H_2O$ is found to be 15 wt%, but it decreases to 9 wt% due to $3H^{\delta+}$ of the NH_3 moiety cannot be recovered. Furthermore, due to the by-products the storage capacity again decreases to 7.8 wt%. Liberation of ammonia is the main problem associated with using NH_3BH_3 in hydrolysis (Andrieux et al. 2012; Marrero-Alfonso et al. 2007; Moussa et al. 2013). If N_2H_4 moiety of the $N_2H_4BH_3$ can be dehydrogenated in the mild conditions in the presence of selective catalyst then $N_2H_4BH_3$–$3H_2O$ has shown the gravimetric capacity of 10 wt%. In the presence of bimetallic nano-catalyst ($Ni_{0.89}Pt_{0.11}$) in excess water and 50 °C, the dehydrogenation of $N_2H_4BH_3$ resulted in 5.8 mol $H_2 + N_2$ per mole which means 9.7 wt% of H_2 release is found. The main problem associated with all the products of boron-based material hydrolysis is their regenerability (Moussa et al. 2013).

Boron Based Hydrides for Thermolysis

$LiBH_4$ is considered as the most important hydride for thermolysis process. The main problem with the $NaBH_4$ is its thermal stability due to which it dehydrogenates at 530 °C (Züttel et al. 2003). The gravimetric hydrogen storage capacity of $LiBH_4$ was reported to be 18.5 wt% by thermolysis (Moussa et al. 2013). Pristine $LiBH_4$ melts and concomitantly decomposes over a temperature of 275 °C. The $LiBH_4$ dehydrogenation is partial

$$LiBH_4 \Rightarrow LiH + B + 3/2\ H_2 \tag{10.18}$$

Therefore, the excess gravimetric heat storage capacity of $LiBH_4$ decreases to 13.9 wt%. The main advantage of using $LiBH_4$ lies in its possibility of reversibility. It was studied that the solid by products B and LiB can be hydrogenated to $LiBH_4$ at a temperature of 600 °C and under a H_2 pressure of 30 MPa (Orimo et al. 2005). To alter the dehydrogenation and re-hydrogenation conditions of the $LiBH_4$ to be milder, three strategies are identified. The first strategy is to destabilize the $LiBH_4$ by incorporating a metal or metalloid element, a hydride or an amide. The second strategy is addition of catalysts such as metal oxides and metal chlorides. For instance, it was reported that by adding 25 wt% TiO_2 the dehydrogenation temperature decreased to 100 °C and H_2 release rate of 3 wt% within a temperature range of 100–300 °C (Au et al. 2008). The third strategy was to decrease the particle size of the $LiBH_4$ to down to nanoscale by nanoconfinement scaffolds. For instance, $LiBH_4$ impregnated into micro-cellular carbon foam dehydrogenates from a temperature of 200 °C so that 4 wt% of H_2 are released at a temperature of 300 °C (Brun et al. 2010).

Magnesium borohydrides $Mg(BH_4)_2$) is found to be the most suitable compound for hydrogen storage. It will liberate all of its hydrogen which is 14.9 wt% when heated upto 600 °C (Matsunaga et al. 2008). Calcium borohydride ($Ca(BH_4)_2$) is the another much investigated borohydride which contains hydrogen capacity of 11.6 wt%. It partly dehydrogenates 9 wt% of H_2 under heating which follows a two-step reaction starts at 360 °C and it finishes at 500 °C (Miwa et al. 2006). Ammoniaborane is also an attractive solution for thermolysis hydrogen storage material as it stores 19.5 wt% H and offers $3H^{\delta+}$ and $3H^{\delta-}$ can also combine together intermolecular and intermolecular interactions to give $3H_2$. At a temperature below −40 °C, NH_3BH_3 exists in a compound form which is richer in hydrogen; the hydrogen form is represented as NH_4BH_4 which is a densest material with gravimetric capacity of 24.3 wt% (Parry et al. 1958).

10.4.1.4 Aluminium Based Hydrogen Storage

Aluminium hydride is considered to be the promising candidate as a hydrogen storage material due to its high energy storage density. Aluminium hydride (AlH_3) exists as metastable, crystalline compound at the room temperature. It has the volumetric hydrogen density of 148 g H_2/L and the gravimetric hydrogen density that exceeds 10 wt%. It exhibits a heat of reaction of 7 kJ/mol of H_2 and at temperatures below 100 °C it has rapid hydrogen evolution rates (Graetz et al. 2011).

α-AlH_3 decomposes to products Al and H_2 by an endothermic reaction:

$$\alpha - AlH_3 \Rightarrow Al + 3/2\, H_2 \tag{10.19}$$

The experimentally determined enthalpy of formation ranges from −6.0 to −7.6 kJ/mol H_2 with entropy of formation of −129 J/k mol H_2 (Graetz et al. 2011). Compared to α-AlH_3 β and γ-AlH_3 structures are less stable. They undergo an exothermic transition to get converted to α at elevated temperatures greater than 100 °C.

Graetz and Reilly (2006) investigated the decomposition of α, β and γ-AlH_3 materials which are prepared form micro crystallization route. The transformation of less stable β and γ to α structure occurs very rapidly at temperatures above 100 °C.

Destabilization of aluminium hydride is extensively investigated for low temperature fuel cell applications where rapid hydrogen evolution is required at temperatures lower than 100 °C. The thermal stability and the decomposition activation energy of the hydride mainly depend upon the material synthesis condition and its purity. It is observed that alane is very sensitive towards light. Even for a short period exposure to the visible light the activation energy decreases by 15–20% when compared to non-irradiated fresh samples. Addition of transition metals and alkali hydrides has also shown the destabilization effect. Additions of transition metals have shown the pronounced effect on the decomposition kinetics when the synthesis of the AlH_3 is done by adding catalyst in the solution. In case of adding Ti the enhancement in the H_2 evolution rate can be seen at certain levels of few parts per million when Ti material was introduced in solution (Parry et al. 1958).

In-spite of all the advantages of the aluminium hydrides, it has a major limitation with bad reversibility of desorption which has hindered its practical applications in hydrogen storage carrier. To make the aluminium hydride a potential material for hydrogen storage, alantes are considered as a promising material. Among them $LiAlH_4$ and $NaAlH_4$ have attracted extensive attention. $LiAlH_4$ have shown the gravimetric capacities of 10.5 wt% H_2 with limitations of poor reversibility and unsatisfactory kinetics of desorption. Besides, $NaAlH_4$ have shown a theoretical gravimetric capacity of 7.4 wt % with excellent reversibility (Kumar et al. 2008).

Research is carried out on nano confinement of $LiAlH_4$ and $NaAlH_4$ to increase their potential in practical applications. Jiao et al. synthesized the $LiAlH_4$ with NiCo-doped nanoalloy which is encapsulated in graphene layers (NiCo@G). He found that the dehydrogenation temperature of $LiAlH_4$-NiCo@G is reduced to 36 °C. In the molecule 5% Fe_2O_3–$LiAlH_4$ the dehydrogenation temperature was found to be reduced to 70 °C and in the first two dehydrogenation process about 7.4 wt% of H_2 can be released (Jiao et al. 2016). Li et al. observed that the 2 mol% of $CoFe_2O_4$ doped $LiAlH_4$ shows good catalytic effect. The first dehydrogenation step occurs at 65 °C and the second occurs at 130 °C. This system can release 7.2 wt% of hydrogen (Li et al. 2014). Some of the hydrogen storage properties of nano confined metal hydrides is shown in Table 10.4.

10.4.2 Equilibrium Pressure for Metal Hydrides

The equilibrium pressure of the metal hydrides at different temperatures plays very important role in designing the hydrogen storage systems. For the calculation of equilibrium pressure, the reaction entropy and enthalpy for the metal hydride is required. The equilibrium pressure based on the entropy and enthalpy and of the reaction is given by:

$$P_{H_2}[\mathrm{atm}] = \exp[(\Delta \mathrm{H} - \mathrm{T}\Delta \mathrm{S})/\mathrm{RT}] \tag{10.20}$$

where, $P_{H_2}[\mathrm{atm}]$ is the equilibrium pressure, T is the temperature, ΔS is the change in entropy and ΔH is the change in enthalpy.

Based on the above equation the equilibrium pressure required can be calculated by taking the entropy and enthalpy of the various metal hydrides can be calculated. In

Table 10.4 Hydrogen storage properties of nano confined metal hydrides (Schneemann et al. 2018)

Hydride	Porous host	D_{avg} (nm)	V_{tot} ($cm^3 g^{-1}$)	Infiltration method	Solvent or atmosphere	H_2 wt%	References
Mg–H_2	RF-aerogel		1.27	$MgBu_2$	170 – 200 °C, 50 – 78 bar H_2, Ar, heptane	0.62	Nielsen et al. (2009)
Mg–H_2	RF-aerogel	27.1	1.32	$MgBu_2$	150 °C, 150 bar H_2	1.8	Huen et al. (2017)
Mg–H_2	Ni-doped hollow carbon	3.4	1.53	$MgBu_2$	$NH_2NH_2 \cdot H_2O$	6.63	Shinde et al. (2017)
AlH_3	HSAG	2–3	0.66	Solvent impregnation	Et_2O	0.25	Wang et al. (2019)
$LiBH_4$	CMK-3	4.51	1.3	Grinding/melt infiltration	310 °C, 60 bar H_2	3.4	Gosalawit-Utke et al. (2013)
$LiBH_4$	Pyrolyzed aerogel	13	0.8	Melt infiltration	280 – 300 °C, Ar	12.5	Gross et al. 2008)
$LiBH_4$	Activated carbon	1.75–3.2	0.87	Solvent impregnation	THF	9	Fang et al. (2008)
NaH	HSAG	2–3	0.66	Physical mixing	Ar	4.4	Li et al. (2016)
$NaBH_4$	HSAG	2–3	0.66	Metal infiltration	520 °C, 5 bar H_2	7.9	Adelhelm et al. (2010)

the equation R is universal gas constant which is taken as 8.314 ($J\,K^{-1}\,mol^{-1}$) (Brooks et al. 2020). Van't Hoff's lines for various metal hydrides is shown in Fig. 10.1.

10.4.3 Thermal Modelling of Metal Hydrides

The thermal properties such as specific heat capacity and thermal conductivity are usually required for analysing the thermodynamics of metal hydride.

Assumptions of the model:

(1) The ideal gas equation of state is appropriate for the hydrogen gas.
(2) Heat transfer through the form of radiation is negligible.
(3) The viscous dissipation and compression work is negligible.
(4) Gases and solids have same temperature.
(5) Pressure variations affect is neglected.
(6) Equilibrium pressure dependence on the hydrogen concentration is neglected.
(7) The dispersion terms and tortuosity can be modelled as diffusion fluxes.

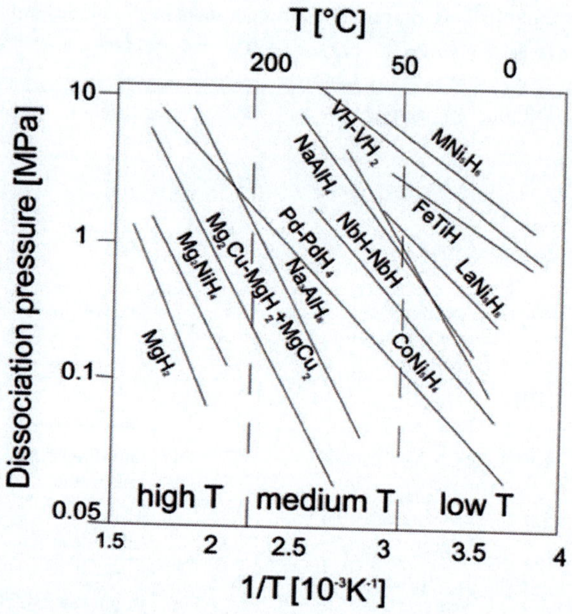

Fig. 10.1 Van't Hoff's lines for various metal hydrides

10.4.3.1 Thermodynamic Equations

As the dehydration and hydration of the metal hydrides is based on the temperature, it is required to analyse the temperature profile.

The general equation for the variation of the internal energy for a porous metal hydride is given by

$$\frac{\partial \rho U}{\partial t} = -(\vec{\nabla} \cdot \rho U \vec{v}) - (\vec{\nabla} \cdot \vec{q}) + Q - P(\vec{\nabla} \cdot \vec{v}) \tag{10.21}$$

In the above equation U is the internal energy between two equilibrium states. $\vec{v}$ is the fluid velocity. Heat added and removed from the system in the form of convection or internal energy is given by Q, ρ is the density and heat transferred by convection is given by q. The term $P(\vec{\nabla} \cdot \vec{v})$ represents the compression work which is usually neglected.

Here, U can be re-written as

$$\mathrm{U} = \mathrm{H} - P/\rho \tag{10.22}$$

Substituting the Eq. (10.3) in Eq. (10.2) and using the substantial derivatives gives

$$\rho \frac{DH}{Dt} = -(\vec{\nabla} \cdot \vec{q}) + \mathrm{Q} + \frac{DP}{Dt}. \tag{10.23}$$

Enthalpy can be expressed as the function of pressure and temperature

$$\frac{DH}{Dt} = C_p \frac{DT}{Dt} + \left(\frac{1}{\rho} + \frac{T}{\rho^2}\frac{\partial \rho}{\partial T}\right)\frac{DP}{Dt} \tag{10.24}$$

In the above equation C_p is the specific heat capacity.
Substituting the Eq. (10.5) in Eq. (10.4) we get

$$\rho C_p\left(\frac{\partial T}{\partial t} + \vec{v} \cdot \nabla T\right) = \lambda \cdot \nabla^2 + Q - \frac{\partial \ln \rho}{\partial \ln T}\left(\frac{\partial P}{\partial t} + \vec{v} \cdot \nabla P\right) \tag{10.25}$$

where for an ideal gas $\frac{\partial \ln \rho}{\partial \ln T} = -1$.
Energy balance equations for the solid and gas combined can be expressed as

$$(\rho C_p)_e \frac{\partial \mathrm{T}}{\partial \mathrm{t}} = \lambda_e \nabla^2 T - \rho_g C_{pg} \vec{v} \cdot \nabla T + \mathrm{m}(\Delta H + T(C_{pg} - C_{ps})) \tag{10.26}$$

$$(\rho C_p)_e = \varepsilon \rho_g^g C_{pg} + (1 - \varepsilon)\rho_s^s C_{ps} \tag{10.27}$$

where, m is the mass of the metal hydride, ε is the porosity of the metal hydride bed. The term g designates the gas phase and s designates the solid phase. λ_e is the effective thermal conductivity of the metal hydride bed, effective thermal conductivity mainly depends on the thermal conductivity of the solid and gas and also on the structural properties of the porous media.

$$\lambda_{\mathrm{e}} = \varepsilon \lambda_{\mathrm{g}} + (1 - \varepsilon)\lambda_{\mathrm{s}} \tag{10.28}$$

where, λ_g is the thermal conductivity of hydrogen gas and λ_s is the thermal conductivity of metal powder.

10.4.3.2 Reaction Kinetics of Metal Hydrides

Rate law of reaction hydrogen uptake and discharge is defined by m. it is the relationship between the reaction rate and species concentration. The value of the rate of reaction is given by the equation as follows

$$\mathrm{m_a} = \mathrm{C_a} \exp\left(-\frac{E_a}{R_g T}\right) \ln\left(\frac{P_g^g}{P_{eqa}}\right)[\rho_{ss} - \rho_s] \tag{10.29}$$

$$\mathrm{m_d} = \mathrm{C_d} \exp\left(-\frac{E_a}{R_g T}\right)\left[\frac{P_g^g - P_{eqd}}{P_{eqd}}\right] \tag{10.30}$$

where, ρ_{ss} is the metal hydride density at the end of the absorption. ρ_s is the density of the metal powder without hydrogen. The reaction constants follow Arrhenius equation for both desorption and absorption process. P_{eq} is the equilibrium pressure of the metal hydride system. The difference between the pressure of the system and the equilibrium pressure is the driving force for the sorption process.

The Van't Hoff equation will give the relation for calculating the equilibrium pressure and temperature of the hydride.

$$\ln P_{\mathrm{eq}} = \frac{\Delta H}{R_{\mathrm{g}\mathrm{T}}} - \frac{\Delta S}{R_{\mathrm{g}}} \tag{10.31}$$

This is same as the equation mentioned above to calculate pressure according to the enthalpy and entropy (Mohammadshahi et al. 2016).

10.5 Requirements of Metal Hydrides for On-Board Applications

The metal hydrides are selected for automobile applications only when then cross the DOE targets. The DOE has set targets on properties such as system system volumetric capacity, gravimetric capacity, durability/operability, storage system cost, charging/discharging rates, fuel quality, environmental, health and safety.

10.5.1 Achieving the Required Pressure

To achieve the proper performance of metal hydrides pressure variations should be achieved inside the storage tank. The dehydrogenation and hydrogenation rates depend on pressure inside the storage tank. As discussed above different metal hydrides require different pressure conditions according to the temperature of operation. The release rate of hydrogen from the metal hydride will also depend on the pressure. Increase in hydrogenation rates of the metal hydrides is observed by creating a vacuum pressure to suck out the air particles from the storage tank.

In-order to achieve the required pressure metal hydride compressors can be used. Metal hydride thermal sorption compression is a reliable efficient method which allows conversion of heat energy into compressed hydrogen gas. For improving the hydrogen storage characteristics, several features are required to be optimized such as pressure plateaus synchronization for multi stage compressors, slope of the isotherm reduction and hysteresis, increase in the life time and cyclic stability and the system design which is associated with the volume expansion of the metal matrix during the hydrogenation (Kumar et al. 2018).

Optimizing the metal hydride compressors mainly depends upon finding a compromise between contradicting factors. The first group A type of factor should assure matching available temperatures and specified pressures, maximum reversible

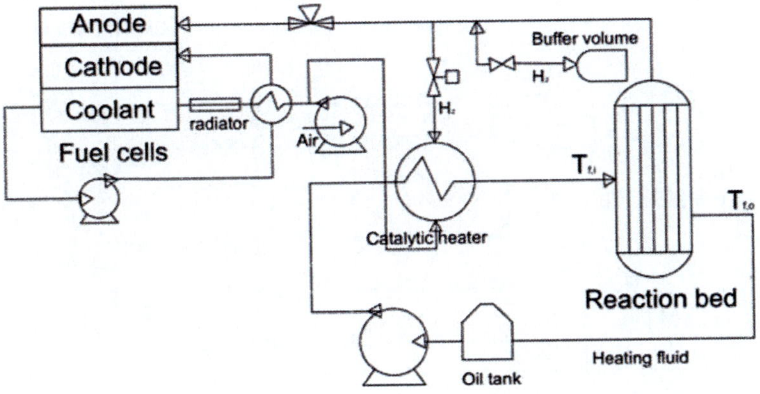

Fig. 10.2 System flow chart for metal hydride storage system coupled with fuel cells

sorption capacity, heat capacity and minimal dilution. The second group B and third group C factors are mainly related to design optimization of metal hydrides compressor, operating conditions and its manufacturing technologies (Lototskyy et al. 2014). HYSTORE technologies Ltd. (with the assistance of other research partners) has developed a small-scale metal hydride compressor demonstrator. The demonstrator operated in 6 stages; each stage deals with the specialized metal hydrides. It is found to be raising the hydrogen gas pressure from <5 bar to more than 200 bar by using water as heat exchanger fluid. The system achieves a stable hydrogen flow of about 3 Nm^3/h (Stamatakis et al. 2018). The system flow chart for metal hydride storage system coupled to fuel cells is as shown in Fig. 10.2.

10.5.2 *Achieving the Required Heat Transfer*

The heat transfer in metal hydride storage systems is the major parameter which controls the storage process. The main factors governing the heat transfer are enhancement of thermal conductivity of the bed by use of metal forms or ENG compacts, reduction in the reactor bed radius, increase in heat transfer area and increase in heat transfer mass flow rate and velocity.

Balzarotti et al. (2020) investigated a method of using pellets of Cu coating and Ni foam for improvement in thermal conductivity. The results showed that the Ni foam is an cost effective method of improving the thermal conductivity. Sanchez et al. investigated the effect of graphite compacts and foam to improve the thermal conductivity and found that the method has shown improvement in heat transfer but the major disadvantage is the compacts has shown to be deteriorated over prolonged usage (Sánchez et al. 2003). Botzung et al. (2008) investigated the effect of Al fins and foam in improving the thermal conductivity and found the significant improvement in thermal conductivity (Wang et al. 2019). Yang et al. (2008), studied the

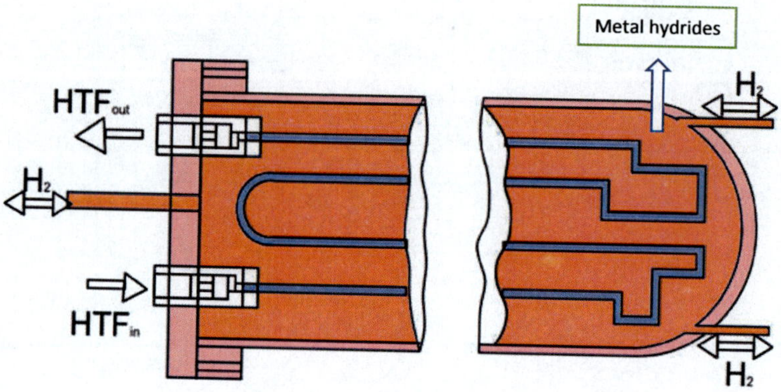

Fig. 10.3 Reactor bed design for the metal hydrides with heat transfer fluid

disc reactor, tubular reactor and annulus disc reactor for improvement in thermal conductivity and heat transfer. The system performance was found to be improved in heat transfer-controlled reaction rate and mass-transfer controlled reaction rate. Ranong et al. (2009) investigated the multiple metal hydride tank and cooling tube in single tank to improve the heat transfer area. It was found that the both designs have shown comparable performances and coiled tubes have shown better gravimetric capacity. Krokos et al. (2009), investigated the geometric optimization and heat transfer velocity optimization for improvement in heat transfer area. The system is observed to capable of an on-board storage for upto 25 km. Dhaou et al. (2010), investigated the double spiral heat exchanger with or without fins and optimized the pitch and fin diameter. The heat exchange performance was increased by 70%.

Chung et al. (2013) investigated the effect of heat pipes with fins which is considered as non-convectional method. It is observed that there is 50% improvement because of the use of heat pipes. The system was not found to be effective for large scale systems. Garrier et al. (2013), investigated the metal hydride compact and phase change material jackets. He found that the PCM system is found to be effective for large scale systems. Mellouli et al. (2016, 2017) investigated the PCM jackets, central tube embedded units with spherical, cylindrical and hexagonal geometries. It is observed that the cylindrical design performs better than the rest of the PCM geometric reactors. Design for the reactor bed with metal hydrides is shown in Fig. 10.3. In the figure HTF_{in} indicates heat transfer fluid inlet and HTF_{out} indicates heat transfer fluid out.

10.5.3 Mass and Volume Considerations

The mass and volume are the major considerations which limited the use of metal hydrides as on-board hydrogen storage systems. Research is directed towards the improvement in volumetric and gravimetric storage densities of the metal hydrides. Many methods such as alloying, nanoconfinement and ball milling are used to improve those parameters. Various techniques and their effect on different metal hydrides are reported in the previous sections.

Apart from the weight of the metal hydrides, the mass and volume of the storage tank also contributes to the major increase mass and volume in storage systems. The material selected for the tank must be of lower density. The thermal conductivity and heat capacity of the storage tank should also be considered. So the material selected for storage tank must be a compromise between the thermal properties and the density considerations. Along with the thermal properties the tank material properties such as corrosion resistance, mechanical strength and should not react with the chemical for the reaction to take place. Based on all these parameters the material should be selected to reduce the mass and volume requirements of the system. The gravimetric and volumetric capacities of the different types of metal hydrides are listed in the above sections.

10.5.4 Recyclability of Metal Hydrides for Many Cycles

The cyclic stability of the metal hydride is also a major parameter which governs the hydrogen storage capacity. Many researchers have tried to improve the cyclic stability of the metal hydrides. According to DOE targets the metal hydride should at least have cyclic stability up to 1500 cycles. Metal hydride composites were prepared which consists of a hydride consisting metal alloy and higher heat conduction material such as expanded natural graphite. The composite has shown a cyclic stability up to 1000 cycles at a constant temperature. The thermal conductivity was found to be decreased at the initial stages but later became constant (Nautiyal et al. 2015). It also showed the higher mechanical integrity. Some metal hydrides are identified as the relevant materials for practical applications. The metal hydrides include the types AB_5 alloys, AB_2 alloys and Mg (Ni). It was observed that the reversible storage capacity of AB_5 alloy decayed during the longer cycles. Even though the material loses its stability, the original stability can be restored by heating the material at 400–500 °C in vacuum conditions (Sakintuna et al. 2007).

Requirements of metal hydrides for on-board applications

Based on the above critical parameters, hydrogen storage systems which are mainly intended for automobile applications are tested and analysed by Giovanni Capurso (Capurso et al. 2018). In the research a metal hydride hydrogen storage tank with 10 cylinders containing 9.5 kg of hydride and the total mass of the system which includes frame, air circulating system, heat exchangers reached to 25 kg. After conducting the required tests simulating the vehicular applications, it was found that hydrogen release in an hour is more than 0.6 m^3 at a flow rate of 25 NL min^{-1}. The gravimetric capacity was found to 1.80 w% of H_2 with H_2 release of 1.65 wt% at the pressure above the fuel cell requirements. In absolute terms, the hydrogen usable is found to be 155 g but the hydrogen stored was found to be in 170 g. The volumetric hydrogen storage capacity was resulted to be 115 g dm^{-3}.

The hydrogen quantity combined with the mass and the dimensions of the storage system yields H_2 system gravimetric capacity of 0.65 wt% and a system volumetric capacity was found to be 6.2 g dm^{-3}. The system is found to be accumulating the energy of more than 22 MJ (6.1 kWh) with a density of 0.88 MJ kg^{-1}. The properties of the hydrogen storage tank were reported to be compared with the specific energies of li-ion batteries, alkaline and lead acid batteries. However, the power density depends on the heat exchanger effectiveness (Du et al. 2019).

HySA systems in collaboration with TF design built a prototype of air heated/cooled cassette-type metal hydride tank which is used for light electric vehicles. The hydrogen storage material used in the tank is multicomponent AB_2 type hydrogen storage alloy. The material has shown a reversible hydrogen storage capacity of 170 NL H_2/kg (1.5 wt% H; T = 20 °C and $P(H_2)$ = 40.1 bar) (Davids et al. 2019).

Mori et al. (2005) developed and tested a new type of high-pressure metal hydride tank for fuel cell vehicles with the collaboration of Toyota motors corporation. The tank is integrated with heat exchanger module and high-pressure cylindrical vessels. With a tank volume of 180 L the hydrogen storage capacity of 7.3 kg is achieved. In addition, the hydrogen is released at 243 K due to the hydrogen desorption alloy with a high dissociation pressure. The cursing range of over 700 km is achieved (Gupta el al. 2018).

10.6 Conclusion

Metal hydrides can be considered as the safest and compact alternative source for hydrogen storage. Over the past decade there was tremendous research going on for optimization of metal hydrides for on-board system requirement.

The present article reviews about the metal hydrides which have a potential to be used as hydrogen storage material. The metal hydrides should possess properties such as high hydrogen storage density, high gravimetric capacity, high cyclability, lower energy loss, light weight, low volumetric density etc. hydrogen should be stored in metal hydrides at moderate temperature and pressure. The research on metal hydrides is mainly focussed on improving its adsorption/desorption properties based on their hydrogen storage capacity, kinetics and thermal properties. Various methods were found to improve their performance. Based on the material properties different techniques are used on different materials. Some of the methods used were highlighted in the present article.

Among all the materials, Mg based metal hydrides are found to be the promising candidates for the hydrogen storage with a gravimetric capacity of 7.6 wt% with high hydrogen desorption capacity temperature of about 300 °C. Efforts are carried out to reduce the desorption temperature and improve the kinetics of the hydride by using appropriate production method, catalysts and nano structuring.

After Mg, the sodium-based hydrides were also identified as the promising candidates for the hydrogen storage with a gravimetric capacity of 4.5 wt% but the reversibility of the hydride posted a major problem. Various doping methods were identified to improve the reversibility and the hydrogen loading capacity of the sodium-based metal hydrides.

The equation required to calculate the equilibrium pressure at various temperatures based on the enthalpy and entropy of the metal hydride chemical reaction is represented in this article. Equations related to the modelling of hydrogen flow and heat transfer analysis based on the heat exchange system and the tank design were also represented in the present work.

The major requirements for the on-board storage system such as methods to achieve the required pressure and temperature for the metal hydride were highlighted. The mass and volume requirements of the metal hydride based on-board storage systems were discussed.

The feasibility of the metal hydrides hydrogen storage tank for the vehicular applications based on the parameters like gravimetric capacity, volumetric capacity, energy density were reported.

References

Adelhelm P, Gao J, Verkuijlen MH, Rongeat C, Herrich M, van Bentum PJM, Gutfleisch O, Kentgens AP, de Jong KP, de Jongh PE (2010) Comprehensive study of melt infiltration for the synthesis of NaAlH4/C nanocomposites. Chem Mater 22(7):2233–2238

Andrieux J, Laversenne L, Krol O, Chiriac R, Bouajila Z, Tenu R, Counioux JJ, Goutaudier C (2012) Revision of the $NaBO_2$–H_2O phase diagram for optimized yield in the H_2 generation through $NaBH_4$ hydrolysis. Int J Hydrogen Energy 37(7):5798–5810

Asano K, Westerwaal RJ, Anastasopol A, Mooij LP, Boelsma C, Ngene P, Schreuders H, Eijt SW, Dam B (2015) Destabilization of Mg hydride by self-organized nanoclusters in the immiscible Mg–Ti system. J Phys Chem C 119(22):12157–12164

Ashby EC, Brendel GJ, Redman HE (1963) Direct synthesis of complex metal hydrides. Inorg Chem 2(3):499–504

Au M, Spencer W, Jurgensen A, Zeigler C (2008) Hydrogen storage properties of modified lithium borohydrides. J Alloy Compd 462(1–2):303–309

Baldé CP, Hereijgers BP, Bitter JH, de Jong KP (2006) Facilitated hydrogen storage in $NaAlH_4$ supported on carbon nanofibers. Angew Chem 118(21):3581–3583

Balzarotti R, Ambrosetti M, Beretta A, Groppi G, Tronconi E (2020) Investigation of packed conductive foams as a novel reactor configuration for methane steam reforming. Chem Eng J 391:123494

Bobet JL, Akiba E, Nakamura Y, Darriet B (2000) Study of Mg-M (M=Co, Ni and Fe) mixture elaborated by reactive mechanical alloying—Hydrogen sorption properties. Int J Hydrogen Energy 25(10):987–996

Bogdanović B, Brand RA, Marjanović A, Schwickardi M, Tölle J (2000) Metal-doped sodium aluminium hydrides as potential new hydrogen storage materials. J Alloy Compd 302(1–2):36–58

Botzung M, Chaudourne S, Gillia O, Perret C, Latroche M, Percheron-Guegan A, Marty P (2008) Simulation and experimental validation of a hydrogen storage tank with metal hydrides. Int J Hydrogen Energy 33(1):98–104

Brooks KP, Sprik SJ, Tamburello DA, Thornton MJ (2020) Design tool for estimating metal hydride storage system characteristics for light-duty hydrogen fuel cell vehicles. Int J Hydrogen Energy 45(46):24917–24927

Brun N, Janot R, Sanchez C, Deleuze H, Gervais C, Morcrette M, Backov R (2010) Preparation of LiBH 4@ carbon micro–macrocellular foams: tuning hydrogen release through varying microporosity. Energy Environ Sci 3(6):824–830

Buchner H (1978) The hydrogen/hydride energy concept. Int J Hydrogen Energy 3(4):385–406

Capurso G, Schiavo B, Jepsen J, Lozano GA, Metz O, Klassen T, Dornheim M (2018) Metal hydride-based hydrogen storage tank coupled with an urban concept fuel cell vehicle: off board tests. Adv Sustain Syst 2(6):1800004

Chung CA, Yang SW, Yang CY, Hsu CW, Chiu PY (2013) Experimental study on the hydrogen charge and discharge rates of metal hydride tanks using heat pipes to enhance heat transfer. Appl Energy 103:581–587

Cracco D, Percheron-Guegan A (1998) Morphology and hydrogen absorption properties of an AB2 type alloy ball milled with Mg_2Ni. J Alloy Compd 268(1–2):248–255

Cui N, He P, Luo JL (1999) Magnesium-based hydrogen storage materials modified by mechanical alloying. Acta Mater 47(14):3737–3743

Cui J, Jonuzaj S, Adjiman CS (2018) A comprehensive approach for the design of solvent-based adhesive products using generalized disjunctive programming. In: Comput Aided Chem Eng, vol 44. Elsevier, pp 427–432

Davids MW, Lototskyy M, Malinowski M, van Schalkwyk D, Parsons A, Pasupathi S, Swanepoel D, van Niekerk T (2019) Metal hydride hydrogen storage tank for light fuel cell vehicle. Int J Hydrogen Energy 44(55):29263–29272

Dehouche Z, Djaozandry R, Goyette J, Bose TK (1999) Evaluation techniques of cycling effect on thermodynamic and crystal structure properties of Mg_2Ni alloy. J Alloy Compd 288(1–2):269–276

Dhaou H, Souahlia A, Mellouli S, Askri F, Jemni A, Nasrallah SB (2010) Experimental study of a metal hydride vessel based on a finned spiral heat exchanger. Int J Hydrogen Energy 35(4):1674–1680

DiPietro, J.P. and Skolnik, E.G., 2000. Sodium hydride-based hydrogen storage system. In *Proceedings of the 2000 DOE hydrogen program review NREL/CP-570–28890*.

Dornheim M (2011) Thermodynamics of metal hydrides: tailoring reaction enthalpies of hydrogen storage materials. IntechOpen

Du J, Lan Z, Zhang H, Lü S, Liu H, Guo J (2019) Catalytic enhanced hydrogen storage properties of Mg-based alloy by the addition of reduced graphene oxide supported V_2O_3 nanocomposite. J Alloy Compd 802:660–667

Fan X, Xiao X, Chen L, Yu K, Wu Z, Li S, Wang Q (2009) Active species of $CeAl_4$ in the $CeCl_3$-doped sodium aluminium hydride and its enhancement on reversible hydrogen storage performance. Chem Commun 44:6857–6859

Fan X, Xiao X, Chen L, Li S, Ge H, Wang Q (2011) Enhanced hydriding—dehydriding performance of $CeAl_2$-doped $NaAlH_4$ and the evolvement of Ce-containing species in the cycling. J Phys Chem C 115(5):2537–2543

Fang ZZ, Wang P, Rufford TE, Kang XD, Lu GQ, Cheng HM (2008) Kinetic-and thermodynamic-based improvements of lithium borohydride incorporated into activated carbon. Acta Mater 56(20):6257–6263

Garrier S, Delhomme B, De Rango P, Marty P, Fruchart D, Miraglia S (2013) A new MgH_2 tank concept using a phase-change material to store the heat of reaction. Int J Hydrogen Energy 38(23):9766–9771

Gizer G, Puszkiel J, Riglos MVC, Pistidda C, Ramallo-López JM, Mizrahi M, Santoru A, Gemming T, Tseng JC, Klassen T, Dornheim M (2020) Improved kinetic behaviour of $Mg(NH_2)_2$–2LiH doped with nanostructured K-modified-$Li_xTi_yO_z$ for hydrogen storage. Sci Rep 10(1):1–12

Gosalawit-Utke R, Milanese C, Javadian P, Jepsen J, Laipple D, Karmi F, Puszkiel J, Jensen TR, Marini A, Klassen T, Dornheim M (2013) Nanoconfined 2LiBH4–MgH2–TiCl3 in carbon aerogel scaffold for reversible hydrogen storage. Int J Hydrogen Energy 38(8):3275–3282

Graetz J, Reilly JJ (2006) Thermodynamics of the α, β and γ polymorphs of AlH3. J Alloy Compd 424(1–2):262–265

Graetz J, Reilly JJ, Yartys VA, Maehlen JP, Bulychev BM, Antonov VE, Tarasov BP, Gabis IE (2011) Aluminum hydride as a hydrogen and energy storage material: past, present and future. J Alloy Compd 509:S517–S528

Gross AF, Vajo JJ, Van Atta SL, Olson GL (2008) Enhanced hydrogen storage kinetics of LiBH4 in nanoporous carbon scaffolds. J Phys Chem C 112(14):5651–5657

Gupta JG, De S, Gautam A, Dhar A, Pandey A (2018) Introduction to sustainable energy, transportation technologies, and policy. In: Sustainable energy and transportation. Springer, Singapore, pp 3–7

Huen P, Paskevicius M, Richter B, Ravnsbæk DB, Jensen TR (2017) Hydrogen storage stability of nanoconfined MgH2 upon cycling. Inorganics 5(3):57

Jepsen LH, Paskevicius M, Jensen TR (2017) Nanostructured and complex hydrides for hydrogen storage. Nanotechnol Energy Sustain, 415–432

Jiao C, Sun L, Xu F, Liu SS, Zhang J, Jiang X, Yang L (2016) NiCo nanoalloy encapsulated in graphene layers for improving hydrogen storage properties of LiAlH 4. Sci Rep 6(1):1–8

Kohno T, Tsuruta S, Kanda M (1996) The hydrogen storage properties of new Mg_2Ni alloy. J Electrochem Soc 143(9):L198

Krokos CA, Nikolic D, Kikkinides ES, Georgiadis MC, Stubos AK (2009) Modeling and optimization of multi-tubular metal hydride beds for efficient hydrogen storage. Int J Hydrogen Energy 34(22):9128–9140

Kumar N (2019) Study of oxygenated ecofuel applications in CI engine, gas turbine, and jet engine. In: Azad A, Rasul M (eds) Advanced biofuels: applications, technologies and environmental sustainability, 1st edn. Elsevier, New York, pp 405–441

Kumar LH, Viswanathan B, Murthy SS (2008) Dehydriding behaviour of LiAlH4—the catalytic role of carbon nanofibres. Int J Hydrogen Energy 33(1):366–373

Kumar P, Srivastava KN, Dhar A (2018) Role of electric vehicles in future road transport. In: Sustainable energy and transportation. Springer, Singapore, pp 43–60

Li L, Peng B, Ji W, Chen J (2009) Studies on the hydrogen storage of magnesium nanowires by density functional theory. J Phys Chem C 113(7):3007–3013

Li L, Qiu F, Wang Y, Liu G, Yan C, An C, Xu Y, Wang Y, Song D, Jiao L, Yuan H (2012) Improved dehydrogenation performances of TiB_2-doped sodium alanate. Mater Chem Phys 134(2–3):1197–1202

Li L, Xu C, Chen C, Wang Y, Jiao L, Yuan H (2013) Sodium alanate system for efficient hydrogen storage. Int J Hydrogen Energy 38(21):8798–8812

Li Z, Zhai F, Wan Q, Liu Z, Shan J, Li P, Volinsky AA, Qu X (2014) Enhanced hydrogen storage properties of $LiAlH_4$ catalyzed by $CoFe_2O_4$ nanoparticles. RSC Adv 4(36):18989–18997

Li Y, Ding X, Zhang Q (2016) Self-printing on graphitic nanosheets with metal borohydride nanodots for hydrogen storage. Sci Rep 6(1):1–10

Liang G, Boily S, Huot J, Van Neste A, Schulz R (1998) Hydrogen absorption properties of a mechanically milled Mg–50 wt.% $LaNi_5$ composite. J Alloys Compd 268(1–2):302–307

Lototskyy MV, Yartys VA, Pollet BG, Bowman RC Jr (2014) Metal hydride hydrogen compressors: a review. Int J Hydrogen Energy 39(11):5818–5851

Lu YS, Wang H, Ouyang LZ, Liu JW, Zhu M (2015) Reversible hydrogen storage and phase transformation with altered desorption pressure in Mg90In5Cd5 ternary alloy. J Alloy Compd 645:S103–S106

Lu C, Zou J, Zeng X, Ding W, Shao H (2019) Enhanced hydrogen sorption properties of core-shell like structured Mg@ $NaBH_4/MgB_2$ composite. J Alloys Compd 810:151763

Marrero-Alfonso EY, Gray JR, Davis TA, Matthews MA (2007) Minimizing water utilization in hydrolysis of sodium borohydride: the role of sodium metaborate hydrates. Int J Hydrogen Energy 32(18):4723–4730
Matsunaga T, Buchter F, Mauron P, Bielman M, Nakamori Y, Orimo S, Ohba N, Miwa K, Towata S, Züttel A (2008) Hydrogen storage properties of $Mg[BH_4]_2$. J Alloy Compd 459(1–2):583–588
Mellouli S, Abhilash E, Askri F, Nasrallah SB (2016) Integration of thermal energy storage unit in a metal hydride hydrogen storage tank. Appl Therm Eng 102:1185–1196
Mellouli S, Askri F, Abhilash E, Nasrallah SB (2017) Impact of using a heat transfer fluid pipe in a metal hydride-phase change material tank. Appl Therm Eng 113:554–565
Miwa K, Aoki M, Noritake T, Ohba N, Nakamori Y, Towata SI, Züttel A, Orimo SI (2006) Thermodynamical stability of calcium borohydride Ca $(BH_4)_2$. Phys Rev B 74(15):155122
Mohammadshahi SS, Gray EM, Webb CJ (2016) A review of mathematical modelling of metal-hydride systems for hydrogen storage applications. Int J Hydrogen Energy 41(5):3470–3484
Mori D, Haraikawa N, Kobayashi N, Kubo H, Toh K, Tsuzuki M, Shinozawa T, Matsunaga T (2005) High-pressure metal hydride tank for fuel cell vehicles. MRS Online Proceedings Library (OPL) 884
Moussa G, Moury R, Demirci UB, Şener T, Miele P (2013) Boron-based hydrides for chemical hydrogen storage. Int J Energy Res 37(8):825–842
Na Ranong C, Hoehne M, Franzen J, Hapke J, Fieg G, Dornheim M, Eigen N, Bellosta von Colbe JM, Metz O (2009) Concept, design and manufacture of a prototype hydrogen storage tank based on sodium alanate. Chem Eng Technol Ind Chem-Plant Equip-Process Eng-Biotechnol 32(8):1154–1163
Nautiyal H, Shree V, Khurana S, Kumar N, Varun (2015) Recycling potential of building materials: a review. In: Muthu SS (ed) Environmental footprints and eco-design of products and processes, 1st edn. Springer, New York, pp 31–50
Nielsen TK, Manickam K, Hirscher M, Besenbacher F, Jensen TR (2009) Confinement of MgH_2 nanoclusters within nanoporous aerogel scaffold materials. ACS Nano 3(11):3521–3528
Nielsen TK, Polanski M, Zasada D, Javadian P, Besenbacher F, Bystrzycki J, Skibsted J, Jensen TR (2011) Improved hydrogen storage kinetics of nanoconfined $NaAlH_4$ catalyzed with $TiCl_3$ nanoparticles. ACS Nano 5(5):4056–4064
Orimo SI, Nakamori Y, Kitahara G, Miwa K, Ohba N, Towata SI, Züttel A (2005) Dehydriding and rehydriding reactions of $LiBH_4$. J Alloy Compd 404:427–430
Orimo SI, Nakamori Y, Eliseo JR, Züttel A, Jensen CM (2007) Complex hydrides for hydrogen storage. Chem Rev 107(10):4111–4132
Ouyang L, Liu F, Wang H, Liu J, Yang XS, Sun L, Zhu M (2020) Magnesium-based hydrogen storage compounds: a review. J Alloys Compd 832:154865
Parry RW, Schultz DR, Girardot PR (1958) The preparation and properties of hexamminecobalt (III) borohydride, hexamminechromium (III) borohydride and ammonium borhydride1. J Am Chem Soc 80(1):1–3
Raman SS, Davidson DJ, Bobet JL, Srivastava ON (2002) Investigations on the synthesis, structural and microstructural characterizations of Mg-based K_2PtCl_6 type (Mg_2FeH_6) hydrogen storage material prepared by mechanical alloying. J Alloy Compd 333(1–2):282–290
Reilly JJ, Wiswall RH (1974) Formation and properties of iron titanium hydride. Inorg Chem 13(1):218–222
Sakintuna B, Lamari-Darkrim F, Hirscher M (2007) Metal hydride materials for solid hydrogen storage: a review. Int J Hydrogen Energy 32(9):1121–1140
Sánchez AR, Klein HP, Groll M (2003) Expanded graphite as heat transfer matrix in metal hydride beds. Int J Hydrogen Energy 28(5):515–527
Saxena V, Kumar N, Saxena V (2021) Combustion, performance and emissions of *Acacia concinna* biodiesel blends in a diesel engine with variable specific heat ratio. J Therm Anal Calorim. https://doi.org/10.1007/s10973-020-10483-7

Schneemann A, White JL, Kang S, Jeong S, Wan LF, Cho ES, Heo TW, Prendergast D, Urban JJ, Wood BC, Allendorf MD (2018) Nanostructured metal hydrides for hydrogen storage. Chem Rev 118(22):10775–10839

Shinde SS, Kim DH, Yu JY, Lee JH (2017) Self-assembled air-stable magnesium hydride embedded in 3-D activated carbon for reversible hydrogen storage. Nanoscale 9(21):7094–7103

Skripnyuk VM, Rabkin E (2012) Mg3Cd: A model alloy for studying the destabilization of magnesium hydride. Int J Hydrogen Energy 37(14):10724–10732

Song W, Dong H, Zhang G, Liu J, Yang G, Liu Y, Li Y, Li J, Shen J, Chen Y, Wei Q (2020) Enhanced hydrogen absorption kinetics by introducing fine eutectic and long-period stacking ordered structure in ternary eutectic Mg–Ni–Y alloy. J Alloys Compd 820:153187

Stamatakis E, Zoulias E, Tzamalis G, Massina Z, Analytis V, Christodoulou C, Stubos A (2018) Metal hydride hydrogen compressors: current developments & early markets. Renew Energy 127:850–862

Suh MP, Park HJ, Prasad TK, Lim DW (2012) Hydrogen storage in metal–organic frameworks. Chem Rev 112(2):782–835

Sun D, Enoki H, Bououdina M, Akiba E (1999) Phase components and hydriding properties of the sintered Mg–xwt.% $LaNi_5$ (x= 20–50) composites. J Alloys Compd 282(1–2):252–257

Wagemans RW, van Lenthe JH, de Jongh PE, Van Dillen AJ, de Jong KP (2005) Hydrogen storage in magnesium clusters: quantum chemical study. J Am Chem Soc 127(47):16675–16680

Wang Y, Aizawa T, Nishimura C (2006) Solid-state synthesis of hydrogen storage Mg_2Co alloys via bulk mechanical alloying. Mater Trans 47(4):1052–1057

Wang J, Li HW, Chen P (2013) Amides and borohydrides for high-capacity solid-state hydrogen storage—materials design and kinetic improvements. MRS Bull 38(6):480–487

Wang H, Zhong H, Ouyang L, Liu J, Sun D, Zhang Q, Zhu M (2014) Fully reversible de/hydriding of Mg base solid solutions with reduced reaction enthalpy and enhanced kinetics. J Phys Chem C 118(23):12087–12096

Wang L, Rawal A, Aguey-Zinsou KF (2019) Hydrogen storage properties of nanoconfined aluminium hydride (AlH3). Chem Eng Sci 194:64–70

Wiberg E, Amberger E (1971) Hydrides of the elements of main groups I-IV. Elsevier Publishing Company

Xia Y, Yang Z, Zhu Y (2013) Porous carbon-based materials for hydrogen storage: advancement and challenges. J Mater Chem A 1(33):9365–9381

Xu C, Lin HJ, Wang Y, Zhang P, Meng Y, Zhang Y, Liu Y, Zhang J, Li L, Shi Q, Li W (2019) Catalytic effect of in situ formed nano-Mg_2Ni and Mg_2Cu on the hydrogen storage properties of Mg-Y hydride composites. J Alloy Compd 782:242–250

Yamaura SI, Kim HY, Kimura H, Inoue A, Arata Y (2002) Electrode properties of rapidly solidified Mg67Ni23Pd10 amorphous alloy. J Alloy Compd 347(1–2):239–243

Yang F, Meng X, Deng J, Wang Y, Zhang Z (2008) Identifying heat and mass transfer characteristics of metal hydride reactor during adsorption—parameter analysis and numerical study. Int J Hydrogen Energy 33(3):1014–1022

Zaluska A, Zaluski L, Ström-Olsen JO (2001) Structure, catalysis and atomic reactions on the nano-scale: a systematic approach to metal hydrides for hydrogen storage. Appl Phys A 72(2):157–165

Zhang J, Zhu Y, Lin H, Liu Y, Zhang Y, Li S, Ma Z, Li L (2017) Metal hydride nanoparticles with ultrahigh structural stability and hydrogen storage activity derived from microencapsulated nanoconfinement. Adv Mater 29(24):1700760

Zhang J, Li Z, Wu Y, Guo X, Ye J, Yuan B, Wang S, Jiang L (2019) Recent advances on the thermal destabilization of Mg-based hydrogen storage materials. RSC Adv 9(1):408–428

Zhao DL, Zhang YH (2014) Research progress in Mg-based hydrogen storage alloys. Rare Met 33(5):499–510

Zheng S, Fang F, Zhou G, Chen G, Ouyang L, Zhu M, Sun D (2008) Hydrogen storage properties of space-confined $NaAlH_4$ nanoparticles in ordered mesoporous silica. Chem Mater 20(12):3954–3958

Zhong HC, Wang H, Liu JW, Sun DL, Zhu M (2011) Altered desorption enthalpy of MgH_2 by the reversible formation of Mg (In) solid solution. Scripta Mater 65(4):285–287
Zhu M, Zhu WH, Chung CY, Che ZX, Li ZX (1999) Microstructure and hydrogen absorption properties of nano-phase composite prepared by mechanical alloying of MmNi5−x (CoAlMn) x and Mg. J Alloy Compd 293:531–535
Zhu M, Wang H, Ouyang LZ, Zeng MQ (2006) Composite structure and hydrogen storage properties in Mg-base alloys. Int J Hydrogen Energy 31(2):251–257
Züttel A, Wenger P, Rentsch S, Sudan P, Mauron P, Emmenegger C (2003) $LiBH_4$ a new hydrogen storage material. J Power Sources 118(1–2):1–7

Chapter 11
Waste Heat Recovery Potential from Internal Combustion Engines Using Organic Rankine Cycle

Krishn Chandra, Avinash Kumar Agarwal, Oronzio Manca, and Andrea Unich

Abstract Increasing fuel prices and stringent emission regulations are forcing improvements in the performance and emissions of internal combustion engines (ICEs). There are several ways to enhance engine performance and emissions. Waste heat recovery has emerged as a practical solution in recent times. The Organic Rankine Cycle (ORC) is a promising technology to recover the waste heat from various sources and potentially enhance the overall performance of ICEs. The ORC system converts waste heat energy into useful power either for ICEs, renewable energy (geothermal, solar and biomass), or industrial waste heat energy. The ORC systems range from a few kW to multi-MW plants. After a long evolution, this technology has matured. The economic performance of an ORC is very important for its further development and wider application. Integration of ORC with ICEs is gaining popularity among researchers. However, it isn't very easy and depends on the physicochemical properties of the working fluid (WF) and the heat source temperature. Several optimization methods could improve its performance and decide the optimum operating conditions for the ORC systems.

Keywords Organic Rankine cycle · Waste heat recovery · Working fluid · Economy

11.1 Introduction

Many industrial processes produce waste heat, typically at lower temperatures, but it can be still recovered to produce useful energy. The waste heat (WH) is the heat dissipated from various energy conversion devices. It is eventually dissipated to

K. Chandra · O. Manca · A. Unich
Department of Engineering, University of Campania Luigi, Vanvitelli, Italy

K. Chandra · A. K. Agarwal (✉)
Engine Research Laboratory, Department of Mechanical Engineering, Indian Institute of Technology Kanpur, Kanpur 208016, India
e-mail: akag@iitk.ac.in

A. K. Agarwal et al. (eds.), *Advanced Combustion for Sustainable Transport*, Energy, Environment, and Sustainability, https://doi.org/10.1007/978-981-16-8418-0_11

the atmosphere at ambient temperature. The waste heat recovery offers the advantage of overall increased efficiency in power generation or could provide auxiliary power used in other applications. The conventional Rankine cycle, often used for primary power generation, uses a high-temperature heat source, typically operating at relatively higher temperatures (200–600 °C). However, the organic Rankine cycle (ORC) uses organic liquid rather than water as a WF, operating at substantially lower temperatures.

The current state-of-the-art ORC systems have several limitations. Many WFs are ozone-depleting or toxic. The components of ORC systems are expensive for the power they produce. For example, a turbine expander is appropriate for the steam Rankine cycle that produces power in the range of MW but is too expensive for a system that produces power in kW. Waste heat recovery (WHR) from various sources has evolved as an effective way to minimize energy losses. Therefore, they reduce greenhouse gas (GHG) emissions. The main sources of waste heat in thermal industrial equipment/systems include exhaust gas, low-quality steam, cooling air, flaring gases, liquids (hot oil and cooling water) and solids (hot steel) (Papapetrou et al. 2018). The conversion of WH into usable energy can reduce energy consumption from the primary sources (Svensson et al. 2019). This leads to enhanced overall system efficiency and reduced emissions.

ICEs are held responsible for anthropogenic emissions of GHG and air pollutants. ICEs are widely used for small-to-medium-large vehicles, gas engines, and ship propulsion, apart from agricultural and industrial applications. The newer emission legislations (such as BS-VI, EURO-VI) for NO_x and Particulate matter (PM) emissions are getting stringent with time. Emission challenges are forcing engine developers to design highly efficient engines. Keeping these challenges in mind, several research groups have turned to WHR for overall engine performance improvement. The strategies for recovering WH are primarily categorized as engine powertrain and bottoming technologies. The engine powertrain technologies employ turbochargers for engine downsizing (Zhang et al. 2013; Innovative and System), which is integrated with various sub-systems such as exhaust gas recirculation (EGR) for NO_X reduction (Squaiella et al. 2013), advanced Miller timing system, and variable valve timing (VVT) system (Codan and Vlaskos 2004). The bottoming technologies include after-treatment systems such as diesel oxidation catalysts (DOC), diesel particulate filters (DPF), and selective catalytic reduction (SCR) to reduce emissions (Johnson 2015) or the WHR systems. The WHR is primarily done using ORC (Lecompte et al. 2015; Quoilin et al. 2013), turbo-compounding (Aghaali and Ångström 2015), and thermoelectric generators (TEGs) (Hamid Elsheikh et al. 2014). The exhaust gas and coolant energy need to be recovered to get the maximum benefit (Chen et al. 2017) of the WHR systems. Heat recovery using the ORC system has been identified as a pioneering technology to reuse the heat from low-to-medium temperature heat sources. Conceptually, the ORC system is similar to the conventional Rankine cycle. The ORC is a closed cycle; therefore, it requires a module similar to the Rankine cycle. However, organic WFs with boiling temperatures lower than water are used for the ORC to recover heat from a low-temperature heat source. The phase change of organic fluid occurs at a substantially lower temperature than boiling water. These

WFs' evaporation pressure and temperature are lower than water in the Rankine cycle (Vanslambrouck et al. 2012). Hence the overall thermal efficiency of ORC systems is not very high. Conversion efficiency enhancement is one of the most significant challenges for ORC implementation in practical applications. The conversion efficiency is quite low for various ORC systems and relies upon the source and sink temperature and the WF properties. The heat sources and corresponding temperature ranges of typical ORC systems are summarized in Table 11.1.

The exhaust gas has higher exergy compared to other potential sources. The temperature and mass flow rate of exhaust gas mainly depends on the engine load. Therefore engine load is a crucial parameter for WHR system design (Hoang 2018). In ICE-WHR prototypes development and investigations, exhaust gas and EGR have been given significant attention due to higher temperature and heat content (Hountalas et al. 2012). Three main components are necessary to recover the waste heat: an available heat source, heat recovery technology, and the use of the recovered energy.

11.1.1 WHR System Evolution and Trends

The ORC is a reliable technology for WHR from geothermal energy sources and CHP using biomass; however, it is rarely used for solar applications. The market of ORC plants has grown exponentially since its first commercial installation in the 1970s. The idea of WHR using an ORC system was put forward by T. Howard in early 1826 (Howard). Afterwards, several engines with naphtha as the WF were designed by Esher Wyss AG and Ofeldt (Wyss 1890). D'Amelio built the first modern ORC system in 1936 (D'Amelio 1936). The same concept was applied to the low-temperature geothermal (GT) plants with a power output of 2.6 kW and 11 kW. In the 1960s, Bronicki and Tabor started to scrutinize the potential fluids on a large scale (Tabor and Bronicki 1964). Prof. Gaia and Angelino Macchi conducted similar tests at Politecnico di Milano (Angelino et al. 1984). Subsequently, numerous manufacturers came forward with their customized technologies. Different combinations of expanders and WF or cycle configurations were made technically feasible.

The establishment of large-scale projects takes more than a year. From 1980 to 2003, the bulk of the research was done for geothermal applications of ORC. 2003 to 2005 saw a sharp expansion in the ORC system market. Their installed capacity increased from 75 to 352 MW in this short period. After 2003, the biomass heat recovery revolution was led by Turboden with the installation of 15–25 units per year. The WHR market shrunk from 2008 to 2013, before the surge again in 2015. However, the market remained stable for ICEs and gas turbines (68%) between 2013 and 2015.

The first implementation of an ORC system in a vehicle was investigated by Doyle and Patel (1976) in 1976. A Mack 676 diesel engine was used to recover heat from the exhaust gas using Fluorinol-50 as the WF. In 2005, El Chammas and Clodic (2005) explored the heat recovery from the exhaust gas and cooling circuits of a 1.4 L hybrid gasoline engine. Arias et al. (2006) simulated the ORC system for the WHR from the

Table 11.1 Heat source and temperature range (Inc 2008; Waste Heat Recovery—Technology Assessment 2016; Heywood 2018)

Category	Heat sources	Temperature range (°C)	Recommended heat recovery technology
High-temperature heat source (>650 °C)	Fume incinerators	600–1450	Steam Rankine cycle Heat exchanger (preheating) Thermal PV Air preheating Steam generation (heating) Thermoelectric
	Glass melting furnace	1000–1550	
	Copper reverberatory furnace	900–1100	
	Zinc refining furnace	760–1100	
	Hydrogen plants	650–1000	
	Solid waste	650–1000	
	Nickel refining furnace	1370–1650	
	Aluminium refining furnaces	650–760	
	Copper refining furnaces	760–815	
	Cement kiln	620–730	
Medium temperature heat source (230–650 °C)	Gas turbine exhaust	370–540	Organic Rankine cycle Air preheating Thermal PV Steam Rankine cycle Heat exchangers (preheating process) Thermoelectric
	Catalytic crackers	425–650	
	Drying and baking ovens	230–600	
	Annealing furnace cooling systems	425–650	
	Steam boiler exhaust	230–480	
	Reciprocating engine exhaust	315–600	
	Heat from EGR	200–750	
Low-temperature heat source (<230 °C)	Processed steam condensate	50–90	Domestic water heating Organic Rankine cycle Absorption and adsorption cooling Piezoelectric Space heating Heat pump Heat exchangers Kalina cycle
	Heat from lubricating oil	80–120	
	Cooling water from:		
	Hot processed liquids and solids	32–232	
	Drying, baking and curing ovens	93–230	
	Bearings	32–88	
	ICEs	66–120	
	Annealing furnaces	66–230	
	Welding and injection moulding machines	32–88	
	Air compressors	27–50	

exhaust gas, coolant, and combined coolant-exhaust of a spark ignition (SI) engine. The WF was preheated in the engine block before being superheated by the exhaust gas.

In 2005, Cummins started investigating the use of the ORC system with its ISX HDDE model for heat recovery. Nelson (2008) reported Cummins ORC activities to recover heat from the EGR and the exhaust gas. Cummins claimed that the total engine efficiency improved by 5–8% using the ORC WHR system. Endo et al. (2007) implemented the ORC to recover heat from the exhaust gas with steam as the WF in a 2.0 L hybrid gasoline engine. In 2008 and 2009, BMW (Drives 2008; Ringler et al. 2009) reported using Turbosteamer (Rankine cycle) for WHR from the exhaust gas and coolant in a passenger car. The WF was water-steam and ethanol in the HT and LT loops, respectively. Eaton (Drive 2017) also tested few new concepts of the WHR using the ORC system in a 13.5 L John Deere heavy-duty diesel engine (HDDE). Bosch (Robert BOSCH GmbH 2014; Seher et al. 2012) explored WHR using an ORC system for commercial vehicle applications. Ricardo plc (Howell and Gibble 2011) implemented an ORC system in a 288 kW Volvo HDDE for WHR from the exhaust gas and EGR with the ethanol–water mixture as the WF. Briggs et al. (2010) implemented the ORC system in a 1.9 L LDDE 4-cylinder engine to recover exhaust gas heat using R-245fa as the WF. This system was equipped with a variable geometry turbocharger and a high-pressure EGR system.

11.1.2 Current State-of-the-Art of the ORC Systems

Figure 11.1 shows the total installed capacity, the total number of plants, and the market share of the different manufacturers in terms of installed capacity and number of plants. The installed capacity of ORC systems worldwide is 2701 MW and in 1754 ORC units. Heat recovery from geothermal sources has been mainly implemented using ORC systems, having a 74.8% share in worldwide installed capacity. However, these applications for geothermal energy recovery need multi-MW plants and massive funding. As a result, only a few companies such as Exergy, ORMAT,

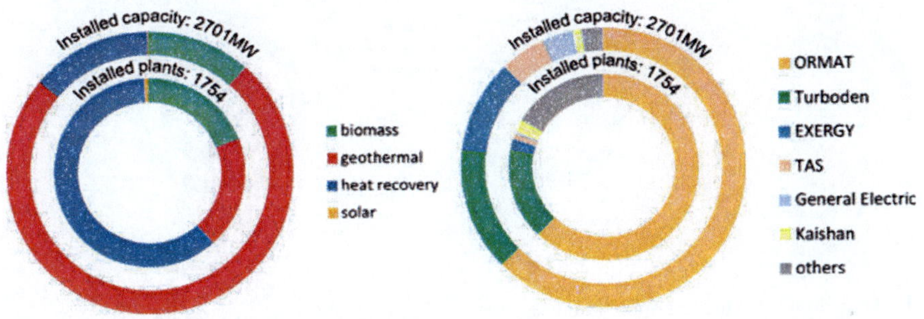

Fig. 11.1 Total installed capacity per application and per manufacturer (Tartière and Astolfi 2017)

Turboden, and Turbine Air System (TAS) are actively pursuing this area. ORMAT is the leader with > 75% of the installed capacity, while ~ 13 and ~ 6% of the market is catered by Exergy and TAS, respectively. Turboden lately entered the geothermal sector with an installed capacity of ~ 2%.

These leading manufacturers in heat recovery systems are actively developing WGR systems for ICEs, other industrial processes, and gas turbines for medium-large plants. Figure 11.2 show the share of installed capacity as per heat recovery applications.

Several manufacturers are exploring small-scale WHR applications ranging from 10 to 150 kWe. This application of WHR covers a 13.9% share of the WHR total market with a sizeable number of ORC plants. Among these, nearly 800 ORC plants are quite small (< 4 kW). Tunoden is the leading player in the Biomass application space with more than 228 plants and an 11% market share. Most of them are combined heat and power (CHP) plants. Several more are under construction. However, their application for solar energy recovery is limited due to higher investment costs. Figure 11.3 shows the major companies active in the market with medium to large plants recovering heat from geothermal, internal combustion engines or industrial processes and biomass. ORMAT is the world leader of ORC units with a total installed

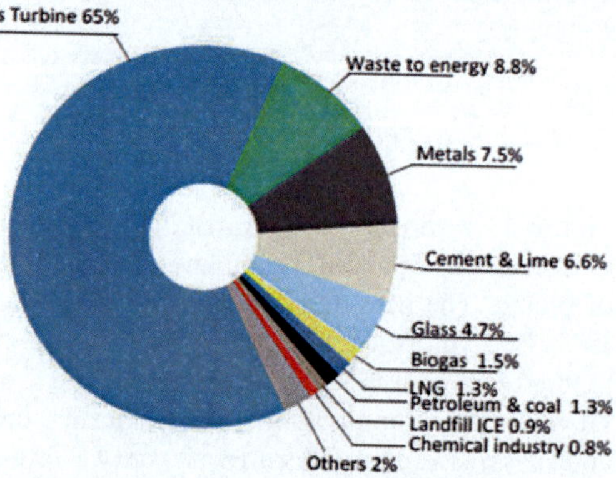

Fig. 11.2 Share of installed capacity per application (Tartière and Astolfi 2017)

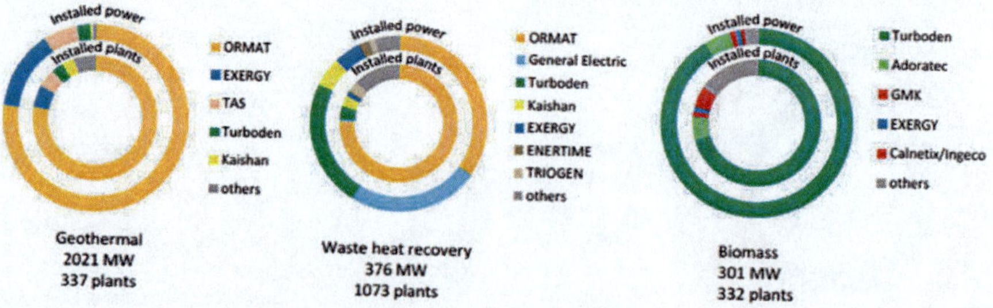

Fig. 11.3 Market share per application and per manufacturer (Tartière and Astolfi 2017)

capacity of 62.9%, followed by Tuboden (13.4%) and exergy (11.1%). Few large-sized plants are installed by General Electric (GE) and TAS. Other companies are also identified with a market share of 3.2% on small-medium size applications.

The manufacturers are producing optimized ORC systems for a definite lower temperature range of waste heat. The three prime producers of ORC units in terms of power and number of installed units are Turboden (45%), ORMAT (24%), and Maxxtec (23%). Table 11.2 shows the manufacturers and ORC systems designed for a wide range of power and temperature. The ORC systems are optimized for particular heat sources and temperature ranges. A small change in working conditions significantly affects their conversion efficiency.

In addition to the above manufacturers, several new companies have entered into the market of lower capacity ORC system manufacturing.

11.2 Fundamentals of Organic Rankine Cycle (ORC)

The conventional Rankine cycle uses heat for the conversion of water into steam. The steam is then expanded in the turbine to produces mechanical power to drive an electricity generator. The boiling point of water is high. However, the WFs in the ORC are organic fluids with lower boiling temperatures and higher molecular weights (Lion et al. 2017). This facilitates the ORC to work with lower temperature heat sources. The most suitable temperature range depends on the WF's thermodynamic properties, which affects the cycle efficiency at different temperatures.

The organic fluids used in an ORC system have higher molecular mass, mass flow rate, and expander efficiency than the steam-operated Rankine cycle plants. Thermodynamically, the efficiency of low-temperature cycles is lower than the high-temperature cycles. Therefore, the efficiency of the ORC systems is much lower than the steam power plant (30–40%). Overall cycle efficiency is lower in the range of ~ 10–20% depending on the evaporator and condenser temperatures. Hence, the economic operation of the ORC-WHR system requires application-specific analysis.

The vehicular integration of the ORC system is rather complex because of the engine's transient and fluctuating operating conditions. Precise control to obtain superior performance, durability, and reliability is required for the practicality of the WHR concept for automotive heat recovery. A simple ORC system layout incorporates preheater, evaporator, expander, regenerator, and condenser (Fig. 11.4). The regenerator enhances the ORC system efficiency by preheating the organic WF.

11.2.1 Thermodynamic Analysis of the ORC System

Vapour form of organic WF from the evaporator expands in the expander to develop mechanical power. Vapours from the expander condense into the liquid form in the condenser. After that, this liquid WF from the condenser is pumped into the

Table 11.2 Leading manufacturers of ORC systems (Quoilin et al. 2013; Anastasovski et al. 2020)

Manufacturer	Capacity (kW)	Applications	Heat source temperature (°C)
ORMAT, US	200–70,000	GE, WHR, SE	150–300
GMK, Germany	50–5000	WHR, GT, B-CHP	120–350
Turboden, Italy (Obernberger and Thek 2008)	≤ 20,000	WHR, SE, BM, GT	≤ 100
Bosch KWK, Germany	65–325	WHR	120–150
Global CSP	75–130	RN	–
Maxxtec, Germany	315–1600	Biomass-CHP	300
Aqylon (Law et al. 2013)	≤ 25,000	GT, BM, WHR, SE, ICI	$HS_{in}/HS_{out} = 305/170$ $CS_{in}/CS_{out} = 35/45$
Cryostar, France	75–20,000	WHR, GT	100–400
Electratherm	≤ 100	GT, LG, STE, BM, FL, WWT	77–122
Opcon, Sweden	350–800	WHR	< 120
Calnetix Power	125	WHR	121
ZE International Group	≤ 130	SG, LG, BM, BME, BG, CH, NG, GI, WHE, CCI, FA, FM, RP, PMP, FA, FL, MP	220–280
Turboden Pure Cycle, US	280	WHR, GT	91–149
SRM	150–2400	RN, WHR	55
Siemens	400–1500	WHR	300
ENER Basque	20	GT, CHP, WHR, BM	80–90
DUR	50–1000	GT, LG, BG, BM, SE, NG, WHR	90
Infinity turbine	10	–	80–120
GE Clean Cycle	125	WHR	> 121
Tri-o-gen (Law et al. 2016), Netherlands	100–170	BM, GE, WHS, OM, DE, FL	350–530
Electratherm, US	50	WHR, Solar	> 93
Exergy	400–650	GT, WHR, CHP, SE, BM	230–315
Enogia (Obernberger et al. 2002)	10–100	GT, WHR, CSP, BM, SM, ICI	70–120
BEP Europe	50	CHP, WHR	85–115
Hydrocurrent	152	EEX, ECA	80–95

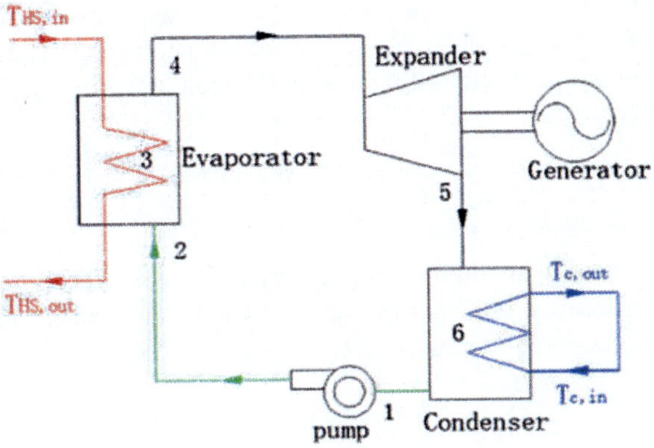

Fig. 11.4 Schematic of a simple ORC system (Song et al. 2015)

evaporator, as shown in Fig. 11.4. The WF is in a saturated or superheated state at the exit of the evaporator. The thermodynamic cycle of the ORC system on a T-s diagram is shown in Fig. 11.5, which is discussed below.

The mass flow rate of the WF can be calculated using Eq. 11.1.

$$\dot{m}_{wf} = \frac{Q_{total} * (1 - \varepsilon)}{h_4 - h_2} \tag{11.1}$$

where Q_{total} is the total heat load of the waste heat source, and ε is the heat loss ratio of the heat exchanger. The total heat (Q_{total}) transferred from the waste heat to WF can be calculated using Eq. 11.2.

$$Q_{total} = \dot{m}_{HS} * C_{p,HS}\left(T_{HS,in} - T_{HS,out}\right) \tag{11.2}$$

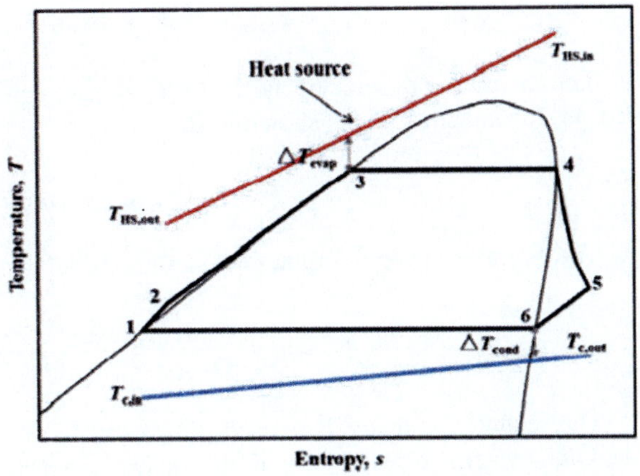

Fig. 11.5 T-s diagram of a simple ORC system (Song et al. 2015)

where, $C_{p,HS}$ is the average specific heat capacity of waste heat source and $T_{HS,in}$ and $T_{HS,out}$ are inlet and outlet temperatures, respectively. The work done on the pump (process 1–2) can be calculated using Eq. 11.3.

$$W_{pump} = \frac{\dot{m}_{wf} * [h_{2s} - h_1]}{\eta_{pump}} \tag{11.3}$$

where h_{2s} is the isentropic enthalpy of the WF after the WF pump and η_{pump} is the efficiency of the pump. Heat transfer from the heat source to the WF in the evaporator (process 2–4) can be calculated by Eq. 11.4.

$$Q_{evaporator} = \dot{m}_{wf} * (h_4 - h_2) \tag{11.4}$$

Work obtain by expander (process 4–5) is calculated by Eq. 11.5

$$W_{\exp} = \dot{m}_{wf} * (h_4 - h_{5s}) * \eta_{\exp} \tag{11.5}$$

where h_{5s} is the isentropic enthalpy of WF at expander outlet, and η_{exp} is the efficiency of the expander. The flow rate of cooling water through the condenser can be calculated by Eq. 11.6.

$$\dot{m}_{cond.} = \frac{\dot{m}_{wf} * (h_5 - h_1)}{(1 - \varepsilon) * C_{p,c} * \left(T_{c,out} - T_{c,in}\right)} \tag{11.6}$$

where $C_{p,c}$ is the average specific heat capacity of cooling water, $T_{c,in}$ and $T_{c,out}$ are inlet and outlet temperature of cooling water, respectively. The heat dissipated from the condenser is calculated by Eq. 11.7.

$$Q_{cond.} = \dot{m}_{wf} * [h_5 - h_1] \tag{11.7}$$

The net power developed by the ORC system can be calculated by substituting Eq. 11.3 from Eq. 11.5, as shown in Eq. 11.8.

$$W_{net} = W_{exp} - W_{pump} \tag{11.8}$$

The overall efficiency of the ORC unit is calculated by Eq. 11.9.

$$\eta_{overall} = \frac{W_{net}}{Q_{total}} = \frac{W_{exp} - W_{pump}}{Q_{total}} \tag{11.9}$$

The second law analysis is used to investigate the irreversibilities and losses in the ORC system corresponding to the exergy destruction rate in every component.

$$I_{pump} = \dot{m}_{wf} * T_0 * (s_2 - s_1) \tag{11.10}$$

$$I_{evap.} = \dot{m}_{wf} * T_0 * \left((s_4 - s_2) - \frac{h_4 - h_2}{T_{evap., avg}} \right) \tag{11.11}$$

$$I_{exp} = \dot{m}_{wf} * T_0 * (s_5 - s_4) \tag{11.12}$$

$$I_{cond.} = \dot{m}_{wf} * T_0 * \left((s_1 - s_5) - \frac{h_1 - h_5}{T_{cond., avg}} \right) \tag{11.13}$$

where T_0 is the ambient temperature, $T_{evap.avg}$ and $T_{cond.,avg}$ are the average temperatures during evaporation and condensation, respectively.

11.2.2 ORC System Components

Pumps

The feed pump in the ORC system is a key component. Particular attention should be given to the feed pump for its sizing and selection. This must satisfy the requirement of the ORC cycle in terms of different parameters such as efficiency, net positive suction head (NPSH), controllability, and tightness.

The pump consumes very low power compared to the output power of the Rankine cycle. However, in an ORC system, the pump's overall efficiency decreases substantially due to irreversibilities. Equation 11.14 shows the Back Work Ratio (BWR), the pump's power consumption, and output power from the expander.

$$BWR = \frac{\dot{W}_{pump}}{\dot{W}_{exp}} \tag{11.14}$$

The BWR is a function of the evaporation temperature of the WF. It is lower at the higher critical temperature of WF. BWR increases with evaporation temperature (T_{ev}) and is significantly higher when the cycle operates near the critical temperature. Therefore, pump efficiency is a critical parameter in the LT and transcritical cycles. Overall isentropic efficiency of 25% has been reported in a 2 kWe capacity ORC system (Quoilin et al. 2010). A pump efficiency of 7% was reported by Reid (2010) in an ORC with HFE-7000 as the WF. The efficiency of 22% with a diaphragm pump was reported by Quoilin (2011) with RHFC-245fa as the WF. The overall efficiency of 20% was reported by Bala et al. (1985) with different types of WFs and sliding vane refrigeration pumps. The efficiency varied between 10 and 20% with a centrifugal pump (Melotte 2012). According to the manufacturer data, the diaphragm's efficiencies were > 60%, and the centrifugal pumps could achieve up to 40–50% efficiencies.

Lower net pressure suction head is a critical parameter for the ORC system design. It is required to set up a strategy to avoid pump cavitation. Here is a standard description of the plan of action to avoid cavitation. The required pressure head can be achieved by installing a lower NPSH pump before the main pump. This approach is selected in Tri-O-Gen ORC plants. In this system, the turbine shaft is directly connected to the main pump. The arrangement is confined in a hermetic container with the generator to avoid the leakage of toluene. To avoid cavitation, the pressure head of the pre-feed pump must be higher than the NPSH of the main pump, given in Eq. 11.15.

$$\Delta P_{pre-feed}(\dot{m}) = NPSH_{pump} \tag{11.15}$$

A static pressure difference supplies the required pressure head. The static pressure difference is due to the vertical distance between the pump and condenser. Turboden used this strategy for large-scale ORC systems. The condition of no cavitation is given in Eq. 11.16.

$$\rho g \Delta h > NPSH_{pump} \tag{11.16}$$

where ρ, g, and Δh are fluid density, gravity, and difference in level, respectively. A small fraction of non-condensable gases (nitrogen) are present in the WF, and they flow through the ORC cycle. However, the remaining fraction of non-condensable gases is in the gaseous state. It accumulates because of condensation of the WF and then transfers by the main flow to the condenser. Therefore, in the liquid receiver and condenser, the concentration of non-condensable gases is higher. Two partial pressures of the pump play an important role in (i) the partial pressure of non-condensing gases in the receiver and condenser ($P_{partial,ncg}$) and (ii) partial pressure of the WF in vapour state at the saturation pressure at a given temperature ($P_{partial,wf}$).

$$P_{total} = P_{partial,ncg} + P_{partial,wf} \tag{11.17}$$

When the WF leaves the gas and liquid interface towards the pump, the pressure is higher than the saturation and subcooled (Schuster et al. 2014). The subcooling in terms of saturation temperature and actual condensing temperature is shown in Eq. 11.18.

$$\Delta T_{sub-cooling} = T_{saturation}(P_{total}) - T_{actual\ condensing} \tag{11.18}$$

For non-cavitation, the NPSH is lower than the partial pressure of non-condensing gases, as shown in Eq. 11.19. This is achieved by providing sufficient subcooling to attain the necessary NPSH by the pump.

$$\Delta P(\Delta T_{sub-cooling}) = P_{total} - P_{partial,wf} = P_{partial,ncg} > NPSH_{pump} \tag{11.19}$$

The thermal subcooling of the WF is used to obtain the required NPSH. This can be achieved in three ways by using a sub-cooler (secondary HE) after the receiver. To avoid additional HE, it uses a sub-cooler directly into the receiver followed by subcooling in the condenser. The third solution is most challenging to control the subcooling in given operating conditions. The temperature of the WF at the sub-cooler outlet ($T_{sub,pump}$) must meet the non-cavitation condition shown in Eq. 11.20.

$$P_{Condenser} - P_{sat}\left(T_{sub,\ pump}\right) > NPSH_{pump} \tag{11.20}$$

The mass flow rate ($\dot{m}$) of the WF is controlled by a pump. An electrical motor is attached to an inverter to allow speed variations. In the case of a positive displacement pump, the flow rate is directly proportional to the speed. At the same time, the pressure head (evaporation and condensing pressure difference) is dependent on the centrifugal pump.

The organic WFs are expensive and can have a higher GWP value, apart from toxic, ODP, and flammable. Hence, it is necessary to make sure that the ORC cycle is completely leak-proof. Instead of the piston-type pumps, diaphragm pumps are selected, which generate pulsed flow. This can create a drawback of pressure fluctuations and errors in flow rate measurement; however, they avoid leakages of the WF by ensuring shaft sealing. The pump pressurizes the WF from the condenser pressure to the evaporator pressure, and they control the mass flow rate of the WF. In the case of a positive displacement pump (diaphragm), the mass flow rate of the WF is proportional to the speed, and direct contact between WF and the pump components is avoided. The only drawback of this type of pump is the pulsed flow. While in the centrifugal pumps, the mass flow rate of the WF depends on the pressure head between the condenser and evaporator pressures. Therefore, it is essential to choose an appropriate pump in terms of cost and efficiency for this application in ORC systems.

Evaporators

The total cost of ORC systems mainly depends on HEs. The optimization of HEs should be performed carefully. The main characteristics of HEs related to heat transfer are pressure drops and efficiency (pinch point), and they are sized according to these two parameters. There are different types of HE for ORC systems, the most familiar shell and tube (for large ORC systems) and plate type HE (for small ORC systems because of compactness). A HE is generally installed on the heat source side and withstand high temperature. In WHR, pressure drop and dimensions are limited due to available space. The acid dew point should avoid the sulfur attack. Hence, flue gasses are not cooled less than 120–180 °C, and this temperature depends on the sulfur content of the exhaust gas. The researchers are working on designing HEs which would be able to resist and evacuate the acid condensate. The heat can be retrieved by assigning direct heat exchange between the heat source and the WF and employing an intermediate heat transfer loop, which transfers the heat from the WH to the evaporator.

A more efficient and straightforward method is direct evaporation. However, at a higher temperature (start-up and transient), the WF can degrade at the maximum temperature of the chemical stability or presence of a hot spot in the HE. The stability and control of the system are also difficult to achieve. The HT loop dampens the fluctuations and permits quieter operation. For engine WHR system using ORC usually utilizes plate, shell and tube, and finned plate HEs. The shell and tube HEs are usually used in stationary and large systems which can tolerate higher pressures. Compact plate-type HEs are preferred to recover heat from the liquids (Quoilin et al. 2013). Plate type HEs withstands temperatures up to 250 °C, which is higher than the shell and tube HEs (Lion et al. 2017). The metal foam HEs are also under investigation and can be used for WHR from the vehicles because of their compact and improved heat transferability (Lin et al. 2011). An intermediate oil circuit is used in a more extensive ORC system, separating the heat source and the WF. This can protect the system against flammability and ensure safety in case of accidental leakage of the WF. For automotive applications, it is necessary to reduce the cost and weight and improve efficiency. The HE must sustain higher temperatures and pressures, fouling, and corrosion (Olofsson 2012).

Organic Expanders

Since the 1970s, the ORC expanders have made evolved significantly. Today, axial and radial expanders above 400 kW are available in the market for the ORC systems. However, for lower capacity ORCs, the expanders are still under development. In recent years, few expanders with lower capacities have come into the market, such as a 165 kW from tri-o-gen, a 50 kW from Electrotherm, a 125 kW from Calnetix (GE), and a 280 kW from UTC power. The macro-expanders are challenging to design. The current trend is to convert the compressors into expanders by altering its operational mode.

The expander is one of the essential components because it determines the efficiency and cost of a lower capacity ORC system (kWe). In this power range, there is no expander currently available in the market at an affordable price. The performance of the expander is directly related to the performance of the ORC system (Lion et al. 2017). The selection of expander depends on the size and operating conditions. The turbo and positive displacement type expander are used in an ORC system. The large-scale ORC turbo-expander shows superior performance at steady-state conditions while reduced performance for off-design operations. It can be a radial or an axial turbine. Axial type expanders are primarily used in big WHR units for stationary applications with a higher mass flow rate of the WF and a lower pressure ratio. It shows a well-defined design in collaboration with the WF of higher molecular mass. The radial expanders are mainly used for lower WF flow rates and higher-pressure ratios (Quoilin et al. 2013).

Turbo-expanders are not satisfactory for small-scale ORC units due to higher rotating speeds, which could potentially cause structural damage. In addition, turbo-expanders do not sustain a higher quantity of liquid during the expansion due to potential damage to the blades. Currently, research and development are underway to reduce these risks and offer an affordable expansion device for automotive ORC systems.

Reciprocating piston, vane, scroll, Wankel, and screw expanders are positive displacement expanders, often used for small-scale WHR (ICE) and CHP applications. These expanders can handle higher pressure ratios due to higher volume ratios (6–14) (Clemente et al. 2011). However, precise balancing, intake and exhaust valve timings, and vibration control are crucial factors. Problems with lubricating oil, sealing, and higher friction are among the challenges faced. Scroll expanders can work on low-pressure ratios due to lower built-in pressure ratios (1.5–4). It undergoes losses of under- or over-expansion. These can reduce the expansion efficiency, and the most common is under-expansion (Lemort et al. 2009). The other losses are leakage, friction, and heat transfer (Declaye et al. 2013). It can sustain temperatures up to 215 °C compared to piston expanders. Commonly, the scroll expanders are retrofitted for automotive applications and are used for low-temperature heat source WHR (Lion et al. 2017). Vane expanders are capable of handling lower power output ORC and can be single-acting or double-acting type. It permits a broad range of vapour quality without any problems. It is suited for mechanical coupling without a gearbox due to its lower speed.

The integration of expanders for automotive applications remains the main issue. There are two options for expander integration: (i) mechanical coupling and (ii) electrical coupling. In mechanical coupling, the net power supplied by the expander is reintroduced to the crankshaft using a gearbox. A clutch can be used to disconnect the expander and the engine. In electrical coupling, the expander shaft is connected to the generator, which produces electrical energy. This electrical energy can be used for auxiliary systems or reintroduce to the powertrain in hybrid vehicles.

11.2.3 Working Fluids for ORC Systems

The WF is the principal part of an ORC system, which decides its economic viability and efficiency. The challenges depend on the commercial availability and selection of the WF. Most WFs are phased out because of international agreements and environmental concerns. Some criteria for selecting suitable WFs are higher thermal stability, compatibility with materials of constructions, lower environmental effects, suitable critical operational parameters, higher performance, and availability at a lower cost. Selecting suitable WFs is a tricky task. The efficiency and performance of the ORC system depend significantly on the WF selection since the thermodynamic conditions of the heat sources are variable (Su et al. 2017; Roedder et al. 2016).

The WF for the ORC systems can be categorized as

1. **Pure WF**: Pure WFs are hydrocarbons (HCs), hydrofluorocarbons (HFCs), perfluorocarbons (PFCs), hydrochlorofluorocarbons (HCFCs), siloxanes, chlorofluorocarbons (CFCs), hydrofluoroethers (HFEs), and hydro-fluoro olefines (HFOs).

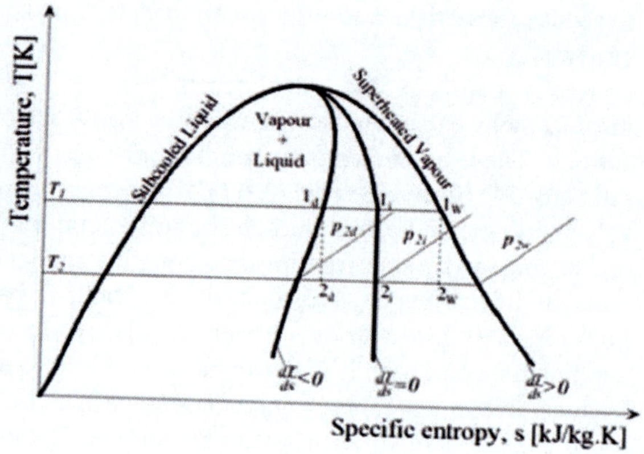

Fig. 11.6 Different WFs on the T-s diagram (Badr et al. 1985)

2. **Zeotropic Mixture**: It is a mixture of WFs with different boiling temperatures. A single substance in a zeotropic mixture does not evaporate or condense at an identical temperature. These zeotropic mixtures exhibit distinct characteristics when used in an ORC system.

Water-steam, hydrocarbons, refrigerants, and alcohols are compatible WFs for the WHR systems. However, hydrocarbons have flammability properties. Meanwhile, refrigerants exhibit more satisfactory abilities for WHR from a low-temperature heat source (Su et al. 2017). The benefits of water/steam are cost, thermodynamic performance, and environment. Alcohols are appropriate for recovering the heat from medium to high-temperature heat sources, but it is flammable. There is a possibility of fire hazards upon exposure to a high-temperature heat source. The WFs can be categorized based on the degree of saturation, as shown in Fig. 11.6.

If $dT/ds < 0$ (negative slope), the WFs are considered 'wet' because WFs fall in the wet zone. The molecular mass of wet WFs is lower (for example, H_2O and ammonia). If $dT/ds = 0$ (infinite slope), WFs are 'isentropic' and of invariable entropy, although the variation of temperature occurs. The isentropic WFs with a saturated vapour curve have medium molecular mass (R134a, R245fa, R123). With $dT/ds > 0$ (positive slope), the WFs are called 'dry'. The WFs with higher molecular mass are considered dry (HFE7000 and HFE7100) (Bao and Zhao 2013; Qiu 2012). Figure 11.7 shows some wet, isentropic, and dry WFs for the ORC system. The wet WFs have a lower molecular mass. Table 11.5 summarizes the WFs from the literature that are used in ORC systems.

The temperature of the heat source affects the selection of WF for an ORC system. Some WFs are suitable for both higher as well as lower temperature heat sources. The selection criteria for a higher temperature heat source includes lower T_c, higher decomposition, and boiling temperature. While for low-temperature heat sources, only T_C is considered as a selection criterion. Several properties should be examined during the selection of WFs for the ORC system. These properties are shown in Tables 11.3 and 11.4.

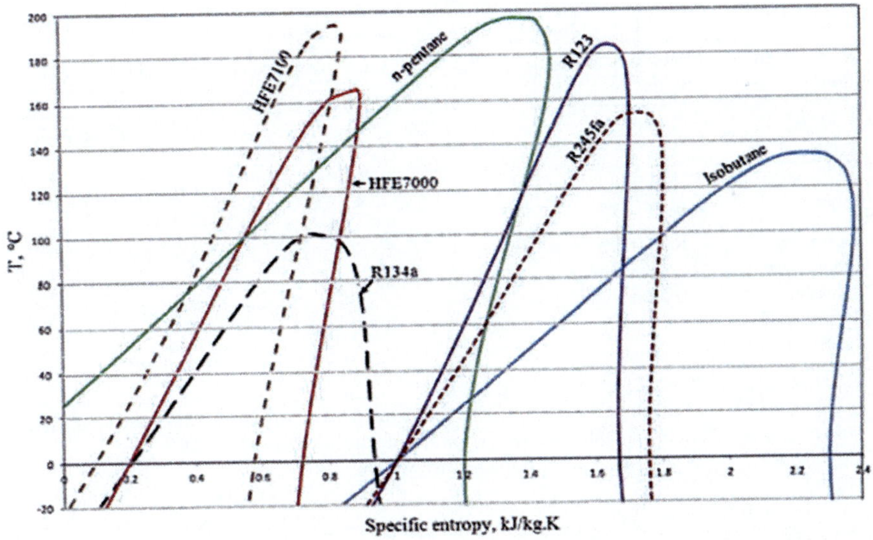

Fig. 11.7 Comparison of *T*–*s* curves for organic WFs used (Qiu 2012)

The selection of WF is a tedious process. The WFs, mainly wet and dry, influence the design and performance of the ORC system. The thermophysical properties of WFs can be utilized as a performance index to compare different WFs. The selection of suitable WF depends on various parameters such as the temperature of the heat source, zeotropic mixtures, operating pressure and temperature, WF properties, flammability, toxicity, chemical instability, GWP, ODP, WF price, material compatibility, and its availability.

WF selection should follow the following criterion:

1. Thermodynamic Performance: The power output and efficiency should be higher at the specified temperature range. This depends on the properties (specific heat, critical point, and acentric factor) of the WFs.
2. Latent heat of vaporization (H_v): The latent heat of vaporization of WFs must be lower (Papadopoulos et al. 2010).
3. Positive or Isentropic Saturation Vapor Curve: This is the dry WF, and a recuperator can increase the overall cycle efficiency.
4. Liquid heat capacity ($C_{p,l}$) must be lower.
5. Density (ρ): The density of WFs must be higher, either in a liquid or vapour state. Higher density results in increased mass flow rate and reduced component size. The lower density of WFs shows a higher volume flow rate. The size of HE must be increased to limit the pressure drop.
6. Viscosity (μ): the outcome of the lower viscosity of WFs in liquid and vapour states is lower losses due to friction and higher heat transfer coefficient in the HEs.

Table 11.3 Properties of different WFs of ORC systems

Considered fluids	Recommended fluids	T_{cd} (°C)	T_{ev} (°C)	Application	Reference
R11, R113, R114	R113	30–50	120	WHR	(Amicabile et al. 2015)
Unconventional WFs	R124 and HCFC-123	35–60	80–110	n/a	(Maizza and Maizza 2001)
Benzene, isopentane, toluene, HCFC-123, HFE 7100 and p-xylene	HCFC-123, benzene and toluene	30	150–200	WHR	(Liu et al. 2004)
R245ca, R152a, water, isopentane, butane, HCFC123, and isobutane	Isopentane, R-245ca and water	55 (100 for water)	60–150 (150–260 for water)	ICE	(Chammas and Clodic 2005)
HCFC123, HFC245fa, R600a and R113	HCFC123 and R113	50	80–220	WHR	(Gu et al. 2009)
Heptane, R12, pentane, ethanol, R365mfc and R141b	Ethanol	50	170	CHP	(Mikielewicz and Mikielewicz 2010)
HFC-134a, R11, benzene	Benzene	35	96–221	ICE WHR	(Vaja and Gambarotta 2010)
n-hexane, neo pentane, n-butane, iso-butane, iso-pentane, RC318, R113, R245fa, R227ea, R236ea, R236fa and C5F12	N hexane	30	50–140	n/a	(Aljundi 2011)
R141b, R113, HFC245fa, HCFC123, R245ca, R114, R236ea, R11 and Butane	HCFC123, R11, R113, R141b, R245ca and HFC245fa	27–87	327	WHR	(Wang et al. 2011)

(continued)

Table 11.3 (continued)

Considered fluids	Recommended fluids	T_{cd} (°C)	T_{ev} (°C)	Application	Reference
HCFC123, R717, R12, and HFC134a	HCFC123	n/a	277	WHR	(Roy et al. 2011)
Butylbenzene, ethylbenzene, toluene, propyl benzene and OMTS	Butylbenzene	90	250–350	CHP	(Drescher and Brüggemann 2007)
n-pentane, HCFC123, ammonia and PF5050	Ammonia	30	70–90	GT	(Madhawa Hettiarachchi et al. 2007)
HCFC123, n-pentane, HFC245fa, and HFC134a	HCFC123 and n-pentane	35	60–100	WHR	(Lemort et al. 2007)
Alkanes, HCFC123, benzene, R141b, R113, HFC245fa, R236ea, R365mfc, R245ca and toluene	Benzene and toluene	40	120	WHR	(Desai and Bandyopadhyay 2009)
Alkanes and fluorinated alkanes, ethers, and fluorinated ethers	RE245, R600, RE134 HFC245fa, R601 and R245ca	30	100	GT	(Saleh et al. 2007)
Ammonia, water, R11, butane, isobutane, R236ea, HCFC123, R245ca, R113, and R141b	R236ea	25	145	WHR	(Dai et al. 2009)
RC318, propylene, R236fa, R227ea, isobutane and HFC245fa	HFC245fa, propylene and R227ea	25	80–115	GT	(Borsukiewicz-Gozdur and Nowak 2007)

(continued)

Table 11.3 (continued)

Considered fluids	Recommended fluids	T_{cd} (°C)	T_{ev} (°C)	Application	Reference
HFE7100, water, n-dodecane, cyclohexane, n-pentane, toluene, isobutane and HFC245fa	n-dodecane	45	120/230	Solar	(Facão et al. 2008)
R113, 123, R245ca, Isobutane	R113	25	100–210	WHR	(Mago et al. 2008)
Refrigerants	R152a, R600, R290	35	60–100	Solar	(Tchanche et al. 2009)

Table 11.4 Properties used in the selection of WFs for ORC systems (Roedder et al. 2016; Papadopoulos et al. 2010)

Thermodynamic properties	Safety	Environmental	Process related parameters
Latent heat of vaporization (H_v) Viscosity (μ) Melting point temperature (T_m) Density (ρ) Liquid heat capacity (C_{pl}) Thermal conductivity (k) Critical temperature (T_c)	Toxicity Flammability	Ozone depletion potential (ODP) Global warming potential (GWP)	Maximum operating pressure (P_{max}) Critical pressure (P_c) Efficiency (η) Mass flow rate ($\dot{m}_f$)

7. The enthalpy and heat transfer coefficient must be higher for system compactness.
8. Higher thermal conductivity (k) to essential to achieve a higher heat transfer coefficient in the condenser and the HE.
9. The melting point should be lower than the ambient operating temperature to ensure that the WF remains in a liquid state.
10. Lower Ozone Depletion Potential (ODP) index and global warming potential (GWP) are preferable to reduce the environmental effect.
11. Acceptable Evaporator Pressure: higher evaporator pressure causes higher cost and complexity.
12. To achieve higher efficiency and evaporating temperature, the critical temperature (T_c) should be > 100 °C.
13. Positive Condensing Gauge Pressure is preferred to avoid air infiltration, and pressure should be higher than the atmospheric pressure.

14. The NBT should be lower to ensure the operating pressure of ORC is higher than the atmospheric pressure.
15. Stability at High Temperature: Thermal stability must be higher to work at a wide range of operating conditions. Organic WFs undergo chemical degradation and separation at higher temperatures. Therefore, source temperature is restricted by the chemical stability of the WF.
16. Safety: Lower toxicity and flammability should be ensured.
17. WF should be easily available at a lower cost.

The selection of WFs in literature covers a wide range of WFs. These WFs are categorised based on critical temperature (Quoilin and Lemort 2009) and are summarized in Table 11.5.

Compliance with the safety factor (SF) and environmental concerns are necessary for the application of WFs. Figure 11.8 presents the selection of WFs based on the NFPA704. The SF mainly depends on ds/dT. Equation 11.21 represents the SF of the WF, which includes Health, Instability, and Flammability factors (Amicabile et al. 2015). According to National Fire Protection Association (NFPA), the WFs are classified based on the health, flammability, and hazards, as given in Eq. 11.21.

$$SF = aH + bI + cF \tag{11.21}$$

where H, I, and F represent the value of health, instability, and flammability factor, respectively, ranging from 0 to 4 based on the recommendations of the National Fire Protection Association (NFPA-704). The constants a (20%), b (35%), and c (45%) are associated with Health, Instability, and flammability factors. The flammability factor is essential to prevent the chances of auto-ignition. The maximum limit of safety factor for each WFs in an ORC system is 2. Any value higher than this is unacceptable.

For safety and economic reasons, the condensation pressure must be 3 bar. The minimum condensation temperature (30 °C) is needed to prevent massive HE and condenser surfaces (Fig. 11.9).

11.3 Economic Analysis of the ORC Systems

The total cost of ORC systems mainly depends on the cost of HEs (up to 80–90%) (Papadopoulos et al. 2010). The naked ORC module cost (C_m) for each part is the product of component cost (C_c) and ORC module cost factor (C_{mcf}).

For pump:

$$C_m = C_c * C_{mcf} = C_c * \left[D_1 + D_2 * F_p * F_m\right] \tag{11.22}$$

$$\log C_c = K_1 + K_2 * \log W_{pump} + K_3\left[\log W_{pump}\right]^2 \tag{11.23}$$

Table 11.5 WFs properties for the ORC systems (Bao and Zhao 2013; Wang et al. 2011; Saleh et al. 2007; Tchanche et al. 2009)

WFs	Tc (K)	Tb (K)	M (kg/kmol)	Pc (MPa)	GWP	ODP	Type	Safety factor
R11	470.96	296.7	137.37	4.407	4750	1	I	A1
R12	385	243.2	120.91	4.14	10,900	0.82	W	A1
R21	451.33	281.8	102.92	5.181	151	0.04	W	B1
R22	369.1	232.2	86.47	4.99	1790	0.04	W	A1
R114	427.7	276.6	170.92	3.257	9180	0.58	D	A1
R123	456.83	301	152.93	3.662	120	0.012	I	n.a
R124	395.3	261	136.48	3.062	619	0.02	D	A1
R134a	374	247	102.03	4.059	1300	0	I	A1
R141b	479.96	305.2	116.95	4.46	700	0.086	I	n.a
142b	410.1	263.9	100.5	4.06	2220	0.06	I	A2
R152a	386.26	259	66.05	4.517	140	0	W	A2
R218	344.95	248.7	188.02	2.671	8830	0	I	A1
R227ea	374.8	256.7	170.03	2.93	3580	0	D	A1
R236fa	397.9	271.6	152.04	3.2	9820	0	D	A1
R245fa	427	288.1	134.05	3.651	1050	0	I	B1
RC318	388.2	267	200.03	2.78	10,300	0	D	A1
R1234ZE	382.4	253.4	114.04	3.63	6	0	I	A2L
HFE7000	164.6	307.9	200.05	2.48	575	0	D	n.a
R1234yf	367.7	243.5	114.04	3.38	4.4	0	I	A2L
HFE7100	195.3	332.6	250	2.23	297	0	D	n.a
HFE7500	261	400.9	414.11	1.555	n.a	0	D	n.a
Butane	425	272.1	58.12	3.8	3	0	D	A3
Hexane	507.7	241.8	86.18	3.06	n.a	n.a	D	n.a
Octane	569.2	398	114.23	2.5	n.a	n.a	D	n.a
Benzene	561.87	353.07	78.11	4.91	n.a	n.a	D	n.a
Propane	369.74	230.9	41.1	4.25	3.3	0	W	A3
Pentane	469.5	308.5	72.15	3.36	4	0	D	n.a
Heptane	540	370.9	100.21	2.73	0	n.a	D	n.a
Propylene	365.4	225	42.08	4.66	2	n.a	W	A3
Toluene	591.6	383.6	92.14	4.126	2.7	n.a	D	n.a

$$\log F_p = C_1 + C_2 * \log P + C_3 * [\log P]^2 \tag{11.24}$$

For heat exchangers:

$$C_m = C_c * C_{mcf} = C_c * [D_1 + D_2 * F_p * F_m] \tag{11.25}$$

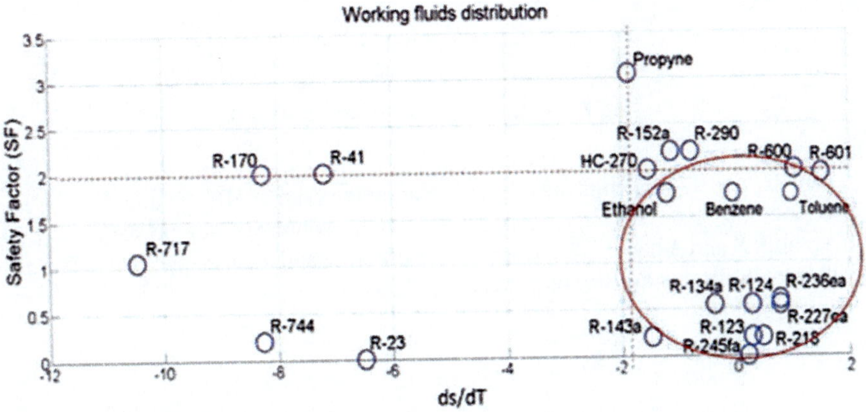

Fig. 11.8 Safety factors of the WFs (Amicabile et al. 2015)

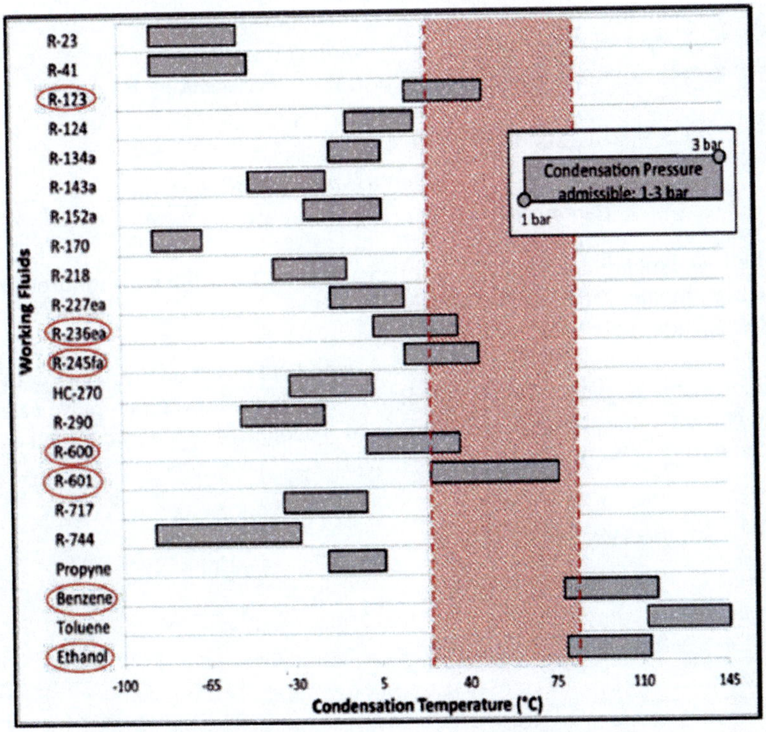

Fig. 11.9 Condensation characteristics of the WFs (Amicabile et al. 2015)

$$\log C_c = K_1 + K_2 * \log A + K_3 * [\log A]^2 \tag{11.26}$$

$$\log F_p = C_1 + C_2 * \log P + C_3 * [\log P]^2 \tag{11.27}$$

where D_1 and D_2 are constants that depend on the type of HE, F_p and F_m represent pressure and material factor, respectively. W_{pump} represents the power required by the pump, P and A denote the component's pressure and area, respectively. The exact values of constants are summarized in Table 11.6.

For expanders:

$$C_m = C_c * C_{mcf} = C_c * [D_2 * F_p * F_m] \tag{11.28}$$

$$\log C_c = K_1 + K_2 * \log W_{exp} + K_3 * [\log W_{exp}]^2 \tag{11.29}$$

where W_{exp} represent the expander power output.

The aggregate expenditures of components are evaluated by Eq. 11.30.

$$C_{total} = [C_{m,pump} + C_{m,HE} + C_{m,exp}]_{2001} \times \frac{[chemical\ engineering\ plant\ cost\ index]_{2018}}{[chemical\ engineering\ plant\ cost\ index]_{2001}} \tag{11.30}$$

The true expenditure for 2018 can be calculated from the expenditure for 2001 announced by the Chemical Engineering Plant Cost Index (CEPCI). The CEPCI for 2018 and 2001 were 648.7 and 397, respectively (Mignard 2014). The economic evaluation of two-stage ORC systems was carried out based on the net present value and depreciated payback period. The net present value showed the future cash flow as the current value using the advantages of discount for expenditure and income induced during operating periods (Eq. 11.31).

$$Net\ present\ value\ (NPV) = -C_{total} + \sum_{t=1}^{n} \left[\frac{CF}{(1+i)^t} \right] \tag{11.31}$$

Table 11.6 Values of different constants for economic analysis

Component	D_1	D_2	K_1	K_2	K_3	F_m	C_1	C_2	C_3
Pump	1.89	1.35	3.3892	0.0536	0.1538	1.6	−0.394	0.396	-0.0023
HE	1.63	1.66	4.325	−0.303	0.1634	1.4	0.0388	−0.1127	0.08183
Expander	0	1	2.705	1.440	−0.178	3.4	0	0	0

where C_{total} and CF represents the total investment cost and net cash flow in n years, respectively, i is the interest rate. The depreciated payback period indicated that the time needed to redeem direct investment cost and could be used to measure how rapidly cash used in investment would be retrieved (Tan et al. 2021; Mohammadzadeh Bina et al. 2017). It is defined in Eq. 11.32.

$$Depreciated\ payback\ period\ (DPP) = \frac{-\ln\left[1 - \frac{k*C_{total}}{CF}\right]}{\ln[1+k]} \tag{11.32}$$

where k represents the discount rate. The capital recovery factor can be calculated by using Eq. 11.33.

$$Capital\ recovery\ factor\ (CRF) = \frac{i*[1+i]^n}{[1+i]^n - 1} \tag{11.33}$$

The production cost of electricity (PCE) constitutes the cost per kWh of electricity supplied by two-stage ORC systems given in Eq. 11.34.

$$Production\ cost\ of\ electricity\ (PEC) = \frac{C_{total}*\left[C_{o,m} + CRF\right]}{W_{net,total}*P_h} \tag{11.34}$$

where $C_{o,m}$ is the operating and maintenance cost, and P_h represents the yearly operating hours.

Table 11.7 shows the installed capital cost, operating and maintenance costs for conventional Rankine cycle and ORC system based on data provided by manufacturers to the ICF. These technologies accounted for almost all waste heat power (WHP) units presently installed. It is supposed to be the governing technologies that

Table 11.7 Costs of waste heat to power technologies (Elson et al. 2015)

Technology	Cost	Electrical energy capacity for WHP technologies				
		50–500 kW	500–1000 kW	1–5 MW	5–20 MW	> 20 MW
Steam Rankine cycle	Installed capital cost, $/kW	3000	2500	1800	1500	1200
	Operating and maintenance cost, $/kWh	0.013	0.009	0.008	0.006	0.005
ORC	Installed capital cost, $/kW	4500	4000	3000	2500	2100
	Operating and maintenance cost, $/kWh	0.02	0.015	0.013	0.012	0.01

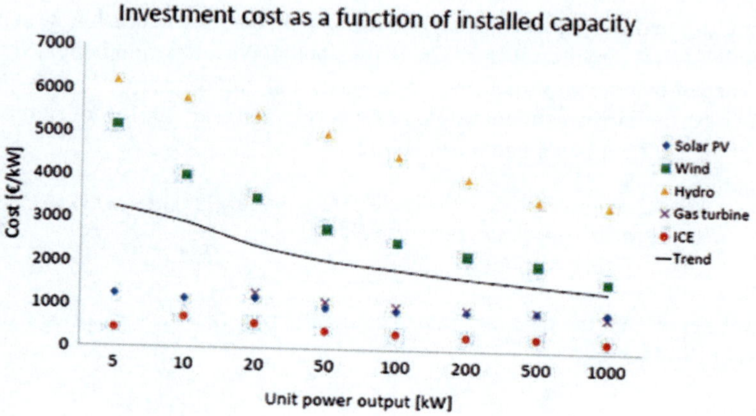

Fig. 11.10 ORC system market scenario (Tocci et al. 2017)

will be established during the next several years. The other WHP systems have not matured yet. Therefore, these technologies are not involved in this cost analysis.

The capital investment costs of the conventional Rankine cycle ranges from $1200 to $3000/kW. The ORC capital investment cost ranges from $2100 to $4500/kW, higher than the conventional Rankine cycle. The operating and maintenance costs range from 0.5 to 1.3 cents/kWh and 1 to 2 cents/kWh for the conventional Rankine cycle and ORC.

The cost of an ORC unit has been evaluated in contrast with the currently available technologies. Figure 11.10 shows the cost of ORC plants installed in a different areas of application.

The cost of an ORC unit should not go over 3500€/kW and 2500€/kW for 5–10 kW and 10–100 kW output power, respectively. In 2016, the total value of the ORC market was $359–402 million per year. Figure 11.11 shows the cost of ORC units for different applications against their size. These costs are indicative, which are collected from scientific publications and the ORC manufacturers (Öhman 2012).

The data scattering is due to the variations in costs for various market policies, manufacturers, and levels of integration. Therefore, it shows the general trends of decreasing cost respective to power. The costs are lower for WHR applications while higher for CHP and geothermal units. The overall cost is two to three times higher than the module cost. This variation is due to buildings, engineering, boiler, and integration processes. This excess cost should never be neglected when assessing the economic feasibility of an ORC plant.

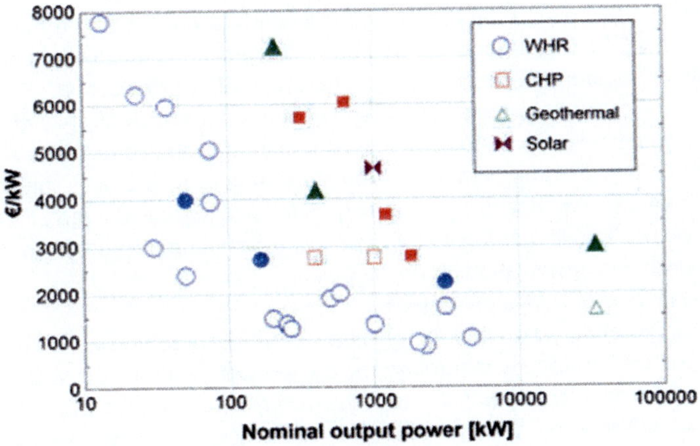

Fig. 11.11 Module and total cost of ORC units (Quoilin et al. 2013)

11.4 WHR System for ICEs: Advantages and Challenges

The WHR technologies extend to yet another market (ICE). This provides another choice to grow the share of renewable energy in the transport sector's energy mix. However, there are some challenges such as cost, application specifics, chemical composition, temperature restriction, and transportability/inaccessibility of heat source. These barriers are required to be tackled for the commercialization of WHR technology in the automotive sector.

1. **Costs**

   The costs of equipment, auxiliaries, and design service lead to a longer payback period in some applications. In addition, various industrial subsectors with higher quality WH sources are notable for small-scale profit margins. The components cost advocate large heat recovery systems but are difficult to implement in small scale systems. Some applications involve costly and advanced materials used in the streams with higher chemical reactivity, high-temperature streams, and exhaust streams that are cooled to a temperature lower than the condensation temperature. The overall cost of material per unit energy heat recovered increases as higher surface areas are needed for a cost-effective and lower temperature heat recovery system. The Scaling, fouling, and corrosion of HE materials causes low profitability and high maintenance costs.
2. **Application-specific constraints**

   The heat recovery can compromise and embarrass the quality and process control systems. The design of equipment are process-specific and needs to adopt the given process.
3. **Temperature constraints**

   Solids and liquids can condense when the hot stream cools down in the recovery system. This can cause fouling and corrosion. The extra cost of materials that can sustain corrosion frequently avoid the low-temperature heat recovery. The materials which maintain chemical and mechanical properties at higher temperatures are expensive. Therefore, the waste heat is dissipated in the air to lower the temperature. This can decrease the standard of available energy for recovery. The variations in heat flow can develop chemical and mechanical stresses in the equipment. The minor differences in the heat source and sink temperatures can decrease the heat transfer rate and need a larger surface area for WHR. The technologies that use lower temperature power production are presently expensive and less developed.

4. **Transportability/ Inaccessibility**

   The gaseous waste heat is released at atmospheric pressure. This can limit the potential to transport through the components without any added energy. It is hard to recover waste heat from unconventional sources and hot equipment surfaces. Numerous locations have limited space for the approach to the waste heat stream.

5. **Chemical composition**

   The accumulation of matter on heat recovery component surfaces will decrease the efficiency and heat transfer rate. The component surfaces will get adversely affected by higher chemical reactivity, which increases the maintenance cost. The WHR from exhaust gas may alter the performance and efficiency of the environmental control devices. The hot gas or liquids with higher chemical reactivity need more advanced materials of construction to withstand corrosion. The waste heat flow is chemically consistent with heat recovery components materials. Exhaust gas may need extra effort to avoid cross-contamination between the fluids.

11.5 Future Directions in WHR Technologies

The ORC systems have been executed commercially since the 1970s, but there are no guidelines for the maintenance of these projects. The matured applications such as biomass and geothermal ORC systems still don't have any guidelines. This may be due to the industries associated with the ORC, who do not wish to reveal desired information. Heat source properties and mass flow rate of heat source affect the selection of the ORC system. The performance of the ORC system depends on the condition of the cooling system. However, there is no standard test for the ORCs. The ORC components are sold based on the design conditions, while they need to work at different conditions due to dynamic variations in the heat source or variations in weather. The waste heat is not in a steady state, which is supplied to the evaporator. The transient conditions are regular for solar and WHR applications.

For this reason, a suitable control system needs to be designed. The waste heat is generally in the form of liquid or exhaust gas in automotive systems, including ICE. They can be retrieved using various types of devices. The HEs appear to be the most accommodated device for transferring heat to the ORC system. But the optimum design of HEs for ICE is necessary, and no regulations exist currently. Every project needs a customised design.

Generally, the environmental effects of the ORC are measured in terms of GWP, ODP, and atmospheric lifetime. However, these are insufficient, and all other factors also need to be considered. The specific cost of an ORC system is higher. That's why ORC technology is not widely used commercially in ICE because they generate very small-scale power.

Nomenclature

WHR	Waste heat recovery
GHG	Green house gas
PM	Particulate matter
WH	Waste heat
EGR	Exhaust gas recirculation
VVT	Variable valve timing
DOC	Diesel oxidation catalyst
DPF	Diesel particulate filter
SCR	Selective catalytic converter
TEG	Thermoelectric generator
ORC	Organic Rankine cycle
CHP	Combined heat and power
GT	Geothermal
WF	Working fluid
EG	Exhaust gas
HT	High temperature
GE	Gas engine
SE	Solar energy
BCHP	Biomass combined heat and power
BM	Biomass
RN	Renewables
LG	Landfill gas
STE	Stationary engines
WWT	Waste water treatment
FL	Flares
SG	Sewage gas
BME	Bio-methane
BG	Biogass
CH	Chemicals
NG	Natural gas
GI	Glass industry
HE	Heat exchanger
CCI	Cement and ceramic industry

FA	Food and animal feedstuff
FM	Furniture making building products
RP	Rubber and plastic
PMP	Paper manufacture and processing
HT	High temperature
MP	Metal production and processing
SF	Safety factor
OM	Off-grid mining
DE	Diesel engine
T_c	Condensing temperature
T_{ev}	Evaporation temperature
EEX	Engine exhaust
ECA	Engine scavenge air
GWP	Global warming potential
ODP	Ozone depletion potential

References

Aghaali H, Ångström HE (2015) A review of turbocompounding as a waste heat recovery system for internal combustion engines. Renew Sustain Energy Rev 49:813–824. https://doi.org/10.1016/j.rser.2015.04.144

Aljundi IH (2011) Effect of dry hydrocarbons and critical point temperature on the efficiencies of organic Rankine cycle. Renew Energy 36:1196–1202. https://doi.org/10.1016/j.renene.2010.09.022

Amicabile S, Lee JI, Kum D (2015) A comprehensive design methodology of organic Rankine cycles for the waste heat recovery of automotive heavy-duty diesel engines. Appl Therm Eng 87:574–585. https://doi.org/10.1016/j.applthermaleng.2015.04.034

Anastasovski A, Rasković P, Guzović Z (2020) A review of heat integration approaches for organic rankine cycle with waste heat in production processes. Energy Convers Manag 221:113175. https://doi.org/10.1016/j.enconman.2020.113175

Angelino G, Gaia M, Macchi E (1984) A review of Italian activity in the field of organic Rankine cycles. In: International VDI ORC HP technology working fluids problems, Zurich, 10–12 Sept, pp 465–482

Arias DA, Shedd TA, Jester RK (2006) Theoretical analysis of waste heat recovery from an internal combustion engine in a hybrid vehicle. SAE Tech Pap 115:777–784. https://doi.org/10.4271/2006-01-1605

Badr O, Probert SD, O'Callaghan PW (1985) Selecting a working fluid for a Rankine-cycle engine. Appl Energy 21:1–42. https://doi.org/10.1016/0306-2619(85)90072-8

Bala EJ, O'Callaghan PW, Probert SD (1985) Influence of organic working fluids on the performance of a positive-displacement pump with sliding vanes. Appl Energy 20:153–159. https://doi.org/10.1016/0306-2619(85)90031-5

Bao J, Zhao L (2013) A review of working fluid and expander selections for organic Rankine cycle. Renew Sustain Energy Rev 24:325–342. https://doi.org/10.1016/J.RSER.2013.03.040

Borsukiewicz-Gozdur A, Nowak W (2007) Comparative analysis of natural and synthetic refrigerants in application to low temperature Clausius-Rankine cycle. Energy 32:344–352. https://doi.org/10.1016/J.ENERGY.2006.07.012

Briggs TE, Wagner R, Edwards KD et al (2010) A waste heat recovery system for light duty diesel engines. SAE Tech Pap. https://doi.org/10.4271/2010-01-2205

Chen T, Zhuge W, Zhang Y, Zhang L (2017) A novel cascade organic Rankine cycle (ORC) system for waste heat recovery of truck diesel engines. Energy Convers Manag 138:210–223. https://doi.org/10.1016/j.enconman.2017.01.056

Clemente S, Micheli D, Reini M, Taccani R (2011) Performance analysis and modeling of different volumetric expanders for small-scale organic rankine cycles. ASME 2011 5th international conference on energy sustainability ES 2011, pp 375–384. https://doi.org/10.1115/ES2011-54302

Codan E, Vlaskos I (2004) Turbocharging medium speed diesel engines with extreme Miller timing. In: Proceedings of the 9th turbocharging conference, Dresden, Germany, 23–24 Sept

D'Amelio (1936) https://www.kaymacor.com/en/rankine-cycle

Dai Y, Wang J, Gao L (2009) Parametric optimization and comparative study of organic Rankine cycle (ORC) for low grade waste heat recovery. Energy Convers Manag 50:576–582. https://doi.org/10.1016/J.ENCONMAN.2008.10.018

Declaye S, Quoilin S, Guillaume L, Lemort V (2013) Experimental study on an open-drive scroll expander integrated into an ORC (Organic Rankine Cycle) system with R245fa as working fluid. Energy 55:173–183. https://doi.org/10.1016/j.energy.2013.04.003

Desai NB, Bandyopadhyay S (2009) Process integration of organic Rankine cycle. Energy 34:1674–1686. https://doi.org/10.1016/J.ENERGY.2009.04.037

Drescher U, Brüggemann D (2007) Fluid selection for the Organic Rankine Cycle (ORC) in biomass power and heat plants. Appl Therm Eng 27:223–228. https://doi.org/10.1016/J.APPLTHERMALENG.2006.04.024

Drive WNF (2017) Final scientific/technical report

Drives A (2008) The turbosteamer: a system introducing the principle of cogeneration in automotive applications, vol 69

El Chammas R, Clodic D (2005) Combined cycle for hybrid vehicles. SAE Tech Pap 2005.https://doi.org/10.4271/2005-01-1171

Elson A, Hampson A, Tidball R (2015) Waste heat to power market assessment

Endo T, Kawajiri S, Kojima Y et al (2007) Study on maximizing exergy in automotive engines. SAE Tech Pap. https://doi.org/10.4271/2007-01-0257

Facão J, Palmero-Marrero A, Oliveira AC (2008) Analysis of a solar assisted micro-cogeneration ORC system. Int J Low-Carbon Technol 3:254–264. https://doi.org/10.1093/ijlct/3.4.254

Gu W, Weng Y, Wang Y, Zheng B (2009) Theoretical and experimental investigation of an organic rankine cycle for a waste heat recovery system. Proc Inst Mech Eng Part A J Power Energy 223:523–533. https://doi.org/10.1243/09576509JPE725

Hamid Elsheikh M, Shnawah DA, Sabri MFM et al (2014) A review on thermoelectric renewable energy: principle parameters that affect their performance. Renew Sustain Energy Rev 30:337–355. https://doi.org/10.1016/j.rser.2013.10.027

Heywood JB (2018) Internal combustion engine fundamentals. McGraw-Hill Education

Hoang AT (2018) Waste heat recovery from diesel engines based on Organic Rankine Cycle. Appl Energy 231:138–166. https://doi.org/10.1016/j.apenergy.2018.09.022

Hountalas DT, Mavropoulos GC, Katsanos C, Knecht W (2012) Improvement of bottoming cycle efficiency and heat rejection for HD truck applications by utilization of EGR and CAC heat. Energy Convers Manag 53:19–32. https://doi.org/10.1016/j.enconman.2011.08.002

Howard T History of ORC. Knowledge Center for Organic Rankine Cycle

Howell T, Gibble J (2011) Development of an ORC system to improve HD truck fuel efficiency. Deer Conf 1:1–21

Inc B (2008) Waste heat recovery: technology opportunities in the US industry. Waste Heat Recover Technol Oppor US Ind 1–112.https://doi.org/10.1017/CBO9781107415324.004

Innovative OS, Systems EC Over story innovative engine systems symbiosis of energy

Johnson TV (2015) Review of vehicular emissions trends. SAE Int J Engines 8.https://doi.org/10.4271/2015-01-0993

Law R, Harvey A, Reay D (2013) Opportunities for low-grade heat recovery in the UK food processing industry. Appl Therm Eng 53:188–196. https://doi.org/10.1016/J.APPLTHERMALENG.2012.03.024

Law R, Harvey A, Reay D (2016) A knowledge-based system for low-grade waste heat recovery in the process industries. Appl Therm Eng 94:590–599. https://doi.org/10.1016/J.APPLTHERMALENG.2015.10.103

Lecompte S, Huisseune H, Van Den Broek M et al (2015) Review of organic Rankine cycle (ORC) architectures for waste heat recovery. Renew Sustain Energy Rev 47:448–461. https://doi.org/10.1016/j.rser.2015.03.089

Lemort V, Cuevas C, Teodorese I, Lebrun J (2007) Contribution à l'étude des cycles de Rankine de récupération de chaleur. VIIIe Colloq Interuniv Fr sur la Therm des Systèmes, pp 305–310

Lemort V, Quoilin S, Cuevas C, Lebrun J (2009) Testing and modeling a scroll expander integrated into an Organic Rankine Cycle. Appl Therm Eng 29:3094–3102. https://doi.org/10.1016/J.APPLTHERMALENG.2009.04.013

Lin W, Yuan J, Sunden B (2011) Review on graphite foam as thermal material for heat exchangers. In: Proceedings—world renewable energy congress—Sweden, 8–13 May 2011, Linköping, Sweden, vol 57, pp 748–755.https://doi.org/10.3384/ecp11057748

Lion S, Michos CN, Vlaskos I et al (2017) A review of waste heat recovery and Organic Rankine Cycles (ORC) in on-off highway vehicle Heavy Duty Diesel Engine applications. Renew Sustain Energy Rev 79:691–708. https://doi.org/10.1016/j.rser.2017.05.082

Liu BT, Chien KH, Wang CC (2004) Effect of working fluids on organic Rankine cycle for waste heat recovery. Energy 29:1207–1217. https://doi.org/10.1016/j.energy.2004.01.004

Madhawa Hettiarachchi HD, Golubovic M, Worek WM, Ikegami Y (2007) Optimum design criteria for an Organic Rankine cycle using low-temperature geothermal heat sources. Energy 32:1698–1706. https://doi.org/10.1016/J.ENERGY.2007.01.005

Mago PJ, Chamra LM, Srinivasan K, Somayaji C (2008) An examination of regenerative organic Rankine cycles using dry fluids. Appl Therm Eng 28:998–1007. https://doi.org/10.1016/J.APPLTHERMALENG.2007.06.025

Maizza V, Maizza A (2001) Unconventional working fluids in organic Rankine-cycles for waste energy recovery systems. Appl Therm Eng 21:381–390. https://doi.org/10.1016/S1359-4311(00)00044-2

Melotte N (2012) Development and optimization of Organic Rankine Cycle control strategies. University of Liège

Mignard D (2014) Correlating the chemical engineering plant cost index with macro-economic indicators. Chem Eng Res Des 92:285–294. https://doi.org/10.1016/J.CHERD.2013.07.022

Mikielewicz D, Mikielewicz J (2010) A thermodynamic criterion for selection of working fluid for subcritical and supercritical domestic micro CHP. Appl Therm Eng 30:2357–2362. https://doi.org/10.1016/j.applthermaleng.2010.05.035

Mohammadzadeh Bina S, Jalilinasrabady S, Fujii H (2017) Energy, economic and environmental (3E) aspects of internal heat exchanger for ORC geothermal power plants. Energy 140:1096–1106

Nelson C (2008) Exhaust energy recovery. In: DEER conference

Obernberger I, Thek G (2008) Combustion and gasification of solid biomass for heat and power production in Europe-State of the art and relevant future development. In: 8th European conference on industrial furnaces and boilers, pp 1–24

Obernberger I, Thonhofer P, Reisenhofer E (2002) Description and evaluation of the new 1,000 kWel Organic Rankine Cycle process integrated in the biomass CHP plant in Lienz, Austria. Euroheat Power 10:1–17

Öhman H (2012) Implementation and evaluation of a low temperature waste heat recovery power cycle using NH_3 in an Organic Rankine Cycle. Energy 48:227–232. https://doi.org/10.1016/j.energy.2012.02.074

Olofsson A (2012) Investigation of materials for use in exhaust gas condensate environment with focus on EGR systems

Papadopoulos AI, Stijepovic M, Linke P (2010) On the systematic design and selection of optimal working fluids for Organic Rankine Cycles. Appl Therm Eng 30:760–769. https://doi.org/10.1016/J.APPLTHERMALENG.2009.12.006

Papapetrou M, Kosmadakis G, Cipollina A et al (2018) Industrial waste heat: Estimation of the technically available resource in the EU per industrial sector, temperature level and country. Appl Therm Eng 138:207–216. https://doi.org/10.1016/J.APPLTHERMALENG.2018.04.043

Patel PS, Doyle EF (1976) Compounding the truck diesel engine with an organic rankine-cycle system. SAE Tech Pap. https://doi.org/10.4271/760343

Qiu G (2012) Selection of working fluids for micro-CHP systems with ORC. Renew Energy 48:565–570. https://doi.org/10.1016/J.RENENE.2012.06.006

Quoilin S (2011) Sustainable energy conversion through the use of Organic Rankine Cycles for waste heat recovery and solar applications. University of Liège, Belgium, Sylvain Quoilin, p 2008

Quoilin S, Lemort V (2009) Technological and economical survey of Organic Rankine Cycle systems. In: 5th European conference on economics and management of energy in industry, vol 278, p 12

Quoilin S, Lemort V, Lebrun J (2010) Experimental study and modeling of an Organic Rankine Cycle using scroll expander. Appl Energy 87:1260–1268. https://doi.org/10.1016/J.APENERGY.2009.06.026

Quoilin S, Van Den BM, Declaye S et al (2013) Techno-economic survey of organic rankine cycle (ORC) systems. Renew Sustain Energy Rev 22:168–186. https://doi.org/10.1016/j.rser.2013.01.028

Reid AD (2010) Low temperature power generation using HFE-7000 in a Rankine cycle. San Diego State University

Ringler J, Seifert M, Guyotot V, Hübner W (2009) Rankine cycle for waste heat recovery of IC engines. SAE Tech Pap 2:67–76. https://doi.org/10.4271/2009-01-0174

Robert BOSCH GmbH (2014) Diesel systems—waste heat recovery system for commercial vehicles—brochure

Roedder M, Neef M, Laux C, Priebe KP (2016) Systematic fluid selection for organic Rankine cycles and performance analysis for a combined high and low temperature cycle. J Eng Gas Turbines Power 138:1–9. https://doi.org/10.1115/1.4031361

Roy JP, Mishra MK, Misra A (2011) Performance analysis of an Organic Rankine Cycle with superheating under different heat source temperature conditions. Appl Energy 88:2995–3004. https://doi.org/10.1016/j.apenergy.2011.02.042

Saleh B, Koglbauer G, Wendland M, Fischer J (2007) Working fluids for low-temperature organic Rankine cycles. Energy 32:1210–1221. https://doi.org/10.1016/J.ENERGY.2006.07.001

Schuster A, Sichert A, Aumann R (2014) Thermodynamic machine and method for the operation thereof

Seher D, Lengenfelder T, Gerhardt J et al (2012) Waste heat recovery for commercial vehicles with a Rankine process. In: 21st Aachen colloquium automobile and engine technology, pp 1–15

Song J, Song Y, Gu Cw (2015) Thermodynamic analysis and performance optimization of an Organic Rankine Cycle (ORC) waste heat recovery system for marine diesel engines. Energy 82:976–985.https://doi.org/10.1016/j.energy.2015.01.108

Squaiella LLF, Martins CA, Lacava PT (2013) Strategies for emission control in diesel engine to meet Euro VI. Fuel 104:183–193. https://doi.org/10.1016/j.fuel.2012.07.027

Su W, Zhao L, Deng S (2017) Simultaneous working fluids design and cycle optimization for Organic Rankine cycle using group contribution model. Appl Energy 202:618–627. https://doi.org/10.1016/J.APENERGY.2017.03.133

Svensson E, Morandin M, Harvey S (2019) Characterization and visualization of industrial excess heat for different levels of on-site process heat recovery. Int J Energy Res 43:7988–8003. https://doi.org/10.1002/er.4787

Tabor H, Bronicki L (1964) Establishing criteria for fluids for small vapor turbines. SAE Tech Pap. https://doi.org/10.4271/640823

Tan H, Cao R, Wang S et al (2021) Proposal and techno-economic analysis of a novel system for waste heat recovery and water saving in coal-fired power plants: a case study. J Clean Prod 281:124372. https://doi.org/10.1016/J.JCLEPRO.2020.124372

Tartière T, Astolfi M (2017) A world overview of the organic rankine cycle market the overview the organic Rankine assessing the feasibility of the heat. Energy Procedia 129:2–9

Tchanche BF, Papadakis G, Lambrinos G, Frangoudakis A (2009) Fluid selection for a low-temperature solar organic Rankine cycle. Appl Therm Eng 29:2468–2476. https://doi.org/10.1016/J.APPLTHERMALENG.2008.12.025

Tocci L, Pal T, Pesmazoglou I, Franchetti B (2017) Small scale Organic Rankine Cycle (ORC): a techno-economic review. Energies 10:1–26. https://doi.org/10.3390/en10040413

Vaja I, Gambarotta A (2010) Internal Combustion Engine (ICE) bottoming with Organic Rankine Cycles (ORCs). Energy 35:1084–1093. https://doi.org/10.1016/j.energy.2009.06.001

Vanslambrouck B, Vankeirsbilck I, Gusev S, De Paepe M (2012) Efficiency comparison between the steam cycle and the organic Rankine cycle for small scale power generation. In: Renewable energy world conference & expo North America, vol 32, p 13

Wang EH, Zhang HG, Fan BY et al (2011) Study of working fluid selection of organic Rankine cycle (ORC) for engine waste heat recovery. Energy 36:3406–3418. https://doi.org/10.1016/j.energy.2011.03.041

Waste Heat Recovery—Technology Assessment (2016). Interni Med pro Praxi 18:163–167

Wyss E (1890) http://douglas-self.com/MUSEUM/POWER/petrol/petrol.htm#ew

Zhang Q, Pennycott A, Brace CJ (2013) A review of parallel and series turbocharging for the diesel engine. Proc Inst Mech Eng Part D J Automob Eng 227:1723–1733. https://doi.org/10.1177/0954407013492108